Advances in Science, Technology & Innovation

IEREK Interdisciplinary Series for Sustainable Development

Advances in Science, Technology & Innovation (ASTI) is a series of peer-reviewed books based on important emerging research that redefines the current disciplinary boundaries in science, technology and innovation (STI) in order to develop integrated concepts for sustainable development. It not only discusses the progress made towards securing more resources, allocating smarter solutions, and rebalancing the relationship between nature and people, but also provides in-depth insights from comprehensive research that addresses the **17 sustainable development goals (SDGs)** as set out by the UN for 2030.

The series draws on the best research papers from various IEREK and other international conferences to promote the creation and development of viable solutions for a **sustainable future and a positive societal** transformation with the help of integrated and innovative science-based approaches. Including interdisciplinary contributions, it presents innovative approaches and highlights how they can best support both economic and sustainable development, through better use of data, more effective institutions, and global, local and individual action, for the welfare of all societies.

The series particularly features conceptual and empirical contributions from various interrelated fields of science, technology and innovation, with an emphasis on digital transformation, that focus on providing practical solutions to **ensure food, water and energy security to achieve the SDGs.** It also presents new case studies offering concrete examples of how to resolve sustainable urbanization and environmental issues in different regions of the world.

The series is intended for professionals in research and teaching, consultancies and industry, and government and international organizations. Published in collaboration with IEREK, the Springer ASTI series will acquaint readers with essential new studies in STI for sustainable development.

ASTI series has now been accepted for Scopus (September 2020). All content published in this series will start appearing on the Scopus site in early 2021.

More information about this series at http://www.springer.com/series/15883

Inamuddin · Rajender Boddula ·
Anish Khan
Editors

Sustainable Production and Applications of Waterborne Polyurethanes

Springer

Editors
Inamuddin
Department of Applied Chemistry
Faculty of Engineering and Technology
Zakir Husain College of Engineering
and Technology
Aligarh Muslim University
Aligarh, India

Rajender Boddula
National Center for Nanoscience and Technology
CAS Key Laboratory of Nanosystem
and Hierarchical Fabrication
Beijing, China

Anish Khan
Department of Chemistry
Faculty of Science
King Abdulaziz University
Jeddah, Saudi Arabia

ISSN 2522-8714 ISSN 2522-8722 (electronic)
Advances in Science, Technology & Innovation
IEREK Interdisciplinary Series for Sustainable Development
ISBN 978-3-030-72868-7 ISBN 978-3-030-72869-4 (eBook)
https://doi.org/10.1007/978-3-030-72869-4

This Springer imprint is published by the registered company Springer Nature Switzerland AG
The registered company address is: Gewerbestrasse 11, 6330 Cham, Switzerland

Contents

Waterborne Polyurethanes for Corrosion Protection

Felipe M. de Souza and Ram K. Gupta

Abstract

Corrosion is a worldwide problem causing the loss of hundreds of thousands of dollars every year. Throughout this chapter, many topics regarding materials, procedures, synthetic routes, and mechanisms of waterborne polyurethane and their composites for corrosion protection are explored. Waterborne polyurethane-based coatings are an extremely versatile group of polymers that can be designed virtually to target any property, for example, thermal, mechanical, abrasion, and, most importantly, corrosion protection in different environments such as against air, moisture, seawater, chemicals, and UV radiation. The preparation of highly stable dispersions along with effective barrier effect for the coatings that include proper packing, high crosslink density, and strong adhesion to substrates are the key factors to develop an effective coating. In this chapter, the most relevant waterborne polyurethane-based composites mentioned in the current literature are discussed in detail to provide a full understanding of how anti-corrosive coatings work and how their properties can be further improved.

Keywords

Polyurethanes • Waterborne polyurethanes • Corrosion • Coatings • Composites • Conducting polymers

1 Introduction

Corrosion is a phenomenon of the deterioration of a material that can take place through chemical, environmental, or biological agents. A material corrodes when it is in a corroding environment and during the corrosion process the material gets oxidized. Since oxidation of the material happens due to the loss of electrons, there must be a path for electron transfer. To better understand the corrosion process, it is important to understand the corrosion cell which consists of two electrodes (anode and cathode), electrolyte, and a conducting path for the electron transport. A classic example of a corrosion cell is Daniel cell where zinc (anode) and copper (cathode) are immersed in a solution and externally connected through a conductive wire (Fig. 1). In this cell arrangement, zinc (Zn) has a higher oxidative potential (E_{ox}) than copper (Cu) which causes the electrons (e^-) to flow naturally through the wire from Zn to Cu. This causes the $Zn^o_{(s)}$ to be converted into $Zn^{2+}_{(aq)}$, which is the oxidation process (corrosion), meanwhile, $Cu^{2+}_{(aq)}$ from the solution reduces to $Cu^o_{(s)}$ on the surface of the metal, which is the reduction process (Ahmad, 2006). This system is spontaneous and in the case of a battery, the aim is to use the electrical energy that is generated when these metals are in contact. However, this process happens mostly in undesirable ways, for example, the corrosion of a metal container due to exposure to air causing a leak, contamination by the oxidized materials in canned food, pipelines that lose their transportation capacity, industrial plants that shut down due to excessive corrosion, and deterioration of the bottom of boats that can cause them to sink. The most common corrosive media are air, moisture, water, brine, gases (H_2S, SO_2, N_xO_y, etc.), acids, and alkalis.

In a real scenario, when a metal is exposed to a corrosive environment such as humid air and ocean water corrosion happens without the need or presence of another metal (like Cu in case of Daniel cell). Figure 2 describes a general corrosion process of a metal exposed to a corrosive

F. M. de Souza
Department of Chemistry, Pittsburg State University, Pittsburg, KS 66762, USA

R. K. Gupta (✉)
Kansas Polymer Research Center, Pittsburg State University, Pittsburg, KS 66762, USA
e-mail: rgupta@pittstate.edu

Inamuddin et al. (eds.), *Sustainable Production and Applications of Waterborne Polyurethanes*, Advances in Science, Technology & Innovation, https://doi.org/10.1007/978-3-030-72869-4_1

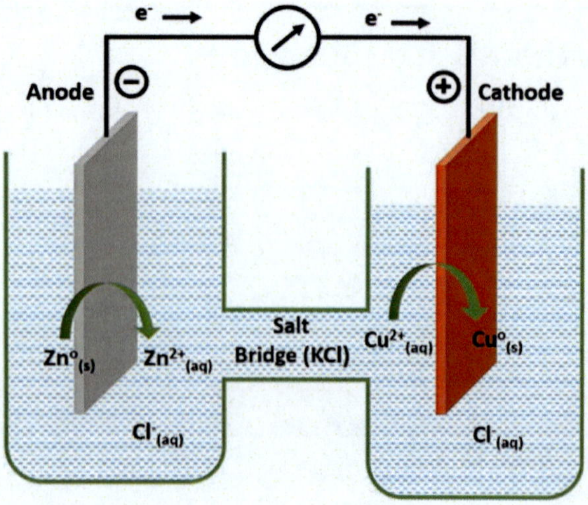

Fig. 1 Scheme of a Daniel cell or galvanic cell

environment such as a metal post on street, metal in constructions, cans having food, etc. The humidity (in case of metal post on street) or difference in concentration of oxygen level (in case of metal in construction) acts as an electrolyte while facilitating the oxidation (corrosion) of the metal (Fig. 2a). A similar situation could arise in pipelines where outside and inside of the pipeline have different environments (Fig. 2b) (Ahmad, 2006).

1.1 Effect of Corrosion on the Economy and Potential Market for Corrosion Inhibitors

Corrosion is a worldwide issue and causes a loss of around USD2.5 trillion per year, which is about 3.4% of the global gross domestic product (GDP) (Koch et al., 2016). The percentage GDP loss of some countries and the rest of the world is shown in Fig. 3. Nevertheless, this scenario can be mitigated by implementing proper ways to prevent corrosion that diminish around 15–35% of the damage. Hence finding ways to prevent corrosion provide a large impact on the economy. The market for corrosion inhibitors reached USD7.4 billion in 2019 and is ascending with a prediction of 3.8% per year up to 2027. The constant increase in the market for corrosion inhibitors is due to the use of more sustainable routes and environmentally friendly products. The development of cost-effective and environment-friendly corrosion protective coatings is much needed to prevent worldwide loss and save the planet. The greatest challenge is to obtain coatings that not only withstand the aggressive environments but also possess an extra set of properties such as high mechanical strength and enhanced thermal stability. Hence, the selection of starting materials to develop such coatings is very crucial and depends on the applications as well as on the materials to be protected.

1.2 Types of Corrosion Inhibitors

Corrosion inhibitors are classified as inorganic, organic, and environmental conditions (Ahmad, 2006). The inorganic inhibitors can be divided into two groups: anodic and cathodic. Environmental conditions are also divided into two groups: Scavengers and Biocides.

The anodic inhibitors, also known as passivation inhibitors, are generally metal oxides that formed on the surface of the anode metal by either induced chemical reactions or natural oxidation with air, the latter called "native oxide layer" that grants protection against corrosion due to their inert nature (McCafferty, 2010). The elements that can be passivated with their respective oxides are Zn, Al, Cu, Mg, Ti, Cd, Sn, Ag, and Si (semi-metal). The thickness of the oxidized layer usually varies between 0.1 and 1.5 nm for effective corrosion protection. The thickness of the oxide layer is one of the very important parameters for effective corrosion protection as a thicker layer causes a crack in the oxide layer which makes the coating ineffective (McCafferty, 2010). Nevertheless, surface passivation has been explored not only for protection but also for other applications. For example, titanium substrate passivated with zinc alloys are used in semiconductors, aluminum oxide and silicon-based coatings are applied in solar cells, porous dielectric coatings are critical in electronic devices (Havemann et al., 1995; Kong et al., 2012; Miyajima et al., 2009).

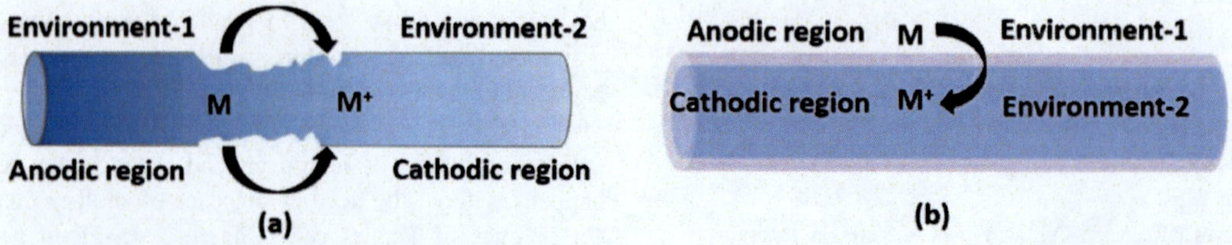

Fig. 2 General schemes for corrosion

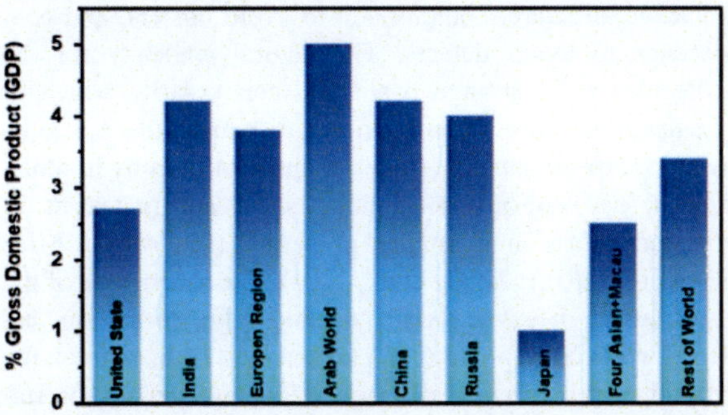

Fig. 3 Cost of corrosion in the percentage of global Gross Domestic Product (GDP) (Koch et al., 2016)

For the cathodic inhibitors, some variations of polyphosphates can be employed. The concept for their adhesion is based on electrostatic attraction. The polyphosphate can be mixed with metal ions such as Ca^{2+} to form a colloidal suspension with the positive charges, which find attractive forces with the electrons on the surface of the cathode to form a protective coating. However, polyphosphates present some disadvantages such as they tend to form complexes with iron that prompts dissolution instead of inhibition. Also, to be functional it must be inserted in a pH between 6.5 and 7.5, otherwise, hydrolysis may take place causing polyphosphate to convert into orthophosphate and lose most of its inhibitor properties (Uhlig et al., 1955). Since the average pH of the ocean is around 8.1, this condition would dismiss its application as a coating for platforms, ships, and related areas. However, to make them suitable for practical applications, other inhibitors such as anodic and/or cathodic can be mixed to combine their properties and optimize the prevention of corrosion through synergistic blending. For example, zinc and chrome-based salts do not form an effective coating but when blended with polyphosphates can improve the overall properties and decrease the quantity of inhibitor that would be normally used. Some inorganic coatings such as chrome-based coatings are harmful to the environment and lose their efficiency under alkaline conditions (Kendig et al., 2001).

Organic inhibitors such as monoamines, diamines, or amides are versatile in the sense that they can be used as both anodic and cathodic inhibitors. The principle of their protection is based on adsorption. For example, amines that present a pair of electrons may interact with the surface of the anode, hence amines are cationic inhibitors, while sulfonates that present a partial positive charge adhere to the surface of the cathode. This process leads to a formation of a bond between the surface of the metal and the organic inhibitor that removes water from that region. This is named as chemisorption, which blocks both anodic and cathodic reactions, hence protecting the metal from oxidization (Sanyal, 1981). Another alternative to prevent oxidation is removing oxygen from the closed systems by using compounds named scavengers. The presence of oxygen speeds up the oxidation reaction rate by reacting with the electrons from the cathode causing a depolarization of the charges and thus causing the corrosion. Ammonium sulfate, sodium sulfite, and hydrazine are the most commonly used scavengers.

1.3 Polymers for Corrosion Inhibitors

As discussed the inhibitors are a group of materials used to prevent corrosion, however, some challenges are required to overcome such as non-environmentally friendly manufacture, costly process, work in specific conditions, and side effects that can emerge during the process. As technology progresses organic coatings appeared as a proper alternative for the coating due to their enhanced anti-corrosion properties against chemicals and water along with high mechanical and thermal stability. Some of the most explored types of organic coatings are epoxy, amine, acrylic, silicone, natural or synthetic rubber, polyvinyl chloride, and urethanes (Czub, 2009; Liu et al., 2013; Meis et al., 2014). Besides used for coatings, these materials also find applications in areas such as paints, adhesives, aerospace, electronics, and biomedical (Bludau et al., 2017; Chen et al., 2007; Ocampo et al., 2005; Osborne et al., 2001; Schwarte et al., 2003; Wu et al., 2014a).

Epoxy resins were synthesized for the first time in 1909 and still have a strong impact on the industries (Bauer, 1979). High mechanical strength, good adhesion to a variety of substrates, chemical and heat resistance allows them to be used as an effective corrosion-resistant coating. The effectiveness of the organic coatings for corrosion inhibition depends on many factors such as molecular weight, chemical structure, curing agent, and processing method. Petrochemicals-based compounds such as bisphenol A (BPA) and novolacs (EPN) are still being used as starting materials for such coatings (Jamali, 2013; Lee, 1967). However, the depletion of hydrocarbon deposits accompanied by economic instability made the current research trend find ways to introduce bio-based materials because of their low cost, effectiveness, trustful source for replenishment and environmentally friendly features. As a result, many bio-based epoxy resins were synthesized using vegetable oils (Fig. 4) (Czub, 2009). The green approach used to obtain epoxidized vegetable oils yielded a variety of materials such as elastomer, foams, adhesives, and coatings. The latter many times surpassing the anti-corrosion properties if compared with petrochemicals (Liu et al., 2005). Hence becoming an attractive alternative path to synthesize environment-friendly materials for effective corrosion protection (Czub, 2009).

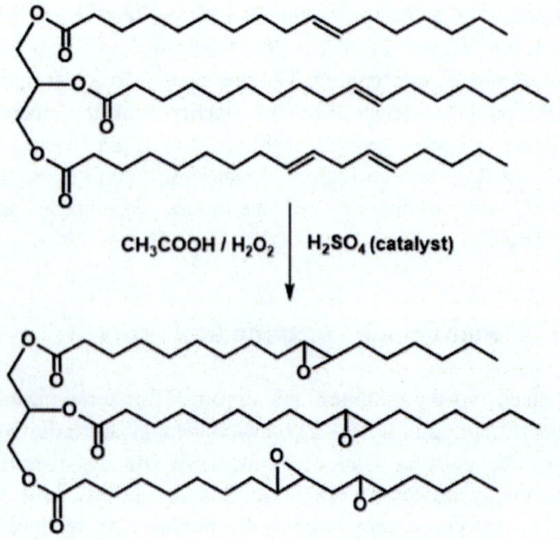

Fig. 4 Epoxidation reaction of vegetable oil

The epoxy resins are the precursor for many reactions filling the role of a prepolymer or starting material that can be further functionalized for coatings. Amine-based compounds can be used for effective corrosion protective coatings, however, the presence of unreacted amines and trapped water in coatings decreases their resistance to corrosion and must be removed. Some approaches such as high temperature curing to remove the water to prevent blister and delamination, higher epoxy/amine ratio, and increase in the coating thickness are applied to make amine-based coatings more effective, however, these approaches increase the cost (Wegmann, 1997). Despite these concerns and lower resistance to water and acids, epoxy/amine resins are still a viable option if compared with chromate-based coatings which are highly toxic and costly. These factors allow epoxy/amine-based coatings to be widely used for civil engineering and aluminum alloys for the aircraft industries (Meis et al., 2014; Wu et al., 2014b).

Acrylics are used to make waterborne dispersions due to the presence of polar groups. Although the presence of polar group is a key factor for proper dispersion of polyurethanes in aqueous media, however, the higher concentration of polar groups allow more permeation of water within the polymeric chains which may create defects in the coatings (Taylor & Winnik, 2004; Wegmann, 1997). The interpenetration of water can happen through physical entrapment and hydrogen-bonding, which is endorsed by the presence of hydrophilic groups (Baukh et al., 2011; Lendvay-Győrik et al., 2007). It may also lead to irreversible swelling of the polymer, which can affect the mechanical property and water resistance of the coating (Cotugno et al., 2001; Han & Drzal, 2003; Tamirisa & Hess, 2006). To avoid such issues, the coating should have a proper interaction of particles to allow a packed structure, compression to avoid blisters, and coalescence to avoid defects. The natural tendency for the molecules is to arrange based on their polarity which in practice may cause a heterogeneous dispersion that has to be averted. For this, the controlled number of both hydrophilic groups, as well as crosslinking sites, are important to enhance corrosion resistance properties (Liu et al., 2013; Tillet et al., 2011; Zhang et al., 2006). The schematics of the mechanisms of waterborne coating film formation are described in Figs. 5 and 6 (Wegmann, 1997). In general, the crosslinking between distinct polar groups leads to the stability of the film by decreasing the interaction of these groups with water in the polymeric structure, which decreases its water uptake (Yang & Zhang, 2005; Zhang et al., 2006). Another important factor for proper film formation is the molecular weight of the resin segment, which is mostly hydrophobic. It can have different effects in a way that a higher molecular weight provides enhanced barrier performance. But the increase of hydrophobic groups causes difficulties for a proper dispersion leading to poor coating and decrease of overall properties. Thus, acrylic groups help define the proper hydrophilic/hydrophobic ratio in the polymeric chain to obtain a stable waterborne dispersion with simultaneous high anti-corrosion performance (Liu et al., 2013).

Silicon-based coatings are a popular class of eco-friendly materials that are used in automotive industries, aerospace, cookware, cable for electronic industry, semiconductors, medical devices, corrosion resistance, etc. (Shit & Shah, 2013). Silicon rubbers present some advantages such as resistance to virtually any environmental conditions such as rain, UV light, and temperature amplitude (Meng et al., 2012). Such coatings also present a high performance when dealing with oils, polar organic compounds, and to store acids and alkalis (Southwart, 1976). The previous report showed that the crosslink density of polysiloxanes along with the introduction of nanoparticles could enhance their properties to make them suitable against strong acids or alkali environments (Zhou et al., 2012). Thus, in general, silicon-based coatings present some advantages such as resistance against water and polar organic substances due to their high hydrophobicity and wide range of work temperatures that prevent the coating from turning into a brittle material even after long exposure. However, they tend to hold dirt particles, do not always present proper adhesion to some substrates, and provide relatively lower tear resistance (Shit & Shah, 2013).

The latex binders are another environmentally friendly option for coatings due to their good adhesion to metal and can be used as an adaptable matrix for composites in several ways to become more resistant against corrosion. In the early development of latex binders as coating, many methods were developed to form stable films without cracking and to

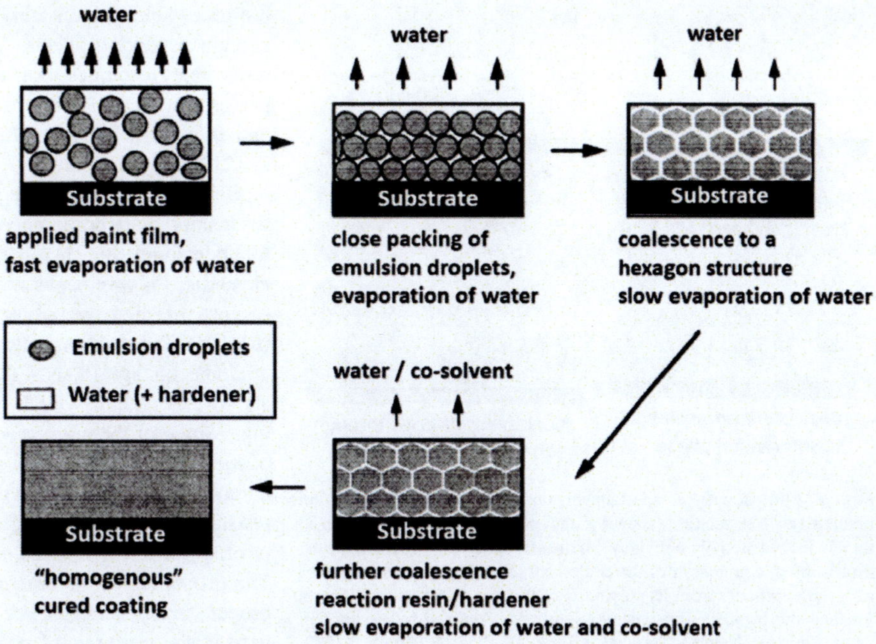

Fig. 5 A mechanism for waterborne coating film formation. Adapted with permission from Wegmann (1997). Copyright (1997) Elsevier

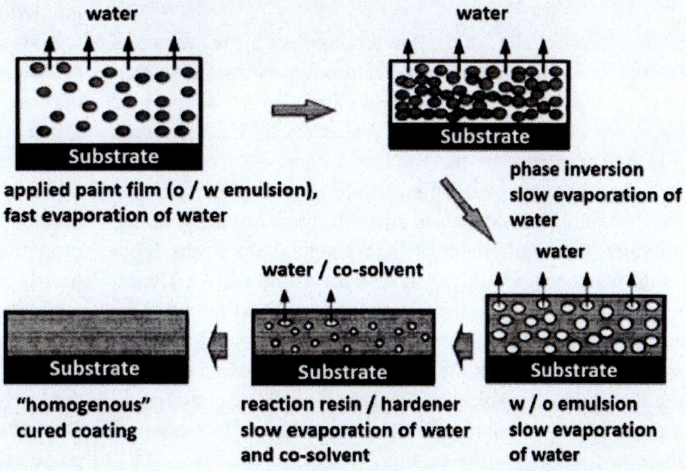

Fig. 6 A mechanism for waterborne film formation through reverse emulsification. Adapted with permission from Wegmann (1997). Copyright (1997) Elsevier

reduce the amount of volatile organic contents (VOCs) in the formulations (Tomba et al., 2008). Latter an interesting approach named "Designed Diffusion™" was developed as a greener alternative for the introduction of water as a dispersing agent without the loss of affecting their coating effectiveness. The schematic for film formation is shown in Fig. 7 (Keddie & Routh, 2010). It consists of a blend of a soft polymer (low T_g) along with a hard polymer (high T_g). The soft polymer ideally should be more hydrophilic to carry out the water and the hard polymer presents higher interaction with the coalescent. Both polymers are mixed to allow percolation of the soft polymer within the polymeric chains of the hard polymer to form a stable dispersion. The homogenous mixture can then be applied to the surface of the substrate. Thermal curing may be required to evaporate the water and coalescent (such as *n*-methyl pyrrolidone) as both tend to migrate to the soft polymer that slowly hardens along with the hard polymer during film formation. This method can reduce VOCs up to 30% (Fu et al., 2017). The concept of diffusion throughout the latexes polymeric matrixes have been employed in different cases such as for dispersion of zinc phosphates, molybdates, and chromates. The study showed that the migration of the pigments into the chains is a decisive factor to enhance corrosion resistance

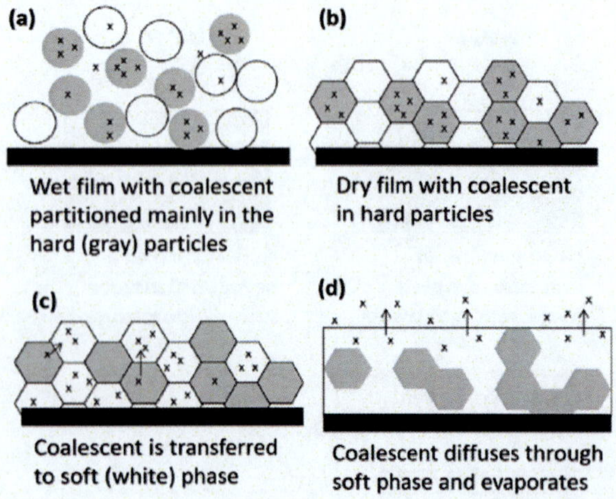

Fig. 7 Film formation mechanism through the design diffusion method. **a** Wet mixture showing the coalescent as "*x*" dispersed mostly in the hard polymer "gray", **b** the settlement of particles after the film is dried, **c** migration of the coalescent to the soft polymer "white" as water evaporates from its polymeric chain, and **d** after evaporation is finished the blended film coating is formed. Adapted by permission from Springer Nature: Springer eBook (Keddie & Routh, 2010), Copyright (2010)

(Grundmeier et al., 2006). Another approach for the use of latexes is as a template for electropolymerization of conductive polymers such as polyaniline (PANI) to provide improved corrosive resistance (Wei et al., 1995; Yeh et al., 2001). The hardships to fabricate a coating of pure PANI include brittleness of the film, insolubility in polar solvents, low solubility in non-polar solvents, non-fusibility of the polymer, the requirement of high potential for electrodeposition which causes oxidation of most of the metals. Hence, it could a challenge to make a coating of PANI by itself. To counter these issues, latexes can be used to form a microsphere nanoparticle as the core of the structure and PANI as the shell (Abu & Aoki, 2005). This structure can adhere to the surface of metal such as iron enabling the formation of the corrosion protective coating.

As the technology evolved many other materials appeared as viable options for coating applications and among them polyurethane-based coatings have received a huge highlight due to its effectiveness, low cost, and variety of synthesis options. The need to find eco-friendly synthetic approaches has pushed the scientific community to develop waterborne polyurethane coatings. Waterborne polyurethane uses eco-friendly solvents and drastically reduces the use of VOCs, which not only reduces the cost by replacing organic solvents with water but also implements non-toxic materials for the synthesis such as soybean, canola, corn oil among many others. Besides the functionalization of the bio-renewable polyunsaturated oils, they can also be used straight as drying agents. Waterborne polyurethanes using renewable resources not only deliver improved coating performance at a lower cost but also provide a green alternative to petrochemicals (Lligadas et al., 2013; Wang et al., 2012; Xue et al., 2014).

Before entering into the procedures and composites used for making waterborne polyurethane coatings, it is important to understand the chemistry and formulations of the key elements. Polyurethanes are obtained from the reaction of two groups: di, tri, or polyisocyanate (R–N=C=O) and polyol (R′–OH). The reaction between these two groups leads to the urethane linkage [R–N–C(O)–O–R′] (Tersac, 2007). There are several features which needed to prepare effective coatings. Some of them are processing methods, the appearance of the coating, mechanical strengths, thermal stability, high durability in radiation, and chemical (water, acid, and alkali) resistance as shown in Fig. 8 (Chen & Wojcik, 2000). The priority may change based on the application of the coatings. The properties of the coating can be tuned by selecting the proper starting reagents. For example, the types of isocyanate used in the synthesis of polyurethanes could alter the properties of the coating. Methylene diisocyanate (MDI) contains aromatic rings which provide a rigid structure to polyurethane and thus improves the mechanical strength of the coating, however, the coating has low UV stability as it becomes yellowish when exposed to sunlight and loses the pigmented color (if added) with time. On the other hand, isophorone diisocyanate (IPDI) presents an aliphatic structure that provides UV stability and homogenous mixture with pigments, but due to the flexible chain, the mechanical properties are relatively lower. Thus, this type of material is mostly used in the automobile industry as coatings with attractive colors, glossiness as well as protection for the coated metal or plastic part against oxidation (Chen & Wojcik, 2000). The chemical structure and functionality of the polyols used for the preparation of polyurethanes also affect the properties of the coatings. For example, polyester-based polyols provide stronger mechanical strength being both flexible and tough while acrylic polyols usually improve the appearance and UV stability (Tersac, 2007). The stoichiometry ratio of diisocyanate and polyol (NCO/OH) is an important factor that could affect the properties of the coating. For example, the excess of polyol (OH groups) in polyurethanes will provide increased flexibility and improved adhesion to surface but will decrease resistance to solvents, moisture, and chemicals. The excess of diisocyanate (NCO groups) will deliver a coating which will take a longer time to dry. Such coating will have lower adhesion to a surface but will have improved mechanical strength and higher resistance to chemicals (Chen & Wojcik, 2000).

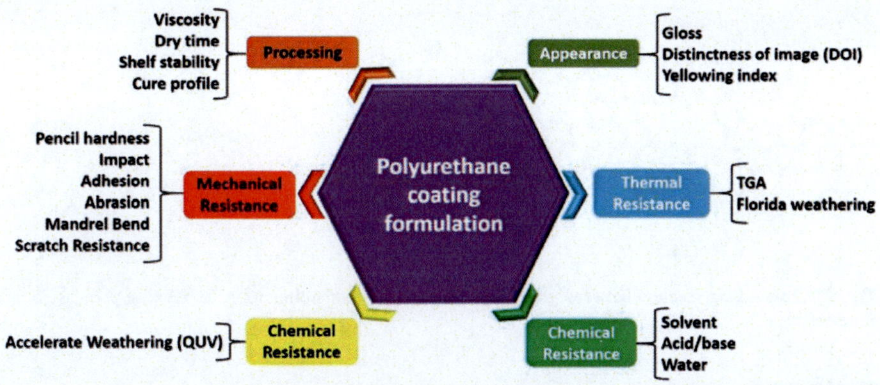

Fig. 8 Polyurethane coating formulation process

2 Waterborne Polyurethanes for Corrosion Protection

Waterborne polyurethanes find their wide applications in corrosion protection on a variety of substrates due to their low VOCs, strong adhesion to the surface, low flammability, high mechanical strength, thermal stability, chemical and weather resistance, etc. In the following sections, we have provided details of waterborne polyurethanes-based compounds for corrosion protection.

2.1 Composite Coatings Based on Waterborne Polyurethanes and Graphene

Waterborne polyurethanes allow great versatility in their synthetic procedure to introduce a variety of materials that provides innovative properties. One green approach is mixing graphene with waterborne polyurethanes to provide a coating with high barrier properties and chemical stability that improves the corrosion resistance. The main challenge with graphene is the lack of dispersibility. To improve the dispersibility of graphene in waterborne polyurethanes, graphene is required to functionalize with hydrophilic groups (Qiu et al., 2017; Xiang et al., 2013). Although the functionalization of graphene improves its dispersibility in water, it also decreases the chemical resistance of the coating against water, solvents, acids, and bases (Liu et al., 2013; Zhu & Ooij, 2004). One of the solutions to this problem is crosslinking the structure of waterborne polyurethanes with graphene. This approach can simultaneously decrease the hydrophilicity of the polar groups and increase the percolation of graphene nanosheets into the waterborne polyurethane's structure, hence, improving the barrier and chemical resistance of the final coating. Recent reports demonstrated an interesting crosslinking procedure by using polycarbodiimide (PCD) as a crosslinker for graphene oxide nanosheets that was later dispersed into a waterborne polyurethane system (Cui et al., 2020; Hesselmans et al., 2006; Posthumus et al., 2007). The chemical reaction between polycarbodiimide and graphene oxide can be seen in Fig. 9. Polycarbodiimide reacted with the carboxyl groups of the waterborne polyurethanes dispersion to create a complete crosslinked structure. The ratio of crosslinker, diisocyanate, and functionalized graphene is crucial to prevent agglomeration and to create a dense coating.

A visual effect of protection against corrosion for these coatings can be seen in Fig. 10. The experiment was performed by applying a salt spray over the surface of metal coated with the synthesized composites. The aspects of the coating were analyzed after 72–120 h of exposure to the corrosive spray. It was notable that the sample WPG8.75 (PCD/GO ratio = 8.75) presented the higher protection against corrosion, barely showing any deterioration. The controlled coating using only waterborne polyurethane (WPU) and waterborne polyurethane with graphene (WGO) showed significant deterioration of the coating after the salt spray test. The highly crosslinked structure using proper composition (WPG8.75) prevented corrosion ions from permeating to the metal, mentioned as barrier effect. On top of that, the relatively inert structure of graphene oxide aids the composite to withstand aggressive environments (Cui et al., 2020).

Many other approaches using graphene oxide were also reported. For example, in a recent study the epoxy, hydroxyl, and carboxyl groups from the graphene oxide were used to chemically bond polydopamine. The dopamine is known as a neurotransmitter in the human body and its polymer is found to be an effective anti-corrosion filler (Singer et al., 2015). The procedure for obtaining a graphene oxide and polydopamine (pDop-rGO) composite consisted of a facile mixing the graphene oxide (GO) dispersion with dopamine hydrochloride at room temperature for 24 h under a buffer system of pH = 8.5. The combination of dopamine with GO created a redox system that enabled simultaneous polymerization of dopamine into polydopamine as well as a reduction of the oxygenated groups from the GO. The reaction

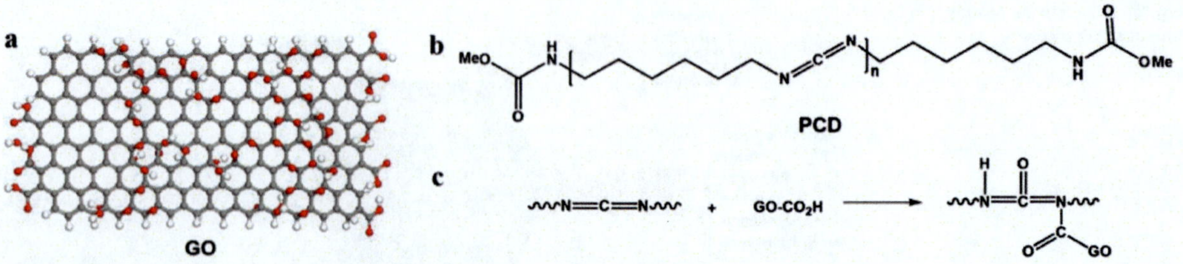

Fig. 9 Crosslinking reaction between graphene oxide and polycarbodiimide. Adapted with permission from Cui et al. (2020). Copyright (2020) Elsevier

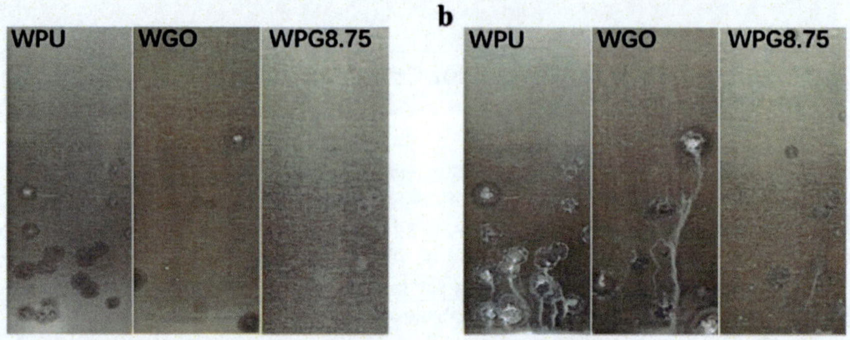

Fig. 10 Salt spray test for the samples after exposure during **a** 72 h and **b** 120 h. Adapted with permission from Cui et al. (2020). Copyright (2020) Elsevier

system allows proper percolation of PDA within the reduced GO (rGO) due to its increased adhesion through π-π stacking. Also, converting GO into reduced GO causes a decrease in hydrophilicity which increases the corrosion resistance in aqueous media (Li et al., 2016). The procedure is described in Fig. 11. Polydopamine plays an important role in improving the dispersibility of the GO and provides strong interfacial adhesion with both organic and inorganic layers (Yang et al., 2013). As previously discussed, one of the factors responsible for improving the anti-corrosion properties is proper dispersion to form a homogenous coating with a regular morphology and proper adhesion to enhance the barrier effect. In addition to these factors, there is also a tendency in the increase of thermal and mechanical properties (Lee et al., 2007; Yang et al., 2013). As discussed, waterborne polyurethanes-graphene composites coatings are promising and environmentally friendly materials that enable many versatile, effective, and facile synthetic routes.

2.2 Composites Based on Waterborne Polyurethanes and Metal Oxides for Corrosion Inhibition

The metal oxides have a long history of being used for corrosion protection through passivation of native metals or by blending with other components to make composite coatings (Hsu et al., 2012; Shimamune & Hosonuma, 1989). The main reason for selecting metal oxides as components for coatings is due to their high stability against most corrosive environments and UV radiation. However, one of the limitations is the concentration of the metal oxides that can be used in the coatings. The higher concentration of the metal oxides in the coating improves the barrier effect but it also causes loss of abrasion, scratch and impact resistance, prompt delamination, decreases flexibility, and increases viscosity. These issues were mostly solved with the introduction of nanostructured metal oxides that have proven to increase the general performance of coatings (Rashvand & Ranjbar, 2013). ZnO, Al_2O_3, TiO_2, $CaCO_3$, and SiO_2 are among the highly explored nanoparticles for corrosion protective coatings (Charpentier et al., 2012; Chen et al., 2014; Yang et al., 2005). Some of the signature properties of the nanoparticles are their high surface area, higher electron density, and smaller size which boost their inherent properties allowing them to disperse evenly in the coating. This improves their packing leading to an exceeding barrier effect compared to macroparticles. Another important factor is that the lower concentrations of nanoparticles can provide the same barrier properties as the macroparticles as well as maintaining the rheological behavior constantly, hence, adding an economical aspect (Chen et al., 2014; Yang et al., 2005). By the fact that corrosion is an electrochemical process, the use of electrochemical characterization methods

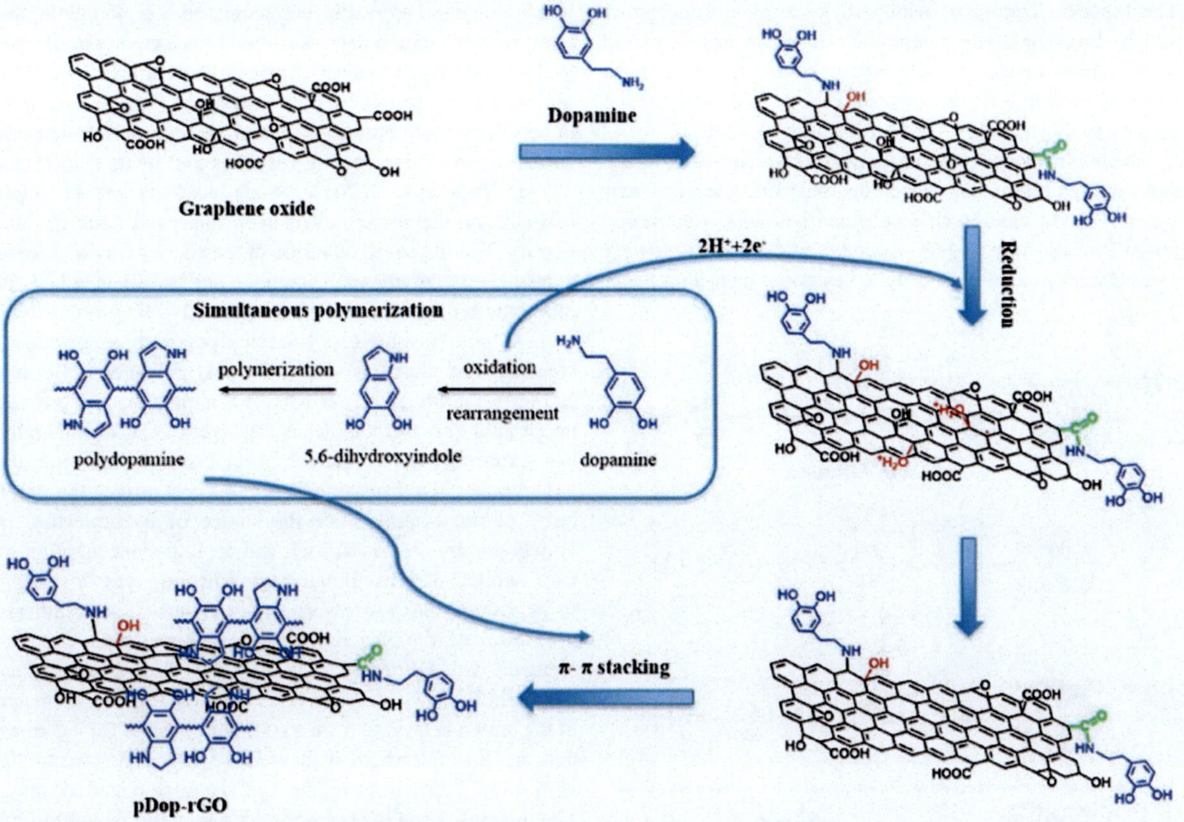

Fig. 11 Schematics for the synthesis of polydopamine and reduced graphene oxide (pDop–rGO) composite. Reproduced with permission from Li et al. (2016). Copyright (2016) American Chemical Society

provides a fundamental understanding of the coatings and their corrosion inhibition mechanism. Among various electrochemical methods, electrochemical impedance spectroscopy (EIS) is a key technique to provide information about corrosion properties such as the adhesion properties of the coating, the area exposed to corrosion, delamination, defects, and resistance of the composite (González et al., 2001; Marcus, 2011; Selman & Varma, 1991).

Aluminum is widely used as a coating material due to its low cost and relatively good stability, however, aluminum as corrosion protective coating may present unsatisfactory properties due to the development of microcracks and pores that accelerate the degradation process (Abedi Esfahani et al., 2012). A way to adequate aluminum for corrosion protective coating is through the synthesis of its oxide nanoparticles (Al_2O_3) that can be easily blended in a highly hydrophobic resin. The strong water-repelling property enables the use of the Al_2O_3 based coating for applications in self-cleaning coatings, anti-icing coatings, coatings to decrease the friction and resistance to corrosion due to the lower interaction of the ions or corrosive species with the coating surface (Dhoke et al., 2009).

Although the nanoparticles in coatings protect from corrosion, they tend to detach from the coating leading to a slow deterioration. They also have a tendency to agglomerate, which decreases the overall properties of the coating. Such issues are mostly arising due to the non-existence of chemical bonding between the nanoparticles and the resin. To avoid such problems, researchers have developed coatings based on nanoparticles that can chemically be bonded to the structure of a waterborne polyurethane (Charpentier et al., 2012; Schaefer & Justice, 2007). Such an approach increases the durability of the coating by forming a strong chemical bonding and homogenously distributing nanoparticles, preventing the agglomeration and detachment of the nanofiller from the coating (Li et al., 2006). Functionalized nanostructured TiO_2 can coordinate with a carboxyl pendant group of a polymer, which is common in waterborne polyurethane since they are responsible for improving the dispersion of the polymeric chain in water (Charpentier et al., 2012).

The functionalization of nanoparticles can be done either with hydroxyl or amine groups that can be chemically reacted with a diisocyanate in waterborne polyurethanes to form urethane or urea linkages, respectively. A schematic for the process is shown in Fig. 12 (Behniafar et al., 2015).

The marine water presents an aggressive corrosive media that requires high performance coatings on objects in seawater. To withstand such harsh environments, waterborne polyurethanes may require specific additives to prevent corrosion. Fortunately, $CaCO_3$, a key component in a bone, is an abundant inorganic compound that is found in the seawater through the decomposition of sea creatures. Due to its high resistance against corrosive environments like the seawater, some researchers have used $CaCO_3$ as a nanofiller in waterborne polyurethanes to improve their performance as corrosion protective coating for ships and boats (Liu et al., 2018a; Tom et al., 2016). Such coatings present high adhesion on the surface along with improved hardness and density. The improved adhesion of the coating on the surface happens through mineralization on the substrate which is promoted with the presence of –OH and –NH groups common in both insoluble polysaccharides such as cellulose, chitosan, and chitin as well as in polyurethanes (Li et al., 2018a; Shahlori et al., 2018). The mineralization process happens in vitro, which means the layer of $CaCO_3$ naturally being formed over the surface of the coatings in a seawater environment. This sustainable approach improves the durability of the coating since the source of its protection is largely present on the sea itself and the high concentration of Ca^{2+} and CO_3^{2-} ions allowing a self-healing type of coating to withstand an aggressive environment. The synthesis procedure of the waterborne polyurethanes was complemented with β-cyclodextrin (β-CD) that presents an amphiphilic crown structure (Hua et al., 2019). The diameter of the crown is used to form a complex with the Ca^{2+} cations that are first adsorbed to the surface and after interacting with CO_3^{2-} anions from the $CaCO_3$ mineralized coating. The mineralization effect for the coating could be noticed by scanning electron microscopy for composite coatings with increasing quantities of β-CD by weight % after exposure to simulated seawater for ten days (Fig. 13).

Silicon is one of the most abundant materials on the planet having similar properties to carbon because of its tetravalence and catenation, which yields many conformations and structures for several applications. This versatility allows silicon to be used as an additive for protective coatings. One of the widely used materials derived from silicon is nano-silica (SiO_2) which is highly hydrophobic, thermally stable, and flexible giving a specific set of properties for coatings (Sangermano et al., 2005). Besides, a small quantity of nano-silica greatly improves the coating properties and it becomes unnoted in the coating due to nanosize which prevents undesired appearance in the coating for specific applications. The size of the nanoparticle is an important factor in avoiding agglomeration and the preferred range of the size is between 5 and 50 nm to disperse evenly into the polymeric structure (Chen et al., 2005). It is found that the SiO_2 polyurethane-based coating can double the corrosion resistance and improve the water contact angle from 62.8 to 85.6° (Liu et al., 2017). A high water contact angle of 151° was reported for nano Al_2O_3 which makes the coating highly hydrophobic. However, silica may be recommended for the scenario in which self-cleaning, aesthetics, mechanical,

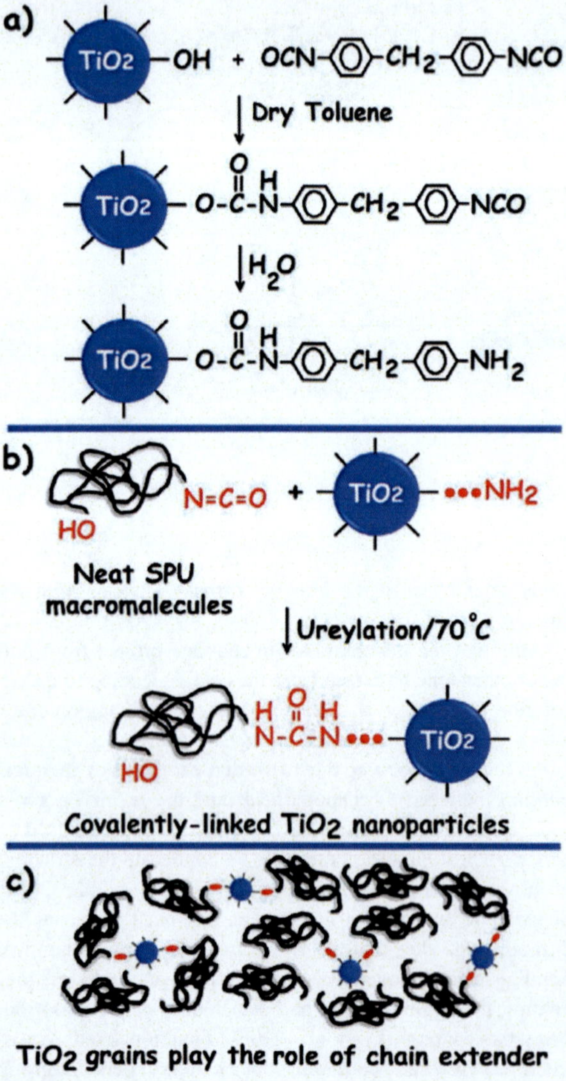

Fig. 12 The reaction of **a** hydroxyl functionalized TiO_2 for reaction with MDI to form urethane linkages, **b** amine-functionalized TiO_2 for urea linkage after reaction with MDI and **c** structure schematics of reactive TiO_2 into polyurethane chain. Adapted with permission from Behniafar et al. (2015). Copyright (2015) Elsevier

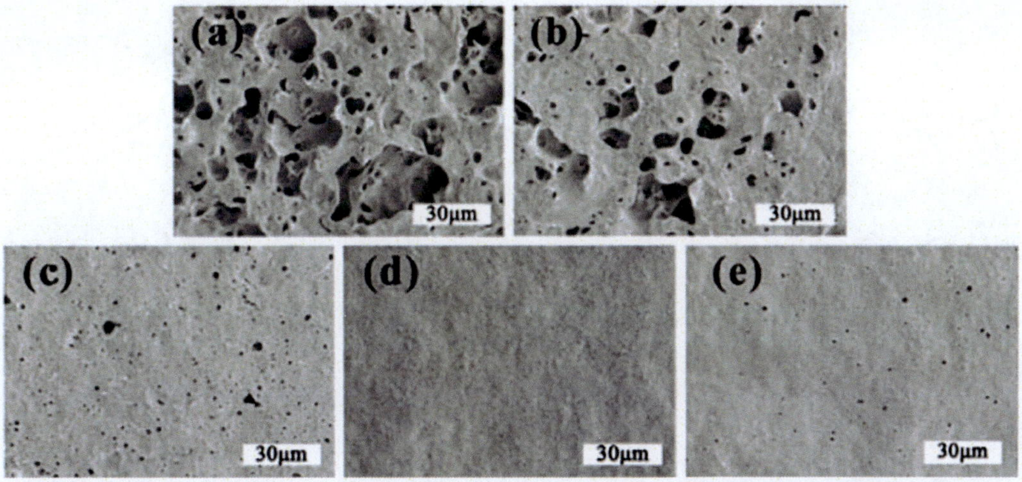

Fig. 13 Micrographs of the mineralized coatings during ten days of exposition in simulated seawater. **a** Neat WPU, **b** 1 β-CD-WPU, **c** 2 β-CD-WPU, **d** 3 β-CD-WPU **e** 4 β-CD-WPU where the number in represents the weight percentage. Adapted with permission from Hua et al. (2019). Copyright (2019) Elsevier

thermal, UV, and moisture resistance are desired (Chen et al., 2005; Liu et al., 2017; Sangermano et al., 2005). It also presents an environmental aspect accompanied by lower cost turning it into a competitive material for the coating market. Because of the toxicity and environmental impact of the materials used in the past, the scientific community had to change their approach focusing on finding new green sources for the production of materials in general. With that perspective in mind, many bio-based materials were introduced to the synthetic approaches producing a green aspect, lower cost, and sometimes even better properties compared with other toxic or petrochemical raw materials. Some examples of green materials are cardanol, cellulose, and lignin's derivatives (Dumas et al., 2015; Rao & Palanisamy, 2011; Sini et al., 2014; Zúñiga et al., 2012).

Vegetable oils are probably the most explored bio-based materials but still have lots of potential applications that require further research. One example of bio-based materials is castor oil, which is a cheap renewable compound that already contains hydroxyl groups and double bonds that can be functionalized using a variety of chemistries. One of the factors that influence the property of coatings in the cross-linked density of the polymeric matrix, since castor oil presents many double bonds as reactive sites it becomes a promising candidate for application in anti-corrosion coatings. Despite the possibility of designing composites to improve the properties, some bio-based coatings can be used as such by performing a copolymerization to obtain a novel structure that yields satisfactory properties. A study developed a castor oil functionalized with a thiol-alkoxysilane that could be easily attached to the double bonds of the castor oil through the thiol-ene coupling reaction (Fu et al., 2014). The increasing content of silicon into the structure of the coating demonstrated an increase in properties such as pendulum hardness, gel content, and tensile strength, and despite the decrease in elongation, water uptake reduces to almost 82% (Fu et al., 2014). This abrupt change happened because of the increase of crosslinked structure and hydrophobic segments that simultaneously difficult the permeation of water and increase the repellence against it. The complete synthetic route can be described in Figs. 14 and 15 (Fu et al., 2014). Despite the large use of petrochemical-based chemicals due to heavy economic features, their slow replacement with bio-based materials is a matter of time, since it has been proven to reduce the harm to the planet and health issues while enabling a sustainable and profitable return for future.

2.3 Coatings Based on Composites of Waterborne Polyurethanes and Conducting Polymers

Conductive polymers have been highlighted along the years due to its tunable conductivity, optical, and chemical properties allowing them to function as a semiconductor or conductive metal through doping. The most known examples for this class of polymers are polyaniline, polypyrrole, polythiophene, polyacetylene, and polyphenylenevinylene which have many applications in electric and electronic fields as supercapacitors, batteries, light-emitting diodes, transistors, photovoltaic cells, electromagnetic interference shielding materials just to name a few. Recently they have also been combined with polyurethanes for various applications including in corrosion protection coatings (Chandrasekhar, 1999).

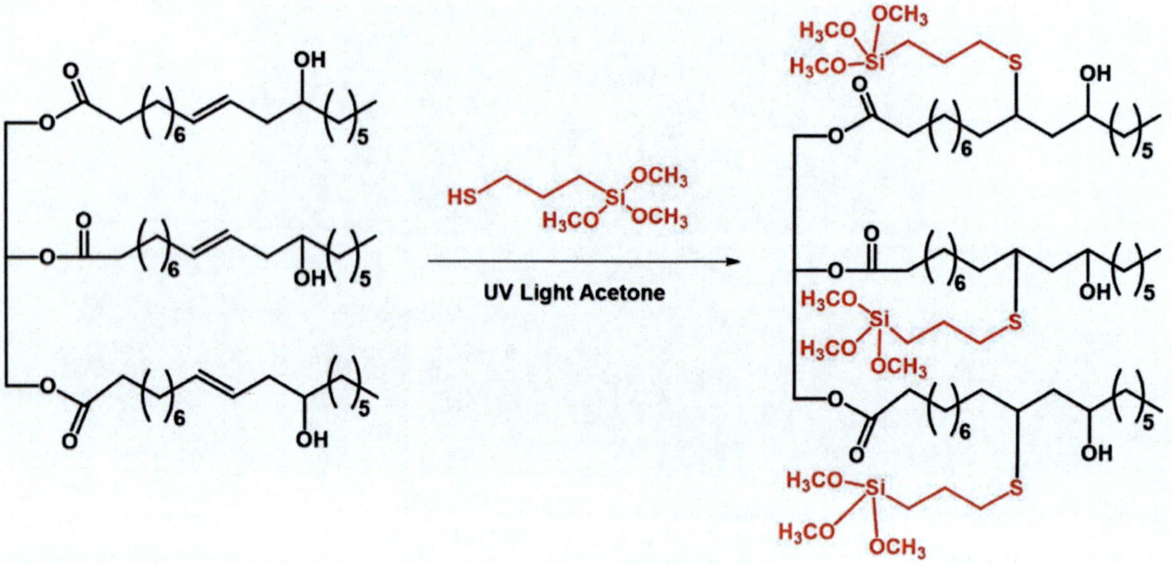

Fig. 14 Synthesis of silanized castor oil through a thiol-ene coupling. Adapted with permission from Fu et al. (2014). Copyright (2014) Elsevier

The main advantage of using conductive polymers is the mixability with the polyurethane matrix, which is due to the attractive intermolecular forces that ease the percolation of chains avoiding agglomerations. A specific approach named interpenetrating polymer networks can be performed for polymeric blending between a conductive polymer and a polyurethane matrix that differentiate them among the other methods (Chou & Lee, 1995). This methodology consists of mixing polymers in which one or more can be crosslinked (Lee et al., 1992). The crosslinking creates a network where the polymers are tightly intertwined within each other. It allows a controlled and homogenous media for the composite as well as improving many of the properties such as mechanical, thermal, barrier effect, and dielectric permittivity. The latter being important to induce a high electrical field that improves the capacitance of the Helmholtz double-layer leading to a stronger corrosion resistance coating (Guiffard et al., 2006; Mishra et al., 2017). Another important factor that arises with the presence of conductive polymers is the free charges that cause the Maxwell–Wagner–Sillars interfacial polarization which increases the dielectric permittivity as well as conductivity (Mishra et al., 2017; Putson et al., 2016).

Polyaniline is a very well-known conductive polymer that can be implemented with a polyurethane matrix due to its unique chemical structure. Polyaniline contains an aromatic ring that adds mechanical strength and rigidity to the coating and the presence of –NH– groups interact with the urethane linkages through hydrogen-bonding improving their high miscibility (El Faydy et al., 2017; Gurunathan et al., 2013; Rodrigues et al., 2005). Although such interactions are strong to provide mechanically robust coating, however, in some cases a phase separation is observed. The chemical grafting of the polymers is one of the solutions to avoid phase separation which is very common in physically blended polymer composites. This procedure provides a homogenous structure that can be synthesized by using an isocyanate-terminated polyurethane prepolymer and polyaniline (Karmakar et al., 2018). The study demonstrated that the addition of –OH groups as in poly(*m*-aminophenol) improved the hydrogen-bonding with the waterborne polyurethanes matrix and yielded better anti-corrosive properties such as corrosion rate of 1.28×10^{-4} mm/year and corrosion potential of − 349.8 mV indicating improved barrier effect and chemical stability. The corrosion protective effect of the coating on mild steel during the salt spray test is shown in Fig. 16 (Karmakar et al., 2018). As seen in the optical images, corrosion started after 72 h for the mild steel coated with pure polyurethanes, however, the corrosion time significantly increased for the mild steel coated with polymer composites.

Polyaniline is reported as a "smart anti-corrosive" coating materials for mild steel as polyaniline undergoes a reduction reaction while the iron experiences oxidation forming a passive layer of iron oxide. This process causes an increase in resistance to corrosion over time by creating a passivation layer underneath the polymeric coating. This process is possible due to one of the signature features of polyaniline, which is its facile redox behavior when submitted to acid or alkaline media (Alam et al., 2008; Tallman et al., 1999). A visual sensor for the corrosion was fabricated using the concept of redox behavior of polyaniline by preparing a

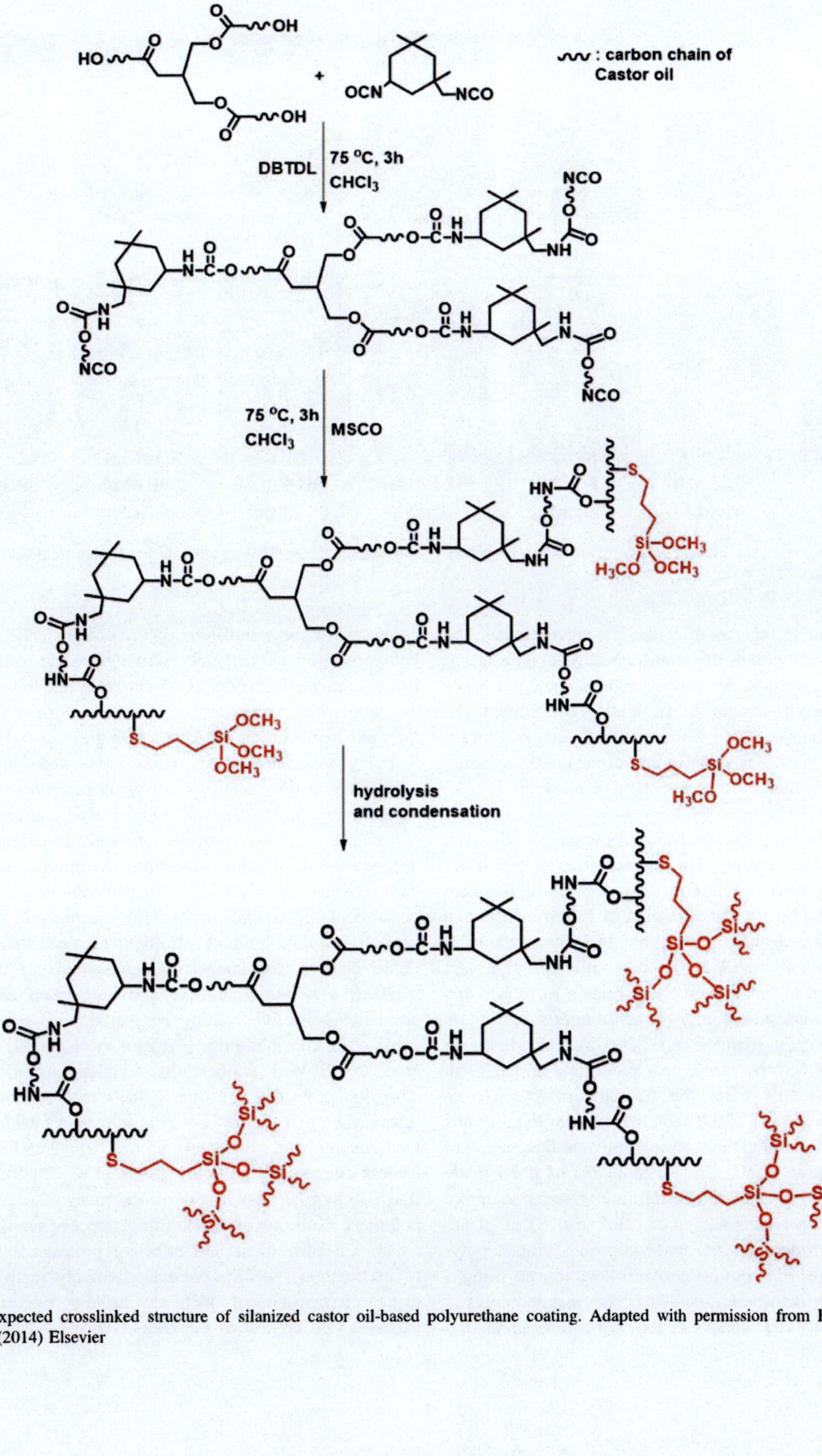

Fig. 15 Expected crosslinked structure of silanized castor oil-based polyurethane coating. Adapted with permission from Fu et al. (2014). Copyright (2014) Elsevier

Fig. 16 Salt spray test over different polymeric coatings on mild steel (Karmakar et al., 2018). Adapted with permission from Karmakar et al. (2018). Copyright (2017) Wiley

composite coating of castor oil-based polyurethane and polyaniline doped with methyl orange. Methyl orange acts as an acid-base sensor and the coating changes the color based on corrosive environments. In an acidic environment, the coating turns into dark red while in an alkaline medium it shows golden color. The visual effects of the color according to the type of corrosive environment are displayed in Fig. 17 (Alam et al., 2008).

Polypyrrole is another conducting polymer mostly used for anti-corrosive coating. The efficiency of polypyrrole-based coatings for corrosion protection is much higher even when it is used in a low concentration of 1% w/w (Iribarren et al., 2006). The corrosion protecting efficiency of the paint with polypyrrole was much higher than without it. The main factor attributed to better results was related to proper dispersion in the paint that allowed a smoother and more homogenous coating (Armelin et al., 2007a). Considering an industrial scenario, the rheological properties impart great importance. In this case, the addition of polypyrrole decreased the viscosity of the paint, but besides that, it must improve the adhesion to the surface to prevent the coating to drip out (Iribarren et al., 2006). Nanowires of polypyrrole and graphene were incorporated into the waterborne epoxy coating for corrosion protection of mild steel (Ding et al., 2019). Polypyrrole/graphene nanocomposites with a polypyrrole diameter of about 50 nm were synthesized using a chemical polymerization method. The nanocomposites showed about 100 times better corrosion protection compared to the waterborne epoxy coating. The improved corrosion protection of polypyrrole/graphene nanocomposites in the waterborne epoxy coating was due to blocking of the pores which decreases the diffusion of corrosive media to the mild steel as seen in Fig. 18 (Ding et al., 2019).

Poly(o-toluidine) along with ZnO were blended in waterborne polyurethane for corrosion protection of carbon steel (Zhang et al., 2019). It was reported that an optimum concentration of the poly(o-toluidine)/ZnO composite was needed for the corrosion protection. The optimum amount of ZnO provided improved corrosion protection on carbon steel because of the creation of passive coating and synergism between poly(o-toluidine) and ZnO which blocked the pores of the coating. Some reports demonstrated that the use of conductive polymers usually provides more satisfactory corrosion protective coating compared with general inorganic corrosion inhibitors (Armelin et al., 2009a; Marathe et al., 2015). Wide applications of conducting polymers for corrosion protection are due to their environment-friendly nature, the need for low concentration and facile synthesis methods to obtain satisfactory properties, which turns it into a more competitive approach (Shao et al., 2009). Polyaniline, polypyrrole, polythiophene, and many other conductive polymers can be synthesized through electrochemical routes, which are convenient and effective procedures to obtain robust coatings. The film of conductive polymer synthesized using electrochemical methods usually presents better adhesion and control of thickness (Armelin et al., 2009a).

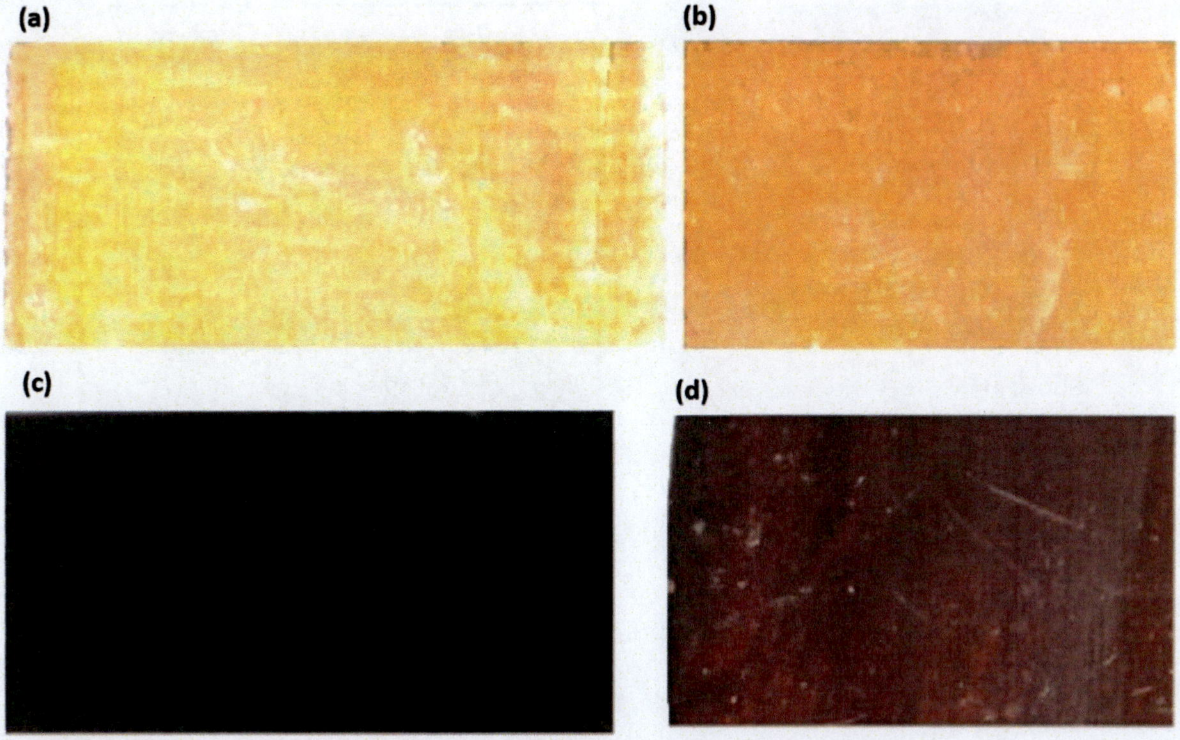

Fig. 17 Digital images of castor oil polyurethane/polyaniline doped with methyl orange composite, **a** neat sample, **b** under 5% NaOH, **c** under 5% HCl, and **d** under 3.5% NaCl. All under 480 h of exposure to the respective corrosive environment. Adapted with permission from Alam et al. (2008). Copyright (2017) Wiley

Coatings based on conductive polymers offers several advantages: (1) low-production cost makes them attractive for large-scale applications (Skotheim, 1997), (2) availability of facile synthesis methods, need of low concentration in coating requirements, and negligible rheological effect (Armelin et al., 2007b), (3) enhanced properties compared with regular inorganic compounds (Armelin et al., 2009a), (4) versatile procedures that allow proper control of the reaction media to design various composites, and (5) eco-friendly features (Armelin et al., 2007b, 2009b; Skotheim, 1997).

2.4 Coatings Based on Nanocomposites of Waterborne Polyurethanes

The inherent properties of the waterborne polyurethane coatings along with the large number of materials used to make composites demonstrate the versatility of the chemistry behind it. Other materials that can also be implemented to enhance the anti-corrosion protection are nano clay, boron nitride, silver, carbon nanotubes, and self-healing coating which will be described in the following section.

Clay is a type of phyllosilicate that is an inorganic polymeric mineral contains Na, Ca, Al, Mg, Si, O, and H_2O which are intercalated within its structure. The most common type of clay is Na^+–montmorillonite (Na^+–MMT) which is a high aspect ratio polymeric mineral presenting plate-like sheets of usually 1×200 nm (thickness $\times$ length) (Fig. 19). The clay presents a sandwiched structure where the edges are silica tetrahedral layers and there is an octahedral structure within the edge that can be made of Mg or Al hydroxide. The presence of Na^+ and Ca^{2+} in between the layers is responsible for improving the hydrophilicity of the clay (Chang et al., 2006). These cations can be replaced by organic cations such as quaternary ammonium cations (Chang et al., 2006; Yeh et al., 2002, 2006). The presence of clay in a coating improves mechanical, thermal, barrier effect along with flame retardant properties (Lan et al., 1994; Tyan et al., 1999; Wang & Pinnavaia, 1998; Yeh et al., 2002). These features in the clay are due to its unique morphology which is formed through layer intercalation or exfoliation of inorganic phases into the organic ones. Such intercalation or exfoliation improves the interaction between them, leading to a better barrier effect and preventing ions to permeate and enter in contact with the

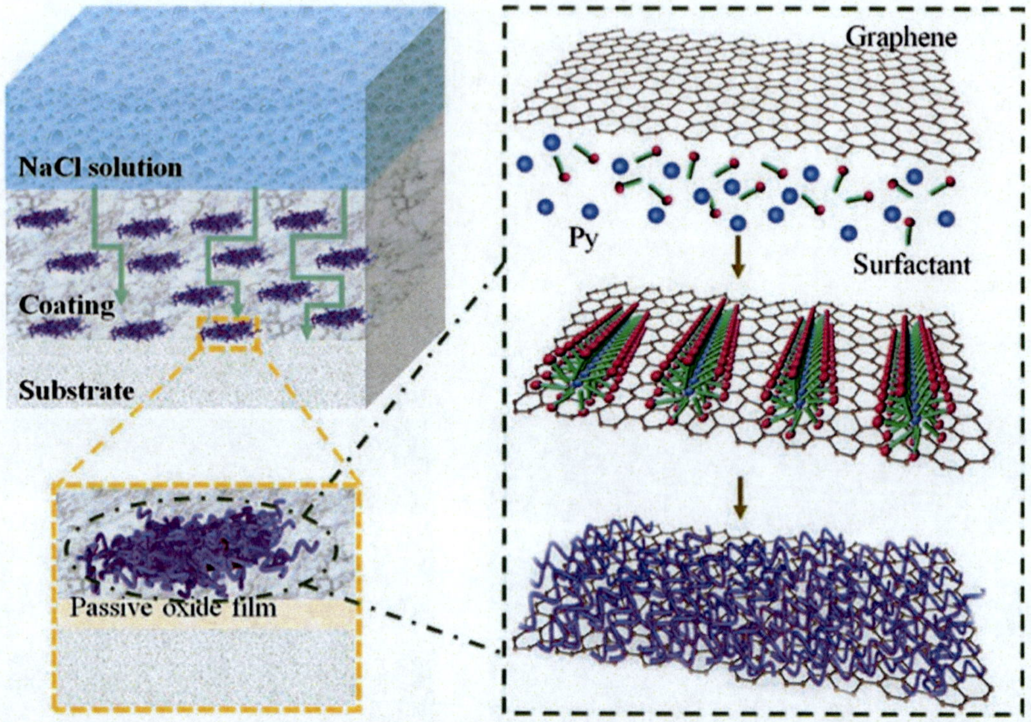

Fig. 18 Schematics of corrosion protection effect of polypyrrole/graphene nanocomposites in waterborne epoxy (Ding et al., 2019)

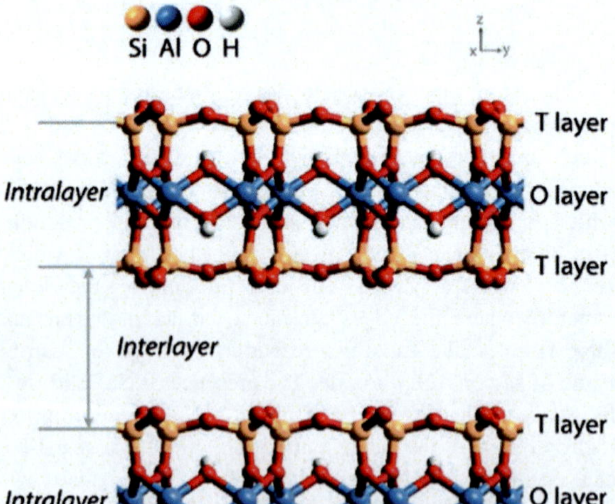

Fig. 19 Chemical structure of montmorillonite (MMT) (Liu et al., 2018b)

surface of the metal (Yeh et al., 2008). The plate shaped of the clay is a significant factor that enhances the impermeability of the coating allowing it to be used to protect against corrosion. The status quo is pushing the scientific community to find waterborne routes for safer and environmentally friendly synthesis. Clay nanocomposites found space on that type of procedure and have been synthesized by suspension and emulsion polymerization (Kim et al., 2002; Meneghetti & Qutubuddin, 2004; Wang et al., 2002).

Even though clay nanocomposites have not been largely studied, some reports have demonstrated that the addition of clay in composite coatings leads to a considerable improvement in anti-corrosion properties by preventing the permeation and diffusion of oxygen and moisture, which is associated with the tortuous surface structure (Chang et al., 2006; Yeh et al., 2008). On top of that, an increase in the working temperature of the coating is also reported, shifting the temperature from around 30–50 °C (Yeh et al., 2001). A previous study demonstrated the facile emulsion polymerization synthesis for a polyaniline doped with clay composite dispersed in water. First, a homogenous dispersion of dodecylbenzene sulfonic acid (DBSA) and aniline in water was prepared under stirring to form a stable dispersion of aniline-DBSA complex. Second, pre-dispersed Na^+–montmorillonite clay in water was added to the aniline-DBSA complex, followed by the addition of an initiator such as ammonium persulfate or iron chlorine to start the emulsion polymerization reaction to obtain the water dispersion composite coating (Yeh et al., 2008). An important aspect of this type of composite is that a low concentration of clay around 0.25–1 wt% was good enough to achieve satisfactory anti-corrosive properties (Yeh et al., 2001, 2008). As noted, clay nanocomposites for coating applications are a viable option due to their capability to enhance anti-corrosion properties even at low quantities in facile synthetic procedures that are simple, efficient, and green.

Boron nitride (BN), a hexagonally structured compound similar to graphene, demonstrates unique properties such as high length over low thickness, good mechanical strength, high barrier effect, and enhanced stability in both acid and alkaline environments which makes them very suitable for coating applications. (Li & Chen, 2016; Lin & Connell, 2012; Meng et al., 2014). It differs from graphene in the sense that BN is an insulator and has a more pronounced thermal stability that can reach around 800 °C (Bernard et al., 2016; Li et al., 2014). Another difference from graphene is the strong intermolecular interaction mostly in a planar axis instead of between layers like in graphene (Gao et al., 2017; Joshi et al., 2017). This planar interaction works as an effective barrier effect, which is one of the reasons to use BN-based coatings for corrosion protection. It has shown to be very effective for corrosion protection on metals like copper and nickel and displays better coating properties than graphene-based coating (Schriver et al., 2013). The BN was also used as a lubricant which imparts a remarkable improvement in the polyurethane coatings to abrasion resistance (Xing et al., 2016). With this set of attributes, the use of BN is targeted to improve anti-corrosion as well as mechanical, thermal, and abrasion properties (Li et al., 2018b). However, due to difficulties in dispersing BN in water, most of the time a chemical modification such as functionalization is required to disperse it in a waterborne polyurethane matrix. A previous report demonstrated the functionalization of BN by introducing hydrophilic groups in its structure through a facile reaction with a base as expressed in Fig. 20 (Li et al., 2018b). The presence of hydroxyl groups in BN facilitates its dispersion in water. The procedure was carried out by simply mixing the functionalized boron nitride with a waterborne polyurethane dispersion. The results showed that a low concentration of functionalized BN (around 2 wt%) was the optimum quantity for coatings due to low interference to the color, proper barrier effect, and remarkable abrasion resistance along with high thermal stability.

Carbon-based compounds such as carbon black, graphene, graphene oxide, reduced graphene oxide, and carbon nanotubes have been used for corrosion protection (Cai et al., 2018; Christopher et al., 2015; Gu et al., 2018; Haeri et al., 2017; Mo et al., 2015; Ramezanzadeh et al., 2015). Carbon nanotubes-based composites are used for corrosion protection due to their stability in most chemical environments, large surface area, high thermal, and mechanical properties (Vosgien Lacombre et al., 2017; Zhao et al., 2018). Some studies have demonstrated that the addition of multi-walled carbon nanotubes (MWCNT) provided a considerable increase in hydrophobicity due to the stack of layers on the nanotubes that inhibited water from permeating within the coating (Asmatulu et al., 2011; Wernik & Meguid, 2014). Also, a low concentration of MWCNT was good enough to provide high mechanical strength, scratch, and UV resistance properties (Asmatulu et al., 2011; Wernik & Meguid, 2014). However, due to high surface energy CNT are prone to agglomerate which deteriorates its effectiveness as corrosion protection. One of the approaches as previously mentioned in this chapter is to use polydopamine as a layer in between the substrate and the coating to enhance the adhesion of CNT to the surface. The poly(2-butyaniline) was also used as a dispersant for CNT to form an even distributed coating (Cui et al., 2018).

A recent work reports on the synthesis of CNT functionalized with polydopamine (PDA) using a facile method (Cai et al., 2018). The functionalized CNT was used as filler in a polyurethane matrix for corrosion inhibitor coating. The schematic for the synthesis is described in Fig. 21 (Cai et al., 2018). The CNT's functionalization with PDA plays an important role in improving the anti-corrosion properties because of its ability to nullify the drawbacks of neat CNT and provide high adhesion to the substrate. The reason for implementing PDA into the CNT's chain was to introduce hydroxyl and amine functionalities to reduces the inherent hydrophobicity of CNT. Simultaneously it also prompts a crosslinked structure that is more effective to block water from permeating to the substrate. Since PDA is an insulator it also prevents the formation of a galvanic system with the substrate, coating, and air. This is an important concern as conductive network of CNT acts as a cathode and the metal substrate behave like anode prompting galvanic corrosion.

Metal nanoparticles present a set of properties and can be used as fillers in a variety of matrixes for corrosion protection. Due to the smaller size of the nanoparticles, they can block the micropores in the coating to enhance the packing and barrier effect preventing the permeation of water and aggressive ions (Shi et al., 2009). Also, nanoparticles can be easily adsorbed in the polymeric matrix due to their high surface area, which helps to decrease the free space between the layers. Hence, they can create a sort of crosslink in the structure of the coating that prevents the permeation of corrosive agents (Dhoke & Khanna, 2009). One specific characteristic of metal nanoparticles in coatings is their antimicrobial properties (Chen et al., 2010; Hsu et al., 2010; Li et al., 2010; Memarzadeh et al., 2012; Palza et al., 2010). Studies demonstrated that waterborne polyurethanes functionalized with silver nanoparticles present one of the best antibacterial properties among other metals along with

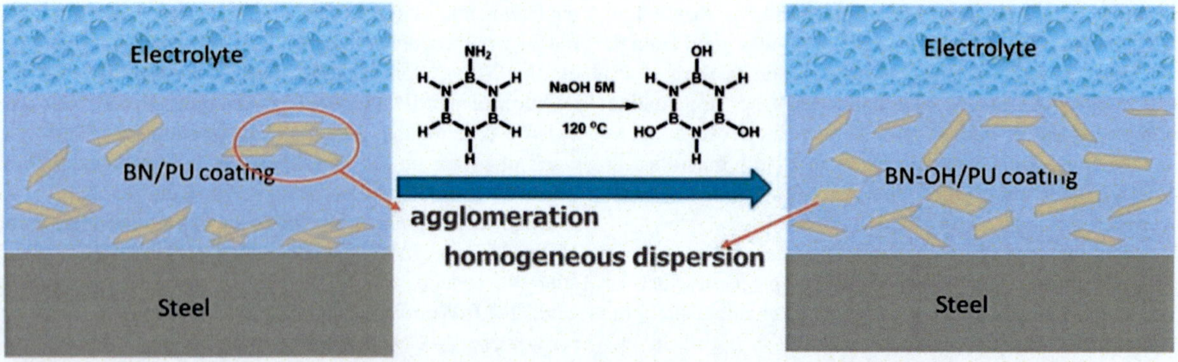

Fig. 20 Hydroxylation procedure of boron nitride. Adapted with permission from Li et al., (2018b). Copyright (2014) Elsevier

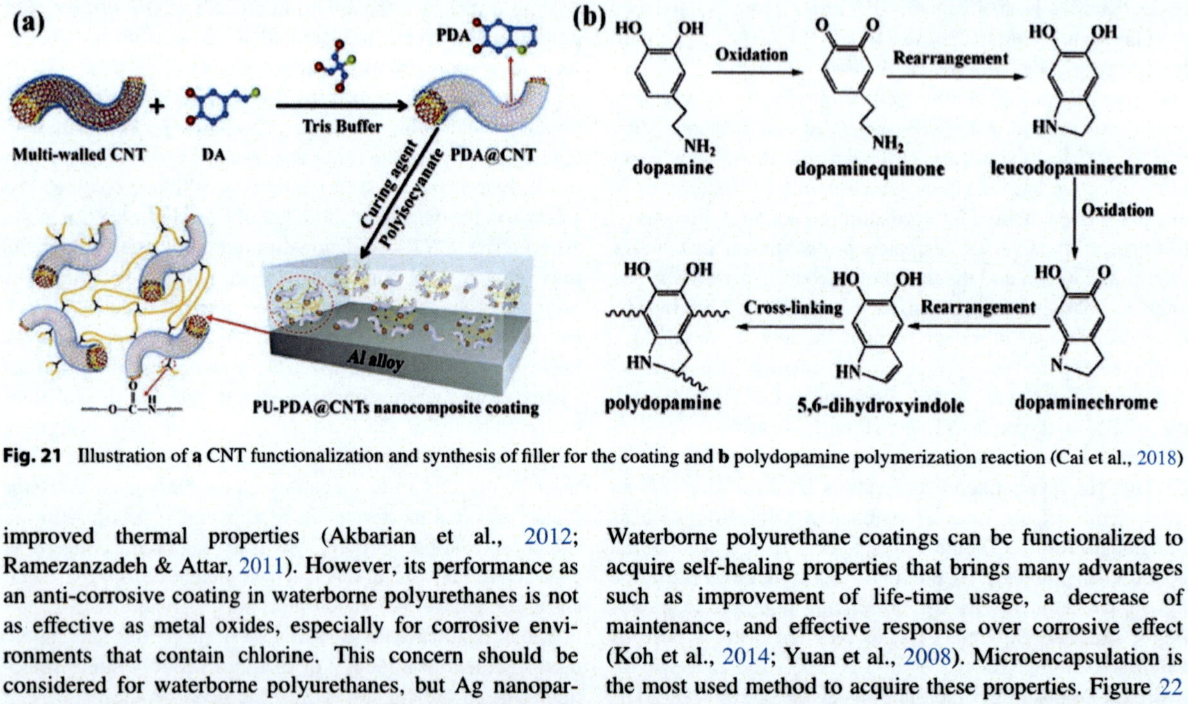

Fig. 21 Illustration of **a** CNT functionalization and synthesis of filler for the coating and **b** polydopamine polymerization reaction (Cai et al., 2018)

improved thermal properties (Akbarian et al., 2012; Ramezanzadeh & Attar, 2011). However, its performance as an anti-corrosive coating in waterborne polyurethanes is not as effective as metal oxides, especially for corrosive environments that contain chlorine. This concern should be considered for waterborne polyurethanes, but Ag nanoparticles blended with high solid polyurethane (HSP) have presented a relatively stable coating. It is believed the interaction between silver cations with carbonyl groups may create stable complexes in a crosslinked-like structure, that are more abundant in HSP, which prevents the interaction with chlorine anions. Nevertheless, the anti-corrosion properties remained similar if compared with the neat polyurethane (Akbarian et al., 2014).

As demonstrated so far that there are many ways to impart corrosion inhibition properties in a waterborne polyurethane coating by selecting the anti-corrosive components and proper design of the experiment to guarantee an even distribution of filler into the polymeric matrix as well as proper packing of the structure to enhance the barrier effect. Waterborne polyurethane coatings can be functionalized to acquire self-healing properties that brings many advantages such as improvement of life-time usage, a decrease of maintenance, and effective response over corrosive effect (Koh et al., 2014; Yuan et al., 2008). Microencapsulation is the most used method to acquire these properties. Figure 22 describes a general example of a self-healing coating mechanism through this approach.

A literature survey suggests that microcapsules such as polyurethane capsules (Hu et al., 2012; Koh et al., 2014; Najjar et al., 2018), phenol, or urea–formaldehyde (Brown et al., 2003; Jadhav et al., 2011; Liu et al., 2009) and silica capsules (Cho et al., 2009; Sondari et al., 2010) are very effective in coating applications particularly for self-healing coatings. Some examples of healing agents are saturated vegetable oils, low viscosity epoxies, catalysts, inorganic inhibitors, and aliphatic diisocyanates (Kim et al., 2016; Suryanarayana et al., 2008; Yuan et al., 2008). The latter can readily react with moisture from the environment to form urea linkages and due to low viscosity, it can flow quickly to

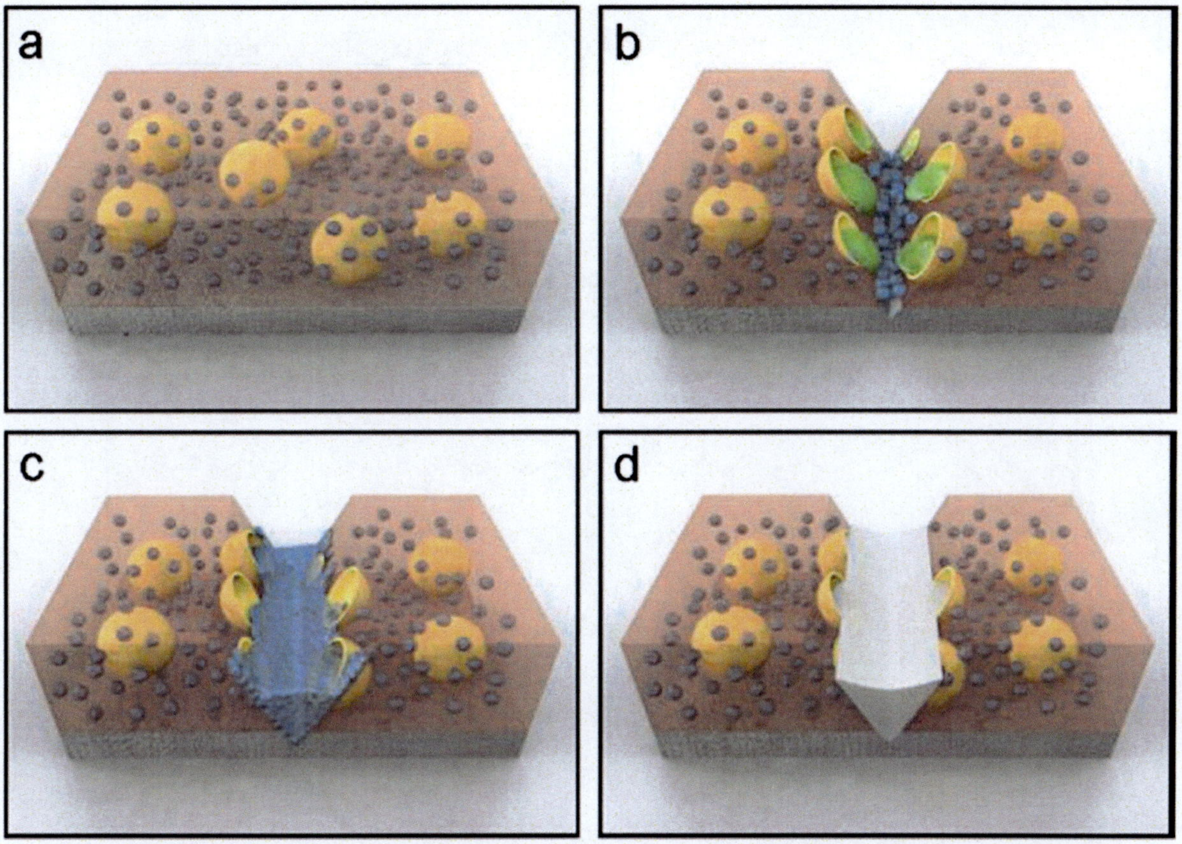

Fig. 22 Self-healing process demonstrating **a** coating before the damage. **b** healing agents (blue) concentrating over the crack along with the release of catalyst (green) from within the microcapsule (yellow). **c** the reaction between the healing agent and catalyst. **d** coating after the healing process. Adapted with permission from Cho et al. (2009). Copyright (2009) Wiley

the damaged area to repair it. The recovery damage effectiveness is the main characteristic of such materials which depends on factors such as rheological behavior, the time required for the action, healing agent/polymer ratio, and environment (Najjar et al., 2018; Yang et al., 2008). A study reports on the synthesis of polyurethane microcapsule based on toluene diisocyanate (TDI) and 1,4 butanediol as starting materials for the prepolymer (Yang et al., 2008). Isophorone diisocyanate (IPDI) was used as an encapsulated healing agent which is an aliphatic isocyanate that reacts with water to form urea linkages and introduces resistance to UV light. The procedure to prepare the microencapsulation consisted of the dissolution of the prepolymer in chlorobenzene around 70 °C followed by the addition of a healing agent. The mixture was then added to an Arabic gum solution where 1,4 butanediol was added as a chain extender to form polyurethane shells. The microsphere's size between 40 and 400 μm was obtained by controlling the stirring rate between 500 and 1500 rpm. The schematics of the reaction are shown in Fig. 23 (Yang et al., 2008). A sequence of images obtained through a scanning electron microscope showed that the higher stirring rate produces smaller microspheres (Fig. 24).

Other study synthesized self-healing coating composed of a waterborne polyurethane as prepolymer, a silica shell, tetraethoxy orthosilicate (TEOS), to improve thermal and adhesion properties, and 2-mercaptobenzothiazole (MBT) as a corrosion inhibitor (Mirmohseni et al., 2019). The implementation of silica also added a pH effect response in a way that upon increasing or decreasing the pH the microsphere disrupts and releases the corrosion inhibitor in the media, providing automatic control. The previous report demonstrated a self-healing eco-friendly polyurethane coating that worked through disulfide bonds in the main chain (Wan & Chen, 2017). The component 2-hydroxyehtyl disulfide was responsible for implementing the self-healing property where the reversibility of S–S bonds provided a 90.5% damage repair after curing at 65 °C for 10 min. The mechanism of the self-healing process is provided in Fig. 25 (Wan & Chen, 2017).

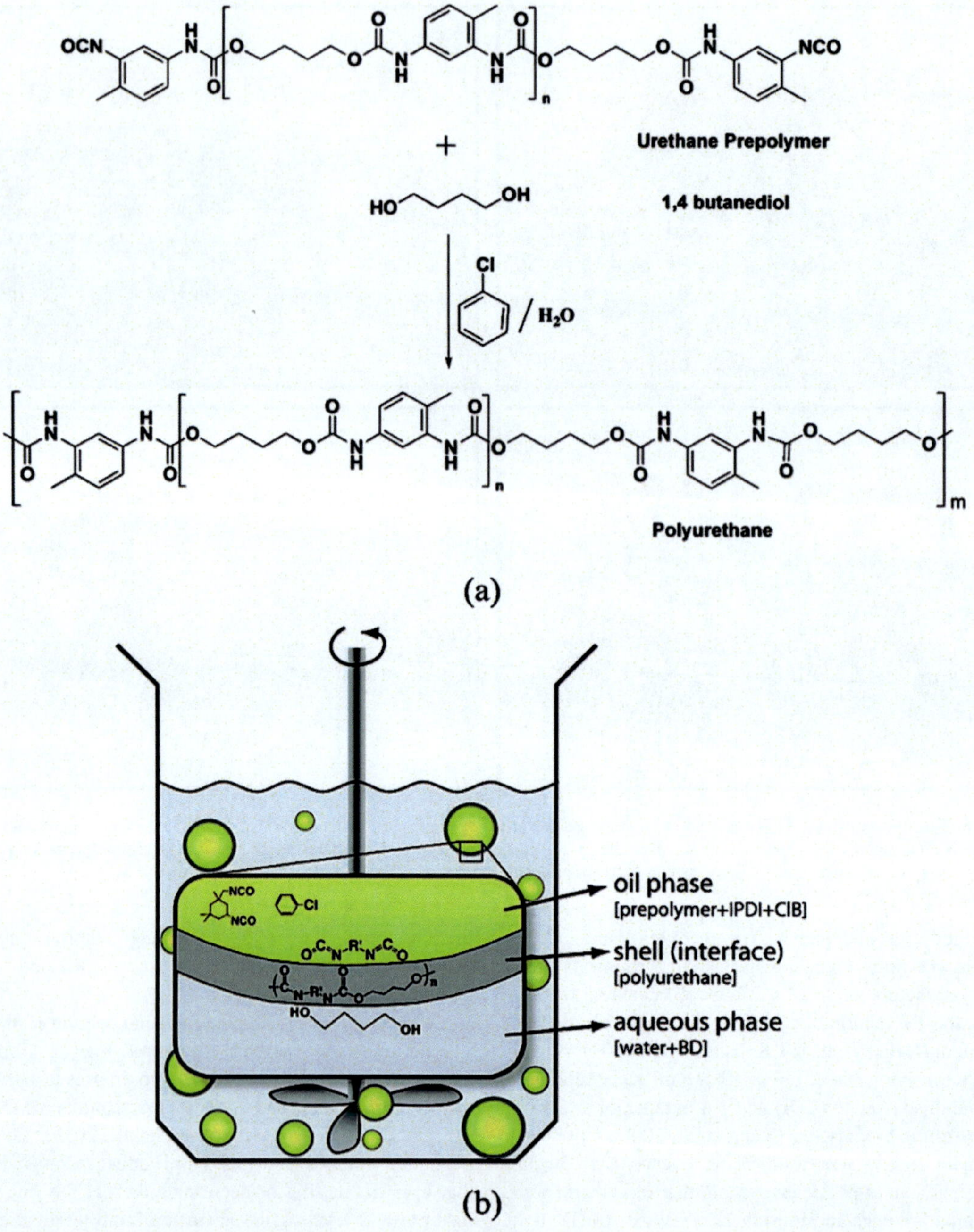

Fig. 23 **a** Reaction for microencapsulation through interfacial polymerization **b** polyurethane shell formation under mechanical stirring. Adapted with permission from Yang et al. (2008). Copyright (2008) American Chemical Society

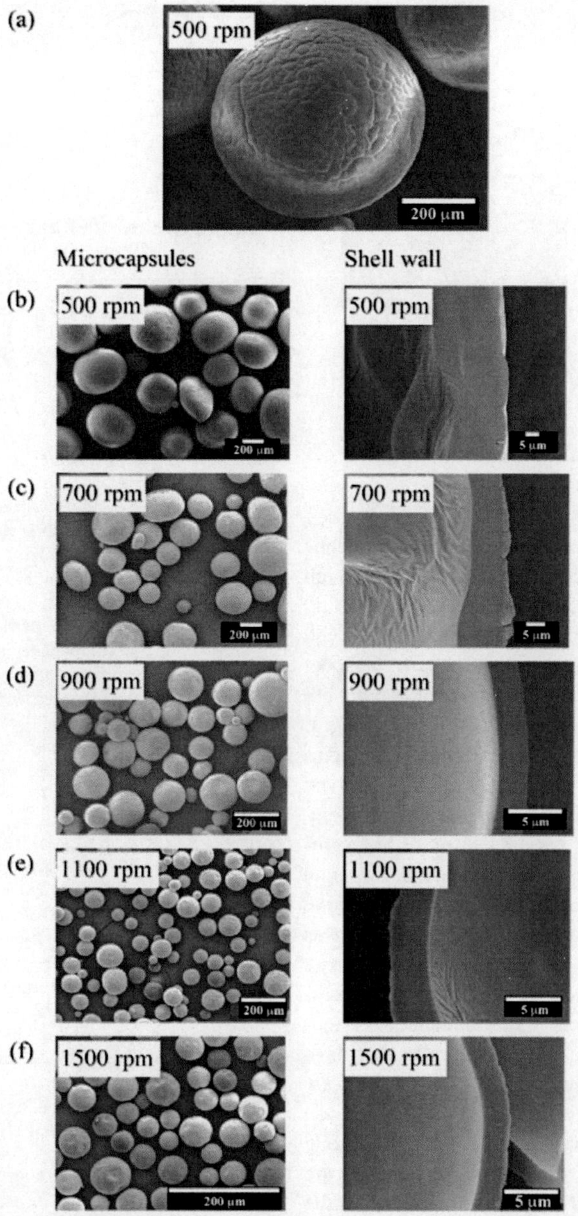

Fig. 24 Micrographs showing the relation between the stirring rate over the diameter of the particle. Adapted with permission from Yang et al. (2008). Copyright (2008) American Chemical Society

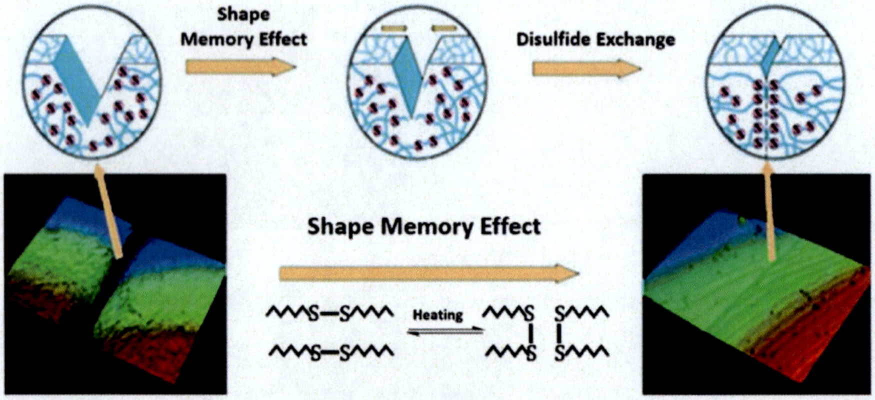

Fig. 25 Self-healing mechanism through disulfide exchange. Adapted with permission from Wan and Chen (2017). Copyright (2016) Springer Nature

3 Conclusion and Future Lookout

In conclusion to this chapter, it is notable that polyurethane coatings present a high demand in the market, which encourages researchers to create novel procedures and find new materials that have not been fully explored yet. Waterborne polyurethane coatings brought a new perspective to the concept of sustainability since they enlarge the life span of most metal surfaces preventing them from going to waste which has a big economic impact. From the beginning, the use of chromium-based coatings partially solved the corrosion problem but are harmful to the environment. The introduction of waterborne polyurethane not only provided a safer route but more enhanced coatings that present a complete set of properties that can be designed in innovative ways. The use of new and eco-friendly materials such as graphene (high hydrophobicity and mechanical property), silicon (thermal stability), conductive polymers (high abrasion and chemical stability), metal oxide nanoparticles (enhanced barrier effect) are some of few examples that have barely scratched the surface of the many possibilities that are yet to be discovered. Therefore, composites based on waterborne polyurethanes are an extremely versatile group of materials that have proven their effectiveness and are among the group of polymers with great importance for the economy.

References

Abedi Esfahani, E., Salimijazi, H., Golozar, M. A., Mostaghimi, J., & Pershin, L. (2012). Study of corrosion behavior of arc sprayed aluminum coating on mild steel. *Journal of Thermal Spray Technology, 21*, 1195–1202. https://doi.org/10.1007/s11666-012-9810-x

Abu, Y. M., & Aoki, K. (2005). Corrosion protection by polyaniline-coated latex microspheres. *Journal of Electroanalytical Chemistry, 583*(1), 133–139. https://doi.org/10.1016/j.jelechem.2005.05.014

Ahmad, Z. (2006). *Principles of corrosion engineering and corrosion control*. Elsevier.

Akbarian, M., Olya, M. E., Ataeefard, M., & Mahdavian, M. (2012). The influence of nanosilver on thermal and antibacterial properties of a 2K waterborne polyurethane coating. *Progress in Organic Coatings, 75*(4), 344–348. https://doi.org/10.1016/j.porgcoat.2012.07.017

Akbarian, M., Olya, M. E., Mahdavian, M., & Ataeefard, M. (2014). Effects of nanoparticulate silver on the corrosion protection performance of polyurethane coatings on mild steel in sodium chloride solution. *Progress in Organic Coatings, 77*(8), 1233–1240. https://doi.org/10.1016/j.porgcoat.2014.03.023

Alam, J., Riaz, U., & Ahmad, S. (2008). Development of nanostructured polyaniline dispersed smart anticorrosive composite coatings. *Polymers for Advanced Technologies, 19*, 882–888. https://doi.org/10.1002/pat.1054

Armelin, E., Oliver, R., Liesa, F., Iribarren, J. I., Estrany, F., & Alemán, C. (2007a). Marine paint fomulations: conducting polymers as anticorrosive additives. *Progress in Organic Coatings, 59*(1), 46–52. https://doi.org/10.1016/j.porgcoat.2007.01.013

Armelin, E., Ocampo, C., Liesa, F., Iribarren, J. I., Ramis, X., & Alemán, C. (2007b). Study of epoxy and alkyd coatings modified with emeraldine base form of polyaniline. *Progress in Organic Coatings, 58*(4), 316–322. https://doi.org/10.1016/j.porgcoat.2007.01.005

Armelin, E., Meneguzzi, Á., Ferreira, C. A., & Alemán, C. (2009a). Polyaniline, polypyrrole and poly(3,4-ethylenedioxythiophene) as additives of organic coatings to prevent corrosion. *Surface and Coatings Technology, 203*(24), 3763–3769. https://doi.org/10.1016/j.surfcoat.2009.06.019

Armelin, E., Alemán, C., & Iribarren, J. I. (2009b). Anticorrosion performances of epoxy coatings modified with polyaniline: A comparison between the emeraldine base and salt forms. *Progress in Organic Coatings, 65*(1), 88–93. https://doi.org/10.1016/j.porgcoat.2008.10.001

Asmatulu, R., Mahmud, G. A., Hille, C., & Misak, H. E. (2011). Effects of UV degradation on surface hydrophobicity, crack, and thickness of MWCNT-based nanocomposite coatings. *Progress in Organic Coatings, 72*(3), 553–561. https://doi.org/10.1016/j.porgcoat.2011.06.015

Bauer, R. S. (1979). *Epoxy resin chemistry*. American Chemical Society.

Baukh, V., Huinink, H. P., Adan, O. C. G., Erich, S. J. F., & van der Ven, L. G. J. (2011). Water-polymer interaction during water uptake. *Macromolecules, 44*, 4863–4871. https://doi.org/10.1021/ma102889u

Behniafar, H., Alimohammadi, M., & Malekshahinezhad, K. (2015). Transparent and flexible films of new segmented polyurethane nanocomposites incorporated by NH2-functionalized TiO2 nanoparticles. *Progress in Organic Coatings, 88*, 150–154. https://doi.org/10.1016/j.porgcoat.2015.06.030

Bernard, S., Salameh, C., & Miele, P. (2016). Boron nitride ceramics from molecular precursors: Synthesis, properties and applications. *Dalt Trans, 45*, 861–873. https://doi.org/10.1039/C5DT03633J

Bludau, H., Czapar, A. E., Pitek, A. S., Shukla, S., Jordan, R., & Steinmetz, N. F. (2017). POxylation as an alternative stealth coating for biomedical applications. *European Polymer Journal, 88*, 679–688. https://doi.org/10.1016/j.eurpolymj.2016.10.041

Brown, E. N., Kessler, M. R., Sottos, N. R., & White, S. R. (2003). In situ poly(urea-formaldehyde) microencapsulation of dicyclopentadiene. *Journal of Microencapsulation, 20*, 719–730. https://doi.org/10.3109/02652040309178083

Cai, G., Hou, J., Jiang, D., & Dong, Z. (2018). Polydopamine-wrapped carbon nanotubes to improve the corrosion barrier of polyurethane coating. *RSC Advances, 8*, 23727–23741. https://doi.org/10.1039/C8RA03267J

Chandrasekhar, P. (1999). *Conducting polymers, fundamentals and applications*. Springer.

Chang, K.-C., Lai, M.-C., Peng, C.-W., Chen, Y.-T., Yeh, J.-M., Lin, C.-L., & Yang, J.-C. (2006). Comparative studies on the corrosion protection effect of DBSA-doped polyaniline prepared from in situ emulsion polymerization in the presence of hydrophilic Na+-MMT and organophilic organo-MMT clay platelets. *Electrochimica Acta, 51*(26), 5645–5653. https://doi.org/10.1016/j.electacta.2006.02.039

Charpentier, P. A., Burgess, K., Wang, L., Chowdhury, R. R., Lotus, A. F., & Moula, G. (2012). Nano-TiO2/polyurethane composites for antibacterial and self-cleaning coatings. *Nanotechnology, 23*, 425606. https://doi.org/10.1088/0957-4484/23/42/425606

Chen, A. T., & Wojcik, R. T. (2000). Polyurethane coatings for metal and plastic substrates. *Metal Finishing, 98*(6), 143–154. https://doi.org/10.1016/S0026-0576(00)80405-5

Chen, Y., Zhou, S., Yang, H., & Wu, L. (2005). Structure and properties of polyurethane/nanosilica composites. *Journal of Applied Polymer Science, 95*, 1032–1039. https://doi.org/10.1002/app.21180

Chen, T.-N., Wuu, D.-S., Wu, C.-C., Chiang, C.-C., Chen, Y.-P., & Horng, R.-H. (2007). Improvements of permeation barrier coatings using encapsulated parylene interlayers for flexible electronic applications. *Plasma Processes and Polymers, 4*, 180–185. https://doi.org/10.1002/ppap.200600158

Chen, Y., Yan, L., Wang, R., Fan, H., & Zhang, Q. (2010). Antimicrobial polyurethane synthetic leather coating with in-situ generated Nano-TiO2. *Fibers and Polymers, 11*, 689–694. https://doi.org/10.1007/s12221-010-0689-1

Chen, X., Yuan, J., Huang, J., Ren, K., Liu, Y., Lu, S., & Li, H. (2014). Large-scale fabrication of superhydrophobic polyurethane/nano-Al2O3 coatings by suspension flame spraying for anti-corrosion applications. *Applied Surface Science, 311*, 864–869. https://doi.org/10.1016/j.apsusc.2014.05.186

Cho, S. H., White, S. R., & Braun, P. V. (2009). Self-healing polymer coatings. *Advanced Materials, 21*, 645–649. https://doi.org/10.1002/adma.200802008

Chou, Y. C., & Lee, L. J. (1995). Mechanical properties of polyurethane-unsaturated polyester interpenetrating polymer networks. *Polymer Engineering and Science, 35*, 976–988. https://doi.org/10.1002/pen.760351204

Christopher, G., Kulandainathan, M. A., & Harichandran, G. (2015). Comparative study of effect of corrosion on mild steel with waterborne polyurethane dispersion containing graphene oxide versus carbon black nanocomposites. *Progress in Organic Coatings, 89*, 199–211. https://doi.org/10.1016/j.porgcoat.2015.09.022

Cotugno, S., Larobina, D., Mensitieri, G., Musto. P., & Ragosta, G. (2001). A novel spectroscopic approach to investigate transport processes in polymers: the case of water–epoxy system. *Polymer (Guildf), 42*, 6431–6438. https://doi.org/10.1016/S0032-3861(01)00096-9

Cui, M., Ren, S., Qiu, S., Zhao, H., Wang, L., & Xue, Q. (2018). Non-covalent functionalized multi-wall carbon nanotubes filled epoxy composites: Effect on corrosion protection and tribological performance. *Surface and Coatings Technology, 340*, 74–85. https://doi.org/10.1016/j.surfcoat.2018.02.045

Cui, J., Xu, J., Li, J., Qiu, H., Zheng, S., & Yang, J. (2020). A crosslinkable graphene oxide in waterborne polyurethane anticorrosive coatings: Experiments and simulation. *Composites Part B Engineering, 188*, 107889. https://doi.org/10.1016/j.compositesb.2020.107889

Czub, P. (2009). Synthesis of high-molecular-weight epoxy resins from modified natural oils and Bisphenol A or Bisphenol A-based epoxy resins. *Polymers for Advanced Technologies, 20*, 194–208. https://doi.org/10.1002/pat.1252

Dhoke, S. K., & Khanna, A. S. (2009). Effect of nano-Fe2O3 particles on the corrosion behavior of alkyd based waterborne coatings. *Corrosion Science, 51*, 6–20. https://doi.org/10.1016/j.corsci.2008.09.028

Dhoke S. K., Khanna, A. S., & Sinha, T. J. M. (2009). Effect of nano-ZnO particles on the corrosion behavior of alkyd-based waterborne coatings. *Progress in Organic Coatings, 64*, 371–382. https://doi.org/10.1016/j.porgcoat.2008.07.023

Ding, Y., Zhong, J., Xie, P., Rong, J., Zhu, H., Zheng, W., Wang, J., Gao, F., Shen, L., He, H., & Cheng, Z. (2019). Protection of mild steel by waterborne epoxy coatings incorporation of polypyrrole nanowires/graphene nanocomposites. *Polymers (Basel), 11*(12). https://doi.org/10.3390/polym11121998

Dumas, L., Bonnaud, L., Olivier, M., Poorteman, M., & Dubois, P. (2015). Eugenol-based benzoxazine: From straight synthesis to taming of the network properties. *Journal of Material Chemistry A, 3*, 6012–6018. https://doi.org/10.1039/C4TA06636G

El Faydy, M., Galai, M., Touhami, M. E., Obot, I. B., Lakhrissi B., & Zarrouk, A. (2017). Anticorrosion potential of some 5-amino-8-hydroxyquinolines derivatives on carbon steel in hydrochloric acid solution: Gravimetric, electrochemical, surface morphological, UV–visible, DFT and Monte Carlo simulations. *Journal of Molecular Liquids, 248*, 1014–1027. https://doi.org/10.1016/j.molliq.2017.10.125

Fu, C., Yang, Z., Zheng, Z., & Shen, L. (2014). Properties of alkoxysilane castor oil synthesized via thiol-ene and its polyurethane/siloxane hybrid coating films. *Progress in Organic Coatings, 77*(8), 1241–1248. https://doi.org/10.1016/j.porgcoat.2014.03.020

Fu, Z., Hejl, A., Swartz, A., Beshah, K., & Dombrowski, G. (2017). Film formation through designed diffusion technology. *Protective Coatings*, 153–168.

Gao, C., Guo, G., Zhang, G., Wang. Q., Wang, T., & Wang, H. (2017). Formation mechanisms and functionality of boundary films derived from water lubricated polyoxymethylene/hexagonal boron nitride nanocomposites. *Material and Design 115*, 276–286. https://doi.org/10.1016/j.matdes.2016.11.016

González, S., Mirza Rosca, I. C., & Souto, R. M. (2001). Investigation of the corrosion resistance characteristics of pigments in alkyd coatings on steel. *Progress in Orgiancs Coatings, 43*(4), 282–285. https://doi.org/10.1016/S0300-9440(01)00210-7

Grundmeier, G., Rossenbeck, B., Roschmann, K. J., Ebbinghaus, P., & Stratmann, M. (2006). Corrosion protection of Zn-phosphate containing water borne dispersion coatings: Part 2: Investigations of the corrosive de-adhesion of model latex coatings on iron.

Corrosion Science, 48(11), 3716–3730. https://doi.org/10.1016/j.corsci.2006.01.007

Gu, B.-E., Huang, C.-Y., Shen, T.-H., & Lee, Y.-L. (2018). Effects of multiwall carbon nanotube addition on the corrosion resistance and underwater acoustic absorption properties of polyurethane coatings. *Progress in Organic Coatings, 121*, 226–235. https://doi.org/10.1016/j.porgcoat.2018.04.033

Guiffard, B., Seveyrat, L., Sebald, G., & Guyomar, D. (2006). Enhanced electric field-induced strain in non-percolative carbon nanopowder/polyurethane composites. *Journal of Physics. D. Applied Physics, 39*, 3053–3057. https://doi.org/10.1088/0022-3727/39/14/027

Gurunathan, T., Rao, C. R. K., Narayan, R., & Raju, K. V. S. N. (2013). Polyurethane conductive blends and composites: Synthesis and applications perspective. *Journal of Materials Science, 48*, 67–80. https://doi.org/10.1007/s10853-012-6658-x

Haeri, S. Z., Asghari, M., & Ramezanzadeh, B. (2017). Enhancement of the mechanical properties of an epoxy composite through inclusion of graphene oxide nanosheets functionalized with silica nanoparticles through one and two steps sol-gel routes. *Progress in Organic Coatings, 111*, 1–12. https://doi.org/10.1016/j.porgcoat.2017.05.003

Han, S. O., & Drzal, L. T. (2003). Water absorption effects on hydrophilic polymer matrix of carboxyl functionalized glucose resin and epoxy resin. *European Polymer of Journal, 39*, 1791–1799. https://doi.org/10.1016/S0014-3057(03)00099-5

Havemann, R. H., Gnade, B. E., & Cho, C.-C. (1995). *Method of fabricating porous dielectric material with a passivation layer for electronics applications*

Hesselmans, L. C. J., Derksen, A. J., & van den Goorbergh, J. A. M. (2006). Polycarbodiimide crosslinkers. *Progress in Organics Coatings, 55*(2), 142–148. https://doi.org/10.1016/j.porgcoat.2005.08.011

Hsu, S., Tseng, H.-J., Lin, Y.-C. (2010). The biocompatibility and antibacterial properties of waterborne polyurethane-silver nanocomposites. *Biomaterials, 31*(26), 6796–6808. https://doi.org/10.1016/j.biomaterials.2010.05.015

Hsu, P.-C., Wu, H., Carney, T. J., McDowell, M. T., Yang, Y., Garnett, E. C., Li, M., Hu, L., & Cui, Y. (2012). Passivation coating on electrospun copper nanofibers for stable transparent electrodes. *ACS Nano, 6*, 5150–5156. https://doi.org/10.1021/nn300844g

Hu, J., Meng, H., Li, G., & Ibekwe, S. I. (2012). A review of stimuli-responsive polymers for smart textile applications. *Smart Materials and Structures, 21*, 53001. https://doi.org/10.1088/0964-1726/21/5/053001

Hua, Y., Li, X., Ma, L., Wang, Y., Fang, H., Wei, H., & Ding, Y. (2019). Self-healing mineralization and enhanced anti-corrosive performance of polyurethane CaCO3 composite film via β-CD induction. *Materials and Design, 177*, 107856. https://doi.org/10.1016/j.matdes.2019.107856

Iribarren, J. I., Armelin, E., Liesa, F., Casanovas, J., & Alemán, C. (2006). On the use of conducting polymers to improve the resistance against corrosion of paints based on polyurethane resins. *Materials and Corrosion, 57*, 683–688. https://doi.org/10.1002/maco.200503952

Jadhav, R. S., Hundiwale, D. G., & Mahulikar, P. P. (2011). Synthesis and characterization of phenol–formaldehyde microcapsules containing linseed oil and its use in epoxy for self-healing and anticorrosive coating. *Journal of Applied Polymer Science, 119*, 2911–2916. https://doi.org/10.1002/app.33010

Jamali, S. S. (2013). *Study on corrosion protection of organic coatings using electrochemical techniques: Developing electrochemical noise method, effective of surface preparation and inhomogeneity of organic coatings.*

Joshi, M. D., Goyal, A., Patil, S. M., & Goyal, R. K. (2017). Tribological and thermal properties of hexagonal boron nitride filled high-performance polymer nanocomposites. *Journal of Applied Polymer Science, 134*(4).

Karmakar, H. S., Arukula, R., Thota, A., Narayan, R., & Rao, C. R. K. (2018). Polyaniline-grafted polyurethane coatings for corrosion protection of mild steel surfaces. *Journal of Applied Polymer Science, 135*, 45806. https://doi.org/10.1002/app.45806

Keddie, J., & Routh, A. F. (2010). *Fundamentals of latex film formation: processes and properties*. Springer Science and Business Media

Kendig, M., Jeanjaquet, S., Addison, R., & Waldrop, J. (2001). Role of hexavalent chromium in the inhibition of corrosion of aluminum alloys. *Surface and Coatings Technologies, 140*(1), 58–66. https://doi.org/10.1016/S0257-8972(01)01099-4

Kim, Y. K., Choi, Y. S., Wang, K. H., & Chung, I. J. (2002). Synthesis of exfoliated PS/Na−MMT nanocomposites via emulsion polymerization. *Chemistry of Materials, 14*, 4990–4995. https://doi.org/10.1021/cm020324h

Kim, S.-R., Getachew, B. A., Park, S.-J., Kwon, O.-S., Ryu, W.-H., Taylor, A. D., Bae, J., & Kim, J.-H. (2016). Toward microcapsule-embedded self-healing membranes. *Environmental Science and Technology Letters, 3*, 216–221. https://doi.org/10.1021/acs.estlett.6b00046

Koch, G., Varney, J., Thopson, N., Moghissi, O., Gould, M., & Payer, J. (2016). *International measures of prevention, application, and economics of corrosion technologies study*. Houston, Texas, USA

Koh, E., Kim, N.-K., Shin, J., & Kim, Y.-W. (2014). Polyurethane microcapsules for self-healing paint coatings. *RSC Advances, 4*, 16214–16223. https://doi.org/10.1039/C4RA00213J

Kong, B., Sun, Z.-W., Lang, C.-I., Tong, J., & Chiang, T., (2012). *Formation of a zinc passivation layer on titanium or titanium alloys used in semiconductor processing*

Lan, T., Kaviratna, P. D., & Pinnavaia, T. J. (1994). On the nature of polyimide-clay hybrid composites. *Chemistry of Materials, 6*, 573–575. https://doi.org/10.1021/cm00041a002

Lee, H., & Neville, K. (1967). Book review-handbook of epoxy resins. *Industrial and Engineering Chemistry, 59*(9), 16–17.

Lee, D. S., Kang, W. K., An, J. H., & Kim, S. C. (1992). Gas transport in polyurethane-polystyrene interpenetrating polymer network membranes. II. Effect of crosslinked state and annealing. *Journal of Membrane Science, 75*, 15–27. https://doi.org/10.1016/0376-7388(92)80003-3

Lee, H., Dellatore, S. M., Miller, W. M., Messersmith, P. B. (2007). Mussel-inspired surface chemistry for multifunctional coatings. *Science, 318*, 426–430. https://doi.org/10.1126/science.1147241

Lendvay-Győrik, G., Pajkossy, T., & Lengyel, B. (2007). Water uptake of water-borne paint resin films as studied by impedance spectroscopy and gravimetry. *Progress in Organics Coatings, 59*(1), 95–99. https://doi.org/10.1016/j.porgcoat.2007.01.015

Li, L. H., & Chen, Y. (2016). Atomically thin boron nitride: Unique properties and applications. *Advanced Functional Materials, 26*, 2594–2608. https://doi.org/10.1002/adfm.201504606

Li, C., Han, J., Ryu, C. Y., & Benicewicz, B. C. (2006). A versatile method to prepare RAFT agent anchored substrates and the preparation of PMMA grafted nanoparticles. *Macromolecules, 39*, 3175–3183. https://doi.org/10.1021/ma051983t

Li, X. H., Xing, Y. G., Li, W. L., Jiang, Y. H., & Ding, Y. L. (2010). Antibacterial and physical properties of poly(vinyl chloride)-based film coated with ZnO nanoparticles. *Food Science and Technology International, 16*, 225–232. https://doi.org/10.1177/1082013209353986

Li, L. H., Cervenka, J., Watanabe, K., Taniguchi, T., & Chen, Y. (2014). Strong oxidation resistance of atomically thin boron nitride nanosheets. *ACS Nano, 8*, 1457–1462. https://doi.org/10.1021/nn500059s

Li, W., Shang, T., Yang, W., Yang, H., Lin, S., Jia, X., Cai, Q., & Yang, X. (2016). Effectively exerting the reinforcement of dopamine reduced graphene oxide on epoxy-based composites via strengthened interfacial bonding. *ACS Applied Materials and Interfaces, 8*, 13037–13050. https://doi.org/10.1021/acsami.6b02496

Li, Z., Zhou, G., Dai, H., Yang, M., Fu, Y., Ying, Y., & Li, Y. (2018a). Biomineralization-mimetic preparation of hybrid membranes with ultra-high loading of pristine metal–organic frameworks grown on silk nanofibers for hazard collection in water. *Journal of Materials Chemistry A, 6*, 3402–3413. https://doi.org/10.1039/C7TA06924C

Li, J., Gan, L., Liu, Y., Mateti, S., Lei, W., Chen, Y., & Yang, J. (2018b). Boron nitride nanosheets reinforced waterborne polyurethane coatings for improving corrosion resistance and antifriction properties. *European Polymer Journal, 104*, 57–63. https://doi.org/10.1016/j.eurpolymj.2018.04.042

Lin, Y., & Connell, J. W. (2012). Advances in 2D boron nitride nanostructures: Nanosheets, nanoribbons, nanomeshes, and hybrids with graphene. *Nanoscale, 4*, 6908–6939. https://doi.org/10.1039/C2NR32201C

Liu, Z., Erhan, S. Z., & Xu, J. (2005). Preparation, characterization and mechanical properties of epoxidized soybean oil/clay nanocomposites. *Polymer (Guildf), 46*, 10119–10127. https://doi.org/10.1016/j.polymer.2005.08.065

Liu, X., Sheng, X., Lee, J. K., & Kessler, M. R. (2009). Synthesis and characterization of melamine-urea-formaldehyde microcapsules containing ENB-based self-healing agents. *Macromolecular Materials and Engineering, 294*, 389–395. https://doi.org/10.1002/mame.200900015

Liu, M., Mao, X., Zhu, H., Lin, A., & Wang, D. (2013). Water and corrosion resistance of epoxy-acrylic-amine waterborne coatings: Effects of resin molecular weight, polar group and hydrophobic segment. *Corrosion Science, 75*, 106–113. https://doi.org/10.1016/j.corsci.2013.05.020

Liu, X., Wang, R., Fan, Y., & Wei, M. (2017). Effect of SiO_2 nanoparticles on the hydrophobic properties of waterborne fluorine-containing epoxy coatings. In *MATEC Web of Conference* (pp. 130).

Liu, X., Zhou, Y., & Pei, C. (2018a). Mimetic biomineralization matrix using bacterial cellulose hydrogel and egg white to prepare various morphologies of CaCO3. *CrystEngComm, 20*, 4536–4540. https://doi.org/10.1039/C8CE00988K

Liu, C., Pidko, E. A., & Hensen, E. J. M. (2018b). Origin of enhanced Brønsted acidity of NiF-modified synthetic mica–montmorillonite clay. *Catalysis Science and Technology, 8*, 244–251. https://doi.org/10.1039/C7CY02053H

Lligadas, G., Ronda, J. C., Galià, M., & Cádiz, V. (2013). Renewable polymeric materials from vegetable oils: A perspective. *Materials Today, 16*(9), 337–343. https://doi.org/10.1016/j.mattod.2013.08.016

Marathe, R. J., Chaudhari, A. B., Hedaoo, R. K., Sohn, D., Chaudhari, V. R., & Gite, V. V. (2015). Urea formaldehyde (UF) microcapsules loaded with corrosion inhibitor for enhancing the anti-corrosive properties of acrylic-based multi-functional PU coatings. *RSC Advances, 5*, 15539–15546. https://doi.org/10.1039/C4RA16327C

Marcus, P. (2011). *Corrosion mechanisms in theory and practice*. CRC Press.

McCafferty, E. (2010). Passivity. In *Introduction to corrosion science* (pp. 209–262). Springer.

Meis, N. N. A. H., van der Ven, L. G. J., van Benthem R. A. T. M., & de With, G. (2014). Extreme wet adhesion of a novel epoxy-amine coating on aluminum alloy 2024-T3. *Progress in Organics Coatings, 77*, 176–183. https://doi.org/10.1016/j.porgcoat.2013.09.001

Memarzadeh, K., Vargas, M., Huang, J., Fan, J., & Allaker, R. P. (2012). Nano metallic-oxides as antimicrobials for implant coatings. *Key Engineering Materials 493*, 489–494. https://doi.org/10.4028/www.scientific.net/KEM.493-494.489

Meneghetti, P., & Qutubuddin, S. (2004). Synthesis of poly(methyl methacrylate) nanocomposites via emulsion polymerization using a zwitterionic surfactant. *Langmuir, 20*, 3424–3430. https://doi.org/10.1021/la0357099

Meng, Y., Wei, Z., Lu, Y. L., & Zhang, L. Q. (2012). Structure, morphology, and mechanical properties of polysiloxane elastomer composites prepared by in situ polymerization of zinc dimethacrylate. *EXPRESS Polymers Letter, 6*(11). https://doi.org/10.3144/expresspolymlett.2012.94

Meng, W., Huang, Y., Fu, Y., Wang, Z., & Zhi, C. (2014). Polymer composites of boron nitride nanotubes and nanosheets. *Journal of Materials Chemistry C, 2*, 10049–10061. https://doi.org/10.1039/C4TC01998A

Mirmohseni, A., Akbari, M., Najjar, R., & Hosseini, M. (2019). Self-healing waterborne polyurethane coating by pH-dependent triggered-release mechanism. *Journal of Applied Polymer Science, 136*, 47082. https://doi.org/10.1002/app.47082

Mishra, R. K., Abraham, J., Kalarikkal, N., Jayanarayanan, K., Joseph, K., & Thomas, S. (2017). Chapter 8-conducting polyurethane blends: Recent advances and perspectives. In S. Thomas, J. Datta, J. T. Haponiuk, & A. B. T-P. P. Reghunadhan (Eds.) (pp 203–231) Elsevier.

Miyajima, S., Irikawa, J., Yamada, A., & Konagai, M. (2009). High quality aluminum oxide passivation layer for crystalline silicon solar cells deposited by parallel-plate plasma-enhanced chemical vapor deposition. *Applied Physics Express, 3*, 12301.

Mo, M., Zhao, W., Chen, Z., Yu, Q., Zeng, Z., Wu, X., & Xue, Q. (2015). Excellent tribological and anti-corrosion performance of polyurethane composite coatings reinforced with functionalized graphene and graphene oxide nanosheets. *RSC Advances, 5*, 56486–56497. https://doi.org/10.1039/C5RA10494G

Najjar, R., Akbari, M., Mirmohseni, A., & Hosseini, M. (2018). Preparation and corrosion performance of healable waterborne polyurethane coatings containing isophoronediisocyanate loaded silica capsules. *Journal of the Taiwan Institute Chemistry Engineering, 93*, 1–10. https://doi.org/10.1016/j.jtice.2018.05.021

Ocampo, C., Armelin, E., Liesa, F., Alemán, C., Ramis, X., & Iribarren, J. I. (2005). Application of a polythiophene derivative as anticorrosive additive for paints. *Progress in Organics Coatings, 53*, 217–224. https://doi.org/10.1016/j.porgcoat.2005.02.009

Osborne, J. H., Blohowiak, K. Y., Taylor, S. R., Hunter, C., Bierwagon, G., Carlson, B., Bernard, D., & Donley, M. S. (2001). Testing and evaluation of nonchromated coating systems for aerospace applications. *Progress in Organics Coatings, 41*(4), 217–225. https://doi.org/10.1016/S0300-9440(01)00132-1

Palza, H., Gutiérrez, S., Delgado, K., Salazar, O., Fuenzalida, V., Avila, J. I., Figueroa, G., & Quijada, R. (2010). Toward tailor-made biocide materials based on poly(propylene)/copper nanoparticles. *Macromolecular Rapid Communications, 31*, 563–567. https://doi.org/10.1002/marc.200900791

Posthumus, W., Derksen, A. J., van den Goorbergh, J. A. M., & Hesselmans, L. C. J. (2007). Crosslinking by polycarbodiimides. *Progress in Organics Coatings, 58*, 231–236. https://doi.org/10.1016/j.porgcoat.2006.09.031

Putson, C., Jaaoh, D., Muensit, N. (2016). Large electromechanical strain at low electric field of modified polyurethane composites for flexible actuators. *Materials Letters, 172*, 27–31. https://doi.org/10.1016/j.matlet.2016.02.131

Qiu, S., Li, W., Zheng, W., Zhao, H., & Wang, L. (2017). Synergistic effect of polypyrrole-intercalated graphene for enhanced corrosion protection of aqueous coating in 3.5% NaCl solution. *ACS Applied Materials and Interfaces, 9*, 34294–34304. https://doi.org/10.1021/acsami.7b08325

Ramezanzadeh, B., & Attar, M. M. (2011). Studying the effects of micro and nano sized ZnO particles on the corrosion resistance and deterioration behavior of an epoxy-polyamide coating on hot-dip galvanized steel. *Progress in Organics Coatings, 71*, 314–328. https://doi.org/10.1016/j.porgcoat.2011.03.026

Ramezanzadeh, B., Ghasemi, E., Mahdavian, M., Changizi, E., & Mohamadzadeh Moghadam, M. H. (2015). Covalently-grafted graphene oxide nanosheets to improve barrier and corrosion protection properties of polyurethane coatings. *Carbon, 93*, 555–573. https://doi.org/10.1016/j.carbon.2015.05.094

Rao, B. S., & Palanisamy, A. (2011). Monofunctional benzoxazine from cardanol for bio-composite applications. *Reactive and Functional Polymer, 71*, 148–154. https://doi.org/10.1016/j.reactfunctpolym.2010.11.025

Rashvand, M., & Ranjbar, Z. (2013). Effect of nano-ZnO particles on the corrosion resistance of polyurethane-based waterborne coatings immersed in sodium chloride solution via EIS technique. *Progress in Organics Coatings, 76*, 1413–1417. https://doi.org/10.1016/j.porgcoat.2013.04.013

Rodrigues, P. C., Lisboa-Filho, P. N., Mangrich, A. S., & Akcelrud, L. (2005). Polyaniline/polyurethane networks. II. A spectroscopic study. *Polymer (Guildf), 46*, 2285–2296. https://doi.org/10.1016/j.polymer.2005.01.020

Sangermano, M., Malucelli, G., Amerio, E., Priola, A., Billi, E., & Rizza, G. (2005). Photopolymerization of epoxy coatings containing silica nanoparticles. *Progress in Organics Coatings, 54*, 134–138. https://doi.org/10.1016/j.porgcoat.2005.05.004

Sanyal, B. (1981). Organic compounds as corrosion inhibitors in different environments—A review. *Progress in Organic Coatings, 9*, 165–236. https://doi.org/10.1016/0033-0655(81)80009-X

Schaefer, D. W., & Justice, R. S. (2007). How nano are nanocomposites? *Macromolecules, 40*, 8501–8517. https://doi.org/10.1021/ma070356w

Schriver, M., Regan, W., Gannett, W. J., Zaniewski, A. M., Crommie, M. F., & Zettl, A. (2013). Graphene as a long-term metal oxidation barrier: Worse than nothing. *ACS Nano, 7*, 5763–5768. https://doi.org/10.1021/nn4014356

Schwarte, S., Wegner, E., & Angermüller, H. (2003). *Self crosslinking polyurethanes and graft copolymers and based on polyurethane, for producing coating materials, adhesives and sealing masses.*

Selman, J. R., & Varma, R. (1991). *Techniques for characterization of electrodes and electrochemical processes*. Wiley.

Shahlori, R., McDougall, D. R., Waterhouse, G. I. N., Yao, F., Mata, J. P., Nelson, A. R. J., & McGillivray, D. J. (2018). Biomineralization of calcium phosphate and calcium carbonate within iridescent chitosan/iota-carrageenan multilayered films. *Langmuir, 34*, 8994–9003. https://doi.org/10.1021/acs.langmuir.8b00434

Shao, Y., Jia, C., Meng, G., Zhang, T., & Wang, F. (2009). The role of a zinc phosphate pigment in the corrosion of scratched epoxy-coated steel. *Corrosion Science, 51*, 371–379. https://doi.org/10.1016/j.corsci.2008.11.015

Shi, X., Nguyen, T. A., Suo, Z., Liu, Y., & Avci, R. (2009). Effect of nanoparticles on the anticorrosion and mechanical properties of epoxy coating. *Surface Coatings Technology, 204*(3), 237–245. https://doi.org/10.1016/j.surfcoat.2009.06.048

Shimamune, T., & Hosonuma, M. (1989). *Metal-metal oxide composites and a coating layer of calcium phosphate on the oxide layer.*

Shit, S. C., & Shah, P. (2013). A review on silicone rubber. *National Academy Science Letter, 36*, 355–365. https://doi.org/10.1007/s40009-013-0150-2

Singer, F., Schlesak, M., Mebert, C., Höhn, S., & Virtanen, S. (2015). Corrosion properties of polydopamine coatings formed in one-step immersion process on magnesium. *ACS Applied Materials and Interfaces, 7*, 26758–26766. https://doi.org/10.1021/acsami.5b08760

Sini, N. K., Bijwe, J., & Varma, I. K. (2014). Renewable benzoxazine monomer from vanillin: Synthesis, characterization, and studies on curing behavior. *Journal of Polymer Science Part A: Polymer Chemistry, 52*, 7–11. https://doi.org/10.1002/pola.26981

Skotheim, T. A. (1997). *Handbook of conducting polymers* (2nd ed.). CRC Press.

Sondari, D., Septevani, A. A., Randy, A., & Triwulandari, E. (2010). Polyurethane microcapsule with glycerol as the polyol component for encapsulated self healing agent. *IJET, 2*, 446–471.

Southwart, D. W. (1976). Comparison of bound rubber and swelling in silicone rubber/silica mixes and in silicone rubber vulcanizates. *Polymer (Guildf), 17*(12), 147–152. https://doi.org/10.1016/0032-3861(76)90085-9

Suryanarayana, C., Rao, K. C., & Kumar, D. (2008). Preparation and characterization of microcapsules containing linseed oil and its use in self-healing coatings. *Progress in Organics Coatings, 63*(1), 72–78. https://doi.org/10.1016/j.porgcoat.2008.04.008

Tallman, D. E., Pae, Y., & Bierwagen, G. P. (1999). Conducting polymers and corrosion: Polyaniline on steel. *Corrosion, 55*, 779–786. https://doi.org/10.5006/1.3284033

Tamirisa, P. A., & Hess, D. W. (2006). Water and moisture uptake by plasma polymerized thermoresponsive hydrogel films. *Macromolecules, 39*, 7092–7097. https://doi.org/10.1021/ma060944u

Taylor, J. W., & Winnik, M. A. (2004). Functional latex and thermoset latex films. *JCT Research, 1*, 163–190. https://doi.org/10.1007/s11998-004-0011-5

Tersac, G. (2007). Chemistry and technology of polyols for polyurethanes. Milhail Ionescu. Rapra Technology, Shrewsbury UK. *Polymer International, 56*, 820–820. https://doi.org/10.1002/pi.2159

Tillet, G., Boutevin, B., & Ameduri, B. (2011). Chemical reactions of polymer crosslinking and post-crosslinking at room and medium temperature. *Progress in Polymer Science, 36*(2), 191–217. https://doi.org/10.1016/j.progpolymsci.2010.08.003

Tom, S., Jin, H.-E., Heo, K., & Lee, S.-W. (2016). Engineered phage films as scaffolds for CaCO3 biomineralization. *Nanoscale, 8*, 15696–15701. https://doi.org/10.1039/C6NR04322D

Tomba, J. P., Ye, X., Li, F., Winnik, M. A., & Lau, W. (2008). Polymer blend latex films: Miscibility and polymer diffusion studied by energy transfer. *Polymer (Guildf), 49*(8), 2055–2064. https://doi.org/10.1016/j.polymer.2008.02.024

Tyan, H.-L., Liu, Y.-C., & Wei, K.-H. (1999). Thermally and mechanically enhanced clay/polyimide nanocomposite via reactive organoclay. *Chemistry of Materials, 11*, 1942–1947. https://doi.org/10.1021/cm990187x

Uhlig, H. H., Triadis, D. N., & Stern, M. (1955). Effect of oxygen, chlorides, and calcium ion on corrosion inhibition of iron by polyphosphates. *Journal of the Electrochemical Society, 102*, 59. https://doi.org/10.1149/1.2429995

Vosgien Lacombre, C., Trinh, D., Bouvet, G., Feaugas, X., Mallarino, S., & Touzain, S. (2017). Influence of pigment on the degradation of anticorrosion polymer coatings using a thermodynamic analysis of electrochemical impedance spectroscopy data. *Electrochimica Acta, 234*, 7–15. https://doi.org/10.1016/j.electacta.2017.03.050

Wan, T., & Chen, D. (2017). Synthesis and properties of self-healing waterborne polyurethanes containing disulfide bonds in the main chain. *Journal of Materials Science, 52*, 197–207. https://doi.org/10.1007/s10853-016-0321-x

Wang, Z., & Pinnavaia, T. J. (1998). Nanolayer reinforcement of elastomeric polyurethane. *Chemistry of Materials, 10*, 3769–3771. https://doi.org/10.1021/cm980448n

Wang, D., Zhu, J., Yao, Q., & Wilkie, C. A. (2002). A Comparison of various methods for the preparation of polystyrene and poly(methyl methacrylate) clay nanocomposites. *Chemistry of Materials, 14*, 3837–3843. https://doi.org/10.1021/cm011656+

Wang, Z., Zhang, X., Wang, R., Kang, H., Qiao, B., Ma, J., Zhang, L., & Wang, H. (2012). Synthesis and characterization of novel soybean-oil-based elastomers with favorable processability and tunable properties. *Macromolecules, 45*, 9010–9019. https://doi.org/10.1021/ma301938a

Wegmann, A. (1997). Chemical resistance of waterborne epoxy/amine coatings. *Progress in Organics Coatings, 32*, 231–239. https://doi.org/10.1016/S0300-9440(97)00062-3

Wei, Y., Wang, J., Jia, X., Yeh, J.-M., & Spellane, P. (1995). Polyaniline as corrosion protection coatings on cold rolled steel. *Polymer (Guildf), 36*(23), 4535–4537. https://doi.org/10.1016/0032-3861(95)96866-7

Wernik, J. M., & Meguid, S. A. (2014). On the mechanical characterization of carbon nanotube reinforced epoxy adhesives. *Materials Design, 59*, 19–32. https://doi.org/10.1016/j.matdes.2014.02.034

Wu, G., Kong, Z., Chen, J., Huo, S., & Liu, G. (2014a). Preparation and properties of waterborne polyurethane/epoxy resin composite coating from anionic terpene-based polyol dispersion. *Progress in Organics Coatings, 77*, 315–321. https://doi.org/10.1016/j.porgcoat.2013.10.005

Wu, T. H., Foyet, A., Kodentsov, A., van der Ven, L. G. J., van Benthem, R. A. T. M., & de With, G. (2014b). Wet adhesion of epoxy–amine coatings on 2024-T3 aluminum alloy. *Materials Chemistry and Physics, 145*(3), 342–349. https://doi.org/10.1016/j.matchemphys.2014.02.022

Xiang, C., Cox, P. J., Kukovecz, A., Genorio, B., Hashim, D. P., Yan, Z., Peng, Z., Hwang, C.-C., Ruan, G., Samuel, E. L. G., Sudeep, P. M., Konya, Z., Vajtai, R., Ajayan, P. M., & Tour, J. M. (2013). Functionalized low defect graphene nanoribbons and polyurethane composite film for improved gas barrier and mechanical performances. *ACS Nano, 7*, 10380–10386. https://doi.org/10.1021/nn404843n

Xing, T., Mateti, S., Li, L. H., Ma, F., Du, A., Gogotsi, Y., & Chen, Y. (2016). Gas protection of two-dimensional nanomaterials from high-energy impacts. *Science and Reports, 6*, 35532. https://doi.org/10.1038/srep35532

Xue, B.-L., Wen, J.-L., & Sun, R.-C. (2014). Lignin-based rigid polyurethane foam reinforced with pulp fiber: Synthesis and characterization. *ACS Sustainable Chemistry and Engineering, 2*, 1474–1480. https://doi.org/10.1021/sc5001226

Yang, M.-J., & Zhang, J.-D. (2005). Design and cross-linking reaction of blend of reactive polymer latex particles. *Journal of Materials Science, 40*, 4403–4405. https://doi.org/10.1007/s10853-005-0467-4

Yang, L. H., Liu, F. C., & Han, E. H. (2005). Effects of P/B on the properties of anticorrosive coatings with different particle size. *Progress in Organics Coatings, 53*(2), 91–98. https://doi.org/10.1016/j.porgcoat.2005.01.003

Yang, J., Keller, M. W., Moore, J. S., White, S. R., & Sottos, N. R. (2008). Microencapsulation of isocyanates for self-healing polymers. *Macromolecules, 41*, 9650–9655. https://doi.org/10.1021/ma801718v

Yang, L., Phua, S. L., Toh, C. L., Zhang, L., Ling, H., Chang, M., Zhou, D., Dong, Y., & Lu, X. (2013). Polydopamine-coated graphene as multifunctional nanofillers in polyurethane. *RSC Advances, 3*, 6377–6385. https://doi.org/10.1039/C3RA23307C

Yeh, J.-M., Liou, S.-J., Lai, C.-Y., Wu, P.-C., & Tsai, T.-Y. (2001). Enhancement of corrosion protection effect in polyaniline via the formation of polyaniline-clay nanocomposite materials. *Chemistry of Materials, 13*, 1131–1136. https://doi.org/10.1021/cm000938r

Yeh, J.-M., Liou, S.-J., Lin, C.-Y., Cheng, C.-Y., Chang, Y.-W., & Lee, K.-R. (2002). Anticorrosively enhanced PMMA−clay nanocomposite materials with quaternary alkylphosphonium salt as an intercalating agent. *Chemistry of Materials, 14*, 154–161. https://doi.org/10.1021/cm010337f

Yeh, J.-M., Huang, H.-Y., Chen, C.-L., Su, W.-F., & Yu, Y.-H. (2006). Siloxane-modified epoxy resin–clay nanocomposite coatings with advanced anticorrosive properties prepared by a solution dispersion approach. *Surface Coatings Technology, 200*(8), 2753–2763. https://doi.org/10.1016/j.surfcoat.2004.11.008

Yeh, J.-M., Yao, C.-T., Hsieh, C.-F., Lin, L.-H., Chen, P.-L., Wu, J.-C., Yang, H.-C., Wu, C.-P. (2008). Preparation, characterization and electrochemical corrosion studies on environmentally friendly waterborne polyurethane/Na+-MMT clay nanocomposite coatings. *European Polymer Journal, 44*, 3046–3056. https://doi.org/10.1016/j.eurpolymj.2008.05.037

Yuan, Y. C., Rong, M. Z., Zhang, M. Q., Chen, J., Yang, G. C., & Li, X. M. (2008). Self-healing polymeric materials using epoxy/mercaptan as the healant. *Macromolecules, 41*, 5197–5202. https://doi.org/10.1021/ma800028d

Zhang, J.-D., Yang, M.-J., Zhu, Y.-R., & Yang, H. (2006). Synthesis and characterization of crosslinkable latex with interpenetrating network structure based on polystyrene and polyacrylate. *Polymer International, 55*, 951–960. https://doi.org/10.1002/pi.2056

Zhang, J., Li, Y., Hu, C., Huang, W., & Su, L. (2019). Anti-corrosive properties of waterborne polyurethane/poly(o-toluidine)-ZnO coatings in NaCl solution. *Journal of Adhesion Science and Technology, 33*, 1047–1065. https://doi.org/10.1080/01694243.2018.1529881

Zhao, Y., Zhao, S., Guo, H., & You, B. (2018). Facile synthesis of phytic acid at attapulgite nanospheres for enhanced anti-corrosion performances of coatings. *Progress in Organics Coatings, 117*, 47–55. https://doi.org/10.1016/j.porgcoat.2018.01.004

Zhou, H., Wang, H., Niu, H., Gestos, A., Wang, X., & Lin, T. (2012). Fluoroalkyl silane modified silicone rubber/nanoparticle composite: A super durable, robust superhydrophobic fabric coating. *Advanced Materials, 24*, 2409–2412. https://doi.org/10.1002/adma.201200184

Zhu, D., & van Ooij, W. J. (2004). Corrosion protection of metals by water-based silane mixtures of bis-[trimethoxysilylpropyl]amine and vinyltriacetoxysilane. *Progress in Organics Coatings, 49*(1), 42–53. https://doi.org/10.1016/j.porgcoat.2003.08.009

Zúñiga, C., Lligadas, G., Ronda, J. C., Galià, M., & Cádiz, V. (2012). Renewable polybenzoxazines based in diphenolic acid. *Polymer (Guildf), 53*(8), 1617–1623. https://doi.org/10.1016/j.polymer.2012.02.026

Waterborne Polyurethane-Polyacrylate Hybrids

Hamidreza Parsimehr and Amir Ershad Langroudi

Abstract

Waterborne Polyurethane-Polyacrylate hybrids (WPUAHs) are prepared via various polymerizations methods such as different types of emulsion polymerization. The WPUAHs have various applications because of their appropriate properties. The WPUAHs have been prepared via separate and in-situ methods. The separate methods provide blends and interpenetrating polymer networks (IPNs). Also, in-situ methods provide blends, IPNs, and copolymers. The morphology, chemical bonds, and interactions between urethane and acrylate segments improve the WPUAHs more than the physical interaction of the blends. Chiefly, the WPUAHs have been used to enhance several types of adhesives and coatings.

Keywords

Waterborne polyurethane • Acrylate • Nanocomposite • IPN • Hybrid • Emulsion • Adhesive • Coating

1 Introduction

The improvement of the polymer properties is considered one of the most important issues in polymer science and technology. Tremendous efforts have been accomplished to reach novel compounds that have better properties. The problem of global warming and ozone layer depletion as two significant crucial issues have been examined carefully. Environmental pollution and green chemistry lead to increase attention to the water-based media instead of the solvent-based media (Goharshenas Moghadam et al., 2020; Kesselmeier & Staudt, 1999). Using deionized water instead of the organic solvents due to a significant reduction of volatile organic compounds (VOC) (Keddie, 1997). The reduction of VOC as a serious issue has been followed by governments, industries, and scientists. Therefore, the study of waterborne polymers is amazingly increased in recent years. The presence of the water as the media in the structure of the waterborne polymers can decrease VOC content drastically. But usually, waterborne polymers need some modifications to compete with solve-based polymers. The high surface tension and also, the high specific heat capacity of water interferes with the water evaporation process than other organic solvents. This interference phenomenon is due to a decrease in the performance of the waterborne polymers (Kast, 1985). Numerous studies have been investigated to resolve these problems (Taylor & Winnik, 2004). The two main methods to enhance the performance of the polymers are chemical modifications of the polymers through mixing with other polymers and adding an optimized content of the nanoparticles. The First method of polymer improvement is the mixing of two or more polymers to fabricate compounds, alloys, hybrids, and interpenetrating polymer networks (IPNs). The type of mixing and the content of interaction between components in the mixture determine the type of the mixture (Sperling, 1994). The second method is the addition of the inorganic nanoparticles to improve several parameters of polymer composites such as thermomechanical (Kim et al., 2003) and corrosion resistance (Abdollahi et al., 2014). But the exceeding the optimized content of inorganic nanoparticles due to dwindling polymer performance because of accumulation, agglomeration, and aggregation of nanoparticles into the polymeric structure (Azadi et al., 2020; Zare, 2016). The nanoparticle agglomeration leads to defects in the morphology of the polymers. These nanoscale defects are due to a significant reduction of several polymer parameters. Hence, nanoparticles must be added at exact optimized content or lower of optimized content into the polymers (McCook et al., 2006). The development of some parameters needs to combine different polymers to

H. Parsimehr · A. E. Langroudi (✉)
Color and Surface Coatings Group, Polymer Processing Department, Iran Polymer and Petrochemical Institute (IPPI), Tehran, Iran
e-mail: a.ershad@ippi.ac.ir

Inamuddin et al. (eds.), *Sustainable Production and Applications of Waterborne Polyurethanes*, Advances in Science, Technology & Innovation, https://doi.org/10.1007/978-3-030-72869-4_2

prepare new compounds. It would not be reached by adding inorganic nanoparticles. Therefore, the second method of enhancement has been developed. One of the most important water-based polymers that attracted scientist's attention is waterborne polyurethane. Polyurethanes have a significant spot in polymer science because of their various applications. The different kinds of polyurethanes have various applications that originate from the varied chemical structure of the polyurethanes. For example, polyurethanes are considered as one of the most important and most useful polymeric foams in materials science (Deng et al., 2003). The polyurethane foams have a rigid structure (Guo et al., 2000) or a flexible structure (Ravey & Pearce, 1997) that depends on the synthesis procedure. The various applications of the polyurethane lead to a significant increase in the polyurethane production from 1947 (Bayer, 1947) up to now. The green chemistry priority leads to increase investigations to prepare polyurethanes from benign and environmentally friendly methods. One of the best solutions is preparing polyurethanes in aqueous media that have been widely examined from 1986 (Rosthauser & Nachtkamp, 1986) up to now. The German chemists have a prominent position in the synthesis of polyurethanes (Bayer, 1947) and waterborne polyurethanes (Rosthauser & Nachtkamp, 1986). The waterborne polyurethanes (WPUs) have various applications that originate from the acceptable performance of the WPUs. WPUs without VOC leads to a significant reduction in environmental pollutions (Li et al., 2019). Emulsion polymerization methods are considered as the most common method to prepare WPUs. The WPUs are prepared via different types of emulsion polymerizations include miniemulsion (Ballard et al., 2013), microemulsion (Wang et al., 2017), and nanoemulsion (Sultan et al., 2014). The simplest difference between these three kinds of emulsions is the particle size that usually depends on the effects of the surfactant. The content of the surfactant and also, the chemical structure of the surfactant have a significant influence on the particle size of the emulsions (Kim et al., 2008). With the increase in the complexity of the demands, the performance of the WPUs must be increased continuously. Numerous studies have been investigated to enhance the performance of the WPUs. The addition of optimized content of inorganic nanoparticles (Liao et al., 2016) and also, mixing with other polymers (Wu et al., 2014) are recognized as two main methods that could have the highest influence to improve the performance of the WPUs. The mixing of the WPUs with other polymers leads to produce novel hybrids and nanocomposites have unique properties that existed in none mixed polymers. The synergic effect of mixing that originates from the morphology of the mixture is responsible for the significant enhancement of the polymeric mixture and polymeric nanocomposites (Kim & Shin, 2002). Another important group of polymers with excellent properties is acrylates. The acrylate polymers are produced on a large scale due to the various applications of the acrylate polymers. Therefore, the mixing of the acrylates and the WPUs could obtain the novel compounds with excellent properties. Numerous studies examined the mixing of polyurethanes and acrylates (Moradi et al., 2018). The interaction of the urethane and acrylate segments in the molecular scale is the main reason for the appropriate performance of Polyurethane-Polyacrylate composites (Zhang et al., 2010). The polyurethane and polyacrylate and their hybrid are illustrated in Fig. 1. The scale of mixing and the morphology of each polymer in the composite determine the type of mixing that occurs in the Polyurethane-Polyacrylate composite.

2 Polyurethane-Polyacrylate Hybrids Preparation

The WPUs are prepared via different types of emulsion polymerization. The main difference between the different types of emulsion polymerization is the particle size of the emulsions (Ruckenstein, 1996). There are two general states that polyurethane and polyacrylates mixed. In the first method, the polyurethane (PU) and Polyacrylate (PA) have been synthesized separately and then, mixed to prepare nanocomposite or interpenetrating polymer networks (IPNs). If these two polymers have some morphologies like sea-island morphology then, recognized as the nanocomposites. The two components are mixed at the nanoscale in nanocomposite structure that one of them is the dominant phase and the other one is the recessive phase. Also, if the two polymers penetrate the structure of the other component then an IPN has been performed. The two polymers are prepared simultaneously in one pot as the second state. The in-situ polymerization of urethane–acrylate leads to prepare hybrid composites. These hybrid composites are obtained in two types of copolymers include random copolymers and graft copolymers. All four probabilities of the mixing Polyurethane-Polyacrylate are illustrated in Fig. 2. The compatibility of polyurethane and Polyacrylate has great importance in the morphology of the component. It could not be categorized easily because of the variety of polyurethanes and polyacrylates.

If the two polymers have been synthesized in two separate reactors and then mixed the final component is a blend (Patrício et al., 2006; Poomalai, 2005) or IPN (Athawale & Kolekar, 1998). Also, if the penetration of the two polymers is completed or partially completed the IPN and semi-IPN (S-IPN) structures have been performed respectively. But if the penetration phenomenon does not occur, the two polymers made a blend. These blends have some common morphologies of the polymeric composites. The nanocomposites

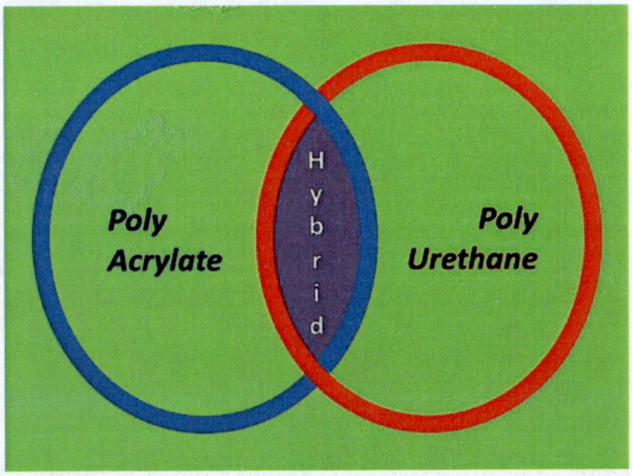

Fig. 1 Polyurethane-Polyacrylate hybrid

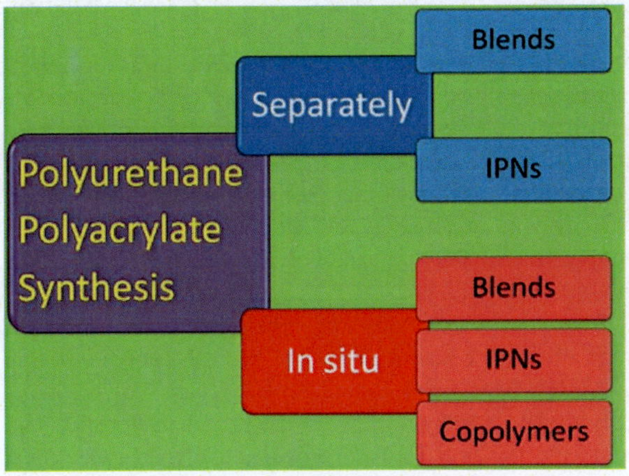

Fig. 2 The mixing states of the polyurethane-polyacrylate

have nanoparticles without accumulation in their structure. In other words, Polyurethane-Polyacrylate blends are recognized as polymer composites unless the nanoparticles are present in the mixture as nanocomposites. The nanocomposite is a polymeric composite with a distribution of the nanoparticles in the polymer structure. The in-situ polymerization of the Polyurethane and Polyacrylate in one reactor leads to prepare IPNs (Widmaier & Bonilla, 2006), blends (Lipatov et al., 2002), and copolymers (Kim et al., 2002). The in-situ formation of blends of two incompatible polymers depends on the kinetics of the polymerization of Polyurethane and Polyacrylate. The kinetics of simultaneously proceeding reactions of radical polymerization of methyl methacrylate and polyaddition reaction of polyurethane formation depended on the ratios of monomeric compounds in the reaction mixture. As distinct from IPNs, phase separation in blends of linear polymers obtained in-situ is not prevented by the formation of crosslink bonds as for IPNs (Lipatov et al., 2002). Also, the in-situ synthesis of polyurethane-Polyacrylate leads to prepare different copolymers including block copolymer (Krol & Chmielarz, 2013) and graft copolymer (Kim et al., 2002). The synthesis conditions and the chemical structure of the monomers determine the type of copolymerization. According to the dynamic mechanical thermal analysis (DMTA) and differential scanning calorimetry (DSC) results the mixture of polyurethane and polyacrylate is characterized. All probabilities of the mixing of polyurethanes and polyacrylates are discussed in the next sections.

2.1 Polyurethane-Polyacrylate Separately Preparation

The simplest method to prepare a polymeric mixture is the preparation of two or more polymers in separated reactors and then, mixing to make a mixture. The result of this mixture could be a blend or an IPN. There is a competition between blend formation and IPN formation that depends on three main parameters include polymer miscibility, mixing conditions, and chemical structure of all mixture ingredients. The mixing conditions include several processing parameters such as mixing temperature, shear rate, impeller geometry, and mixing time. The alteration of each parameter has a significant influence on the morphology of the mixture. Another important parameter is the chemical structure of mixture ingredients. The presence of the solvent, plasticizer, and small molecules has an important influence on the morphology of the Polyurethane-Polyacrylate blends (Mishra et al., 1996). The blends of Polyurethane-Polyacrylate are

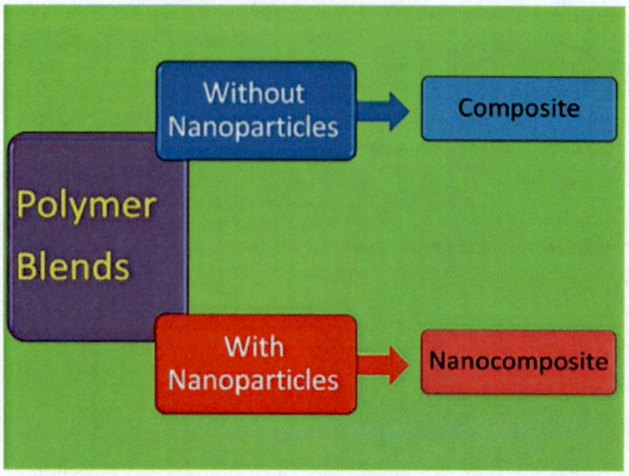

Fig. 3 The polymer blends classifications

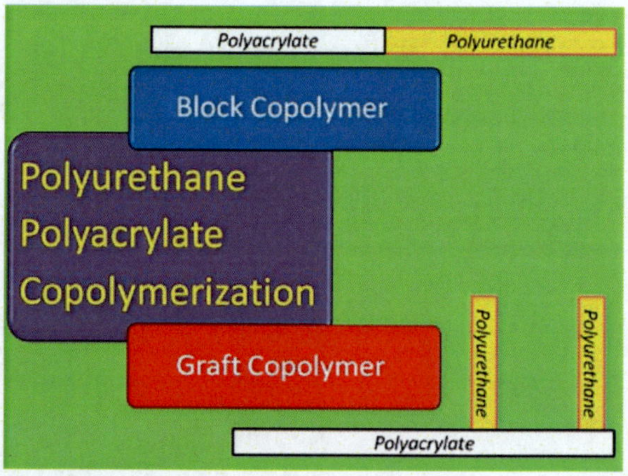

Fig. 4 Polyurethane-polyacrylate copolymers

divided into composite and nanocomposite. The morphology and the miscibility of these two polymers indicate the shape of the polymeric chains in the structure. The Polyurethane-Polyacrylate composites do not have nanoparticles and also, the Polyurethane-Polyacrylate nanocomposites have nanoparticles. The classification of the polymeric blends is illustrated in Fig. 3.

The presence of the nanoparticles due to improving several different properties of the Polyurethane-Polyacrylate nanocomposites such as acoustical properties (Moradi et al., 2020), mechanical resistance (Mo Jeong & Hun Lee, 2003; Yu et al., 2014), rheological properties (Barick & Tripathy, 2010), dielectric behavior (Chilaka & Ghosh, 2014), and thermal resistance (Mo Jeong & Hun Lee, 2003).

2.2 Polyurethane-Polyacrylate In-Situ Preparation

The simultaneous synthesis of Polyurethane and Polyacrylate in one reactor is considered as in-situ preparation. The products of the in-situ preparation are more than separate preparation of Polyurethane-Polyacrylate mixture. The copolymerization of Polyurethane and Polyacrylate is only performed at in-situ preparation. The copolymerization is a phenomenon that occurs at the chain propagation of the polymers. Therefore, the Polyurethane-Polyacrylate copolymers could not be obtained through the mixing of the Polyurethane and Polyacrylate. The in-situ copolymerizations of Polyurethane and Polyacrylate lead to performed two types of copolymers: block copolymer (Krol & Chmielarz, 2013) and graft copolymer (Ma et al., 2002; Mehravar et al., 2019a). It indicates the reactivity ratio of the polymerization reaction between urethane and acrylate is lower than the reactions of the same monomers together. Therefore, Polyurethane-Polyacrylate copolymers with high reactivity ratio between acrylate and urethane include random copolymer and alternative copolymer have not been prepared. The essential reason to prevent the formation of random copolymer and graft copolymer is the considerable tendency of the acrylate radical monomers to react with each other. Also, the structure of the radical polymerization and polyaddition is different that prevents obtaining copolymers like random copolymer and alternative copolymer. The copolymers of Polyurethane-Polyacrylate are illustrated in Fig. 4.

The novel methods of polymerizations such as surface-initiated atom transfer radical polymerization (s-ATRP) have been examined to prepare Polyurethane-Polyacrylate copolymers with specific characteristics (Jin et al., 2010). These methods provide copolymers with novel structures and appropriate properties.

The in-situ synthesis of Polyurethane and Polyacrylate can produce composites and IPNs. Polyurethane-Polyacrylate copolymers do not prepared if Polyacrylate chains and Polyurethane chains do not participate in their chain reactions. In this case, the mixture is a composite or an IPN. The miscibility of two polymers has a significant role in the competition between composite formation and IPN formation (Kausar, 2019). The lower miscibility values lead to performed composites. The mean miscibility values lead to performed semi-IPNs (S-IPN). Also, higher miscibility values lead to performed IPNs. The probabilities of the in-situ reaction of the urethane polymerization and acrylate polymerization are illustrated in Figs. 5 and 6.

Also, other parameters such as the chemical structure of the components, the presence of small molecules like a solvent, and mixing conditions influence the in-situ polymerizations of Polyurethane and Polyacrylate. But the effect of the miscibility has the most influence in this reaction.

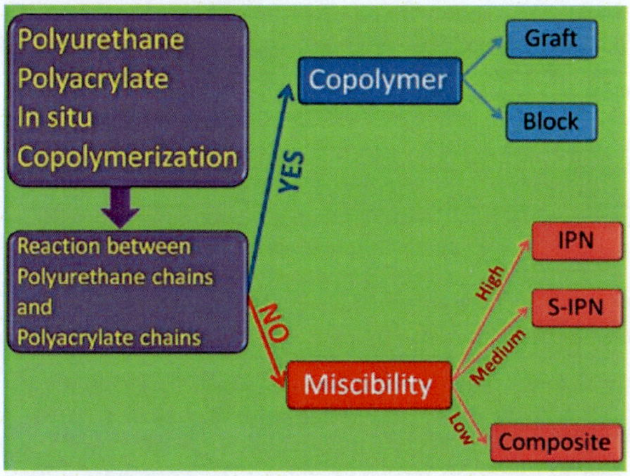

Fig. 5 The products of polyurethane-polyacrylate in-situ polymerization

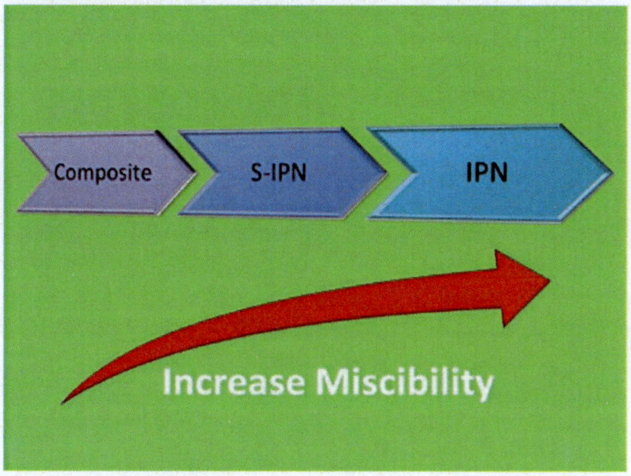

Fig. 6 The miscibility effect on the structure of the in-situ polymerization of polyurethane-polyacrylate

The miscibility parameter determines the structure of the mixture of Polyurethane and Polyacrylate.

3 Waterborne Polyurethane-Polyacrylate Hybrids (WPUAHs) Preparation

The preparation of Polyurethane-Polyacrylate hybrids is examined in the previous section. The importance of waterborne Polyurethanes (WPUs) leads to a significant increase in investigations about WPUs (Zhou et al., 2015). The WPUs are synthesized through different types of emulsion polymerization that are illustrated in Fig. 7. The type of emulsion polymerization (Deng et al., 2020) and the kinetic of polymerization (Kukanja et al., 2002) have a significant influence on the properties of the WPUAHs. Even

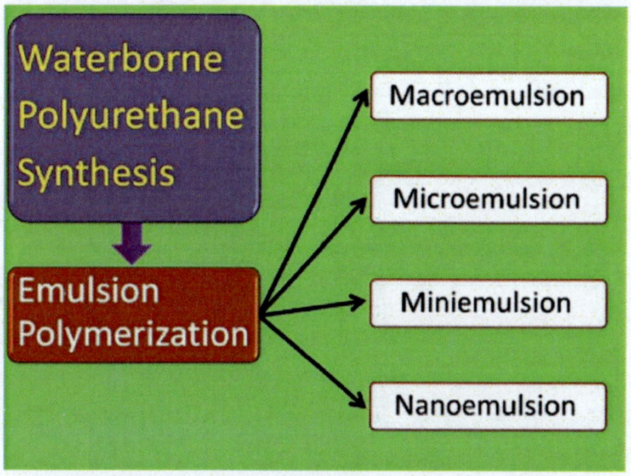

Fig. 7 Conventional methods of waterborne polyurethane preparation

using different feeding methods include semibatch and batch processes have considerable influence on the WPUAHs (Sebenik et al., 2003). Meanwhile, atom transfer radical polymerization (ATRP) and addition-fragmentation chain transfer (RAFT) polymerization processes provide an opportunity to control the chain sequence of WPUs and obtain products with the desired performance (Zhou et al., 2015).

Apart from these methods, several new processes have been developed for the synthesis of novel WPUs include addition-fragmentation chain transfer (RAFT) polymerization (Che et al., 2010), atom transfer radical polymerization (ATRP), and homogeneous solution polymerization (HSP) (Wang et al., 2008). The numerous studies prepared waterborne Polyurethane-Polyacrylate hybrids (WPUAHs) via microemulsion (Dong et al., 1999a), miniemulsion (Ballard et al., 2013), and nanoemulsion (Sultan et al., 2014) polymerizations. Also, the novel studies prepared WPUAHs through state-of-the-art methods of polymerizations such as RAFT (Yu et al., 2015) and ATRP (Krol & Chmielarz, 2013) polymerizations. The investigations about the optimized preparation of WPUAHs still ongoing because of the great importance of the effect of the preparation method on the properties of the WPUAHs (Deng et al., 2020). The WPUAHs have various structures such as core–shell, crosslinked, emulsifier-free, and solvent-free that listed in Table 1. The structure of the particles has a significant role in the morphology and properties of the WPUAHs (Lin et al., 2019). Also, the morphology of WPUAHs has a great influence on the performance of the WPUAHs (Athawale & Kulkarni, 2010). The stability of the latex particles and include acrylic (Pedraza & Soucek, 2005) and urethane particles and also the miscibility of the segments (Chen et al., 2018; Kong et al., 2010) have great importance in the performance of the emulsions. The particle stability of

Table 1 The specific structures of the WPUAHs

Structure	Subject	Method	Year	References
Core–shell	Core–shell Polyurethane-Acrylate hybrid aqueous emulsion	Emulsion	1997	Hirose et al. (1997)
	Core–shell Polyurethane-Acrylate microspheres	Emulsion	1999	Dong et al. (1999b)
	Polyurethane-Acrylate hybrid emulsion	Emulsion	2000	Hirose et al. (2000)
	Fluorinated core–shell Polyurethane-Acrylate hybrid latex	Self-emulsification	2007	Jiang et al. (2007)
	Fluorinated acrylic and siliconated core–shell Polyurethane-Acrylate hybrid	Emulsion	2007	Dai et al. (2007)
	Nanograde Core–shell Polyurethane-Polyacrylate composite	Nanoemulsion	2008	Chai and Tan (2008)
	Core–shell hybrid waterborne Polyurethanes-Polyacrylate latex	Seeded emulsion	2011	Lu et al. (2011)
	Crosslinked core–shell waterborne Polyurethane-Acrylate copolymer	Emulsion	2013	Zhang et al. (2013)
	Crosslinked core–shell waterborne fluorinated Polyurethane-Acrylate	Emulsion	2014	Luo et al. (2014)
	Waterborne epoxy-modified Polyurethane-Acrylate core–shell structure	Nanoemulsion	2017	Sun et al. (2017)
Crosslinked	Two-component waterborne Polyurethane-Acrylate resin	Emulsion	2006	Zhou et al. (2006)
	Waterborne Crosslinked Polyurethane-Acrylate	Emulsion	2012	Xu et al. (2012)
	Crosslinked core–shell waterborne Polyurethane-Acrylate copolymer	Emulsion	2013	Zhang et al. (2013)
	Waterborne Polyurethane-Acrylate hybrid emulsion	Semibatch emulsion	2014	Ou et al. (2014)
	Crosslinked core–shell waterborne fluorinated Polyurethane-Acrylate	Emulsion	2014	Luo et al. (2014)
	Waterborne Polyurethane/Acrylate emulsions in two-component coating	Emulsion	2015	Ma et al. (2015a)
	Waterborne Crosslinked Polyurethane-Acrylate	Emulsion	2016	HyeLin et al. (2016)
	Emulsifier-free crosslinkable waterborne Polyurethane-Acrylic hybrid	Emulsion	2016	Cheon et al. (2016)
	Emulsifier-free crosslinkable waterborne Polyurethane-Acrylate	Emulsion	2016	Lee et al. (2016)
	Solvent-/emulsifier-free crosslinkable waterborne Polyurethane-Acrylate	Emulsion	2017	Che et al. (2017)
Emulsifier-free	Nanograde core–shell Polyurethane/Polyacrylate composite emulsion	Nanoemulsion	2008	Chai and Tan (2008)
	Polyurethane-fluorinated Acrylate hybrid latex	Emulsion	2011	Xin et al. (2011a)
	Core–shell hybrid waterborne Polyurethanes-Polyacrylate latex	Seeded emulsion	2011	Lu et al. (2011)
	Solvent-/emulsifier-free waterborne fluorinated Polyurethane-Acrylate	Emulsion	2013	Shin et al. (2013)
	Solvent/emulsifier-free waterborne Polyurethane/self crosslinkable fluorinated Acrylate copolymer hybrid emulsion	Emulsion	2015	Park et al. (2015a)
	Emulsifier-free crosslinkable waterborne Polyurethane-Acrylate hybrid	Emulsion	2016	Cheon et al. (2016)
	Emulsifier-free crosslinkable waterborne Polyurethane-Acrylate	Emulsion	2016	Lee et al. (2016)
	Solvent-/emulsifier-free waterborne Polyurethane-Acrylate	Emulsion	2017	Ryu et al. (2017)
	Solvent-/emulsifier-free crosslinkable waterborne Polyurethane-Acrylate	Emulsion	2017	Che et al. (2017)
	Emulsifier-free waterborne Polyurethane-Acrylate hybrid lattice	Emulsion	2018	Jiang et al. (2018)
	Emulsifier-free waterborne Polyurethane-Acrylate	Emulsion	2019	Wu et al. (2019)

(continued)

Table 1 (continued)

Structure	Subject	Method	Year	References
Graft copolymer	Seeded emulsion polymerization of methyl methacrylate–Polyurethane	Emulsion	2002	Kim et al. (2002)
	Waterborne Polyurethane-Acrylate hybrid emulsion	Emulsion	2005	Šebenik and Krajnc (2005)
	Soybean oil-Based waterborne Polyurethane-Acrylate emulsion	Emulsion	2007	Lu and Larock (2007)
	Nanograde core–shell Polyurethane-Polyacrylate composite	Nanoemulsion	2008	Chai and Tan (2008)
	Polyurethane-Acrylate hybrid emulsion	Emulsion	2010	Jiang et al. (2010)
	Waterborne Polyurethane-Acrylate hybrid	Miniemulsion	2011	Degrandi-Contraires et al. (2011)
	Waterborne Polyurethane-Acrylate crosslinked core–shell copolymer	Copolymerization	2013	Zhang et al. (2013)
	Waterborne Polyurethane-Acrylate adhesive	Miniemulsion	2013	Degrandi-Contraires et al. (2013)
	Waterborne fluorinated Polyurethane-Acrylate/SiO_2 hybrid	Sol–Gel	2013	Yu et al. (2013)
	Waterborne fluorinated Polyurethane-Acrylate hybrid emulsion	Emulsion	2014	Zhao et al. (2014a)
	Polyurethane-Acrylate hybrid	Copolymerization	2015	Ma et al. (2015b)
	Waterborne Polyurethane-Acrylate leather coating	Emulsion	2016	Bahadur et al. (2016)
	Waterborne Polyurethane-Acrylate hybrid emulsion	Emulsion	2017	Yi et al. (2017)
	Low viscosity Waterborne Polyurethane-Acrylate emulsion	Emulsion	2018	Zhu et al. (2018)
	Hybrid waterborne Polyurethane-Acrylate	Self-emulsification	2018	Alvarez et al. (2018)
	Waterborne Polyurethane-Acrylate hybrid emulsion	Emulsion	2019	Lin et al. (2019)
	Waterborne Polyurethane-Acrylate hybrid	Emulsion	2019	Honarkar (2019)
	Waterborne fluoro-silicon Polyurethane-Polyacrylate	Emulsion	2019	Zhou et al. (2019)
Solvent-free	PDMS modified waterborne Polyurethane-Acrylate hybrid emulsion	Emulsion	2008	Zhang et al. (2008)
	Solvent-free waterborne Polyurethane-Polyacrylate composite	Emulsion	2013	Shi et al. (2013)
	Solvent-/emulsifier-free waterborne fluorinated Polyurethane-Acrylate	Emulsion	2013	Shin et al. (2013)
	Solvent/emulsifier-free Waterborne Polyurethane/self crosslinkable fluorinated acrylic copolymer hybrid emulsions	Emulsion	2015	Park et al. (2015a)
	Solvent-/emulsifier-free waterborne Polyurethane-Acrylate	Emulsion	2017	Ryu et al. (2017)
	Solvent-/emulsifier-free crosslinkable waterborne Polyurethane-Acrylate	Emulsion	2017	Che et al. (2017)
Reactive emulsifier	Low viscosity waterborne Polyurethane-Acrylate emulsion	Emulsion	2018	Zhu et al. (2018)

WPUAHs as an important parameter should be improved to reach optimized properties (Santamaria-Echart et al., 2015).

The WPUAHs with chemical bonds between Polyurethane and Polyacrylate chains have better mechanical properties than WPUAHs with physical interactions between Polyurethane and Polyacrylate chains. It indicates the chemical bonds such as the crosslink reaction (HyeLin et al., 2016) between Polyurethane and Polyacrylate segments have a significant role to enhance the performance of the WPUAHs (Kukanja et al., 2000; Peruzzo et al., 2011). The main difference between the blends and the hybrid coating with the same overall composition was the higher degree of mixing between Polyurethane and Polyacrylate in the case of hybrids as well as a more evenly distributed Polyacrylate phase in hybrids (Chen et al., 2019). The domain size of Polyacrylate in hybrids is about 5–10 times smaller when compared to that in blends, resulting in a much larger interface between the two materials (Chen et al., 2019). Therefore, the copolymers (block copolymer and graft copolymer) and IPNs of WPUAHs have better properties than the physical blends of WPUAHs (Adler et al., 2001; Brown et al., 2005; Kukanja et al., 2000). Various organic and inorganic substances have been examined into the structure of WPUAHs to enhance their performance that

Table 2 The enhancement of WPUAHs via different materials

Substance	Subject	Method	Year	References
Epoxy	Waterborne epoxy-modified Polyurethane-Acrylate core–shell	Nanoemulsion	2017	Sun et al. (2017)
Fluorinated	Fluorinated Polyurethane-Acrylate Hybrid	Copolymerization	2005	Li et al. (2005a)
	Fluorinated core–shell Polyurethane-Acrylate hybrid latex	Self-emulsification	2007	Jiang et al. (2007)
	Fluorinated siliconated core–shell Polyurethane-Acrylate hybrid	Emulsion	2007	Dai et al. (2007)
	Polyurethane-fluorinated acrylic hybrid latex	Emulsion	2011	Xin et al. (2011a)
	Polyurethane-fluorinated Acrylate hybrid latex	Emulsion	2011	Xin et al. (2011b)
	Solvent-/emulsifier-free waterborne fluorinated Polyurethane-Acrylate	Emulsion	2013	Shin et al. (2013)
	Waterborne fluorinated Polyurethane-Acrylate/SiO_2 hybrid	Sol–Gel	2013	Yu et al. (2013)
	Waterborne Polyurethane-Acrylate hybrid emulsion	Emulsion	2013	Lee et al. (2013)
	Waterborne fluorinated Polyurethane-Acrylate hybrid emulsion	Emulsion	2013	Wang et al. (2013)
	Waterborne fluorinated Polyurethane-Acrylate	Emulsion	2014	Wang et al. (2014)
	Crosslinked core–shell waterborne fluorinated Polyurethane-Acrylate	Emulsion	2014	Luo et al. (2014)
	Waterborne fluorinated Polyurethane-Acrylate hybrid emulsion	Emulsion	2014	Zhao et al. (2014a)
	Waterborne Polyurethane-fluorinated Acrylate emulsion	Emulsion	2014	Zhao et al. (2014b)
	Solvent/emulsifier-free waterborne Polyurethane/self crosslinkable fluorinated Acrylate copolymer hybrid emulsion	Emulsion	2015	Park et al. (2015a)
	Fluorinated UV-curable Polyurethane-Acrylate	Emulsion	2015	Park et al. (2015b)
	Waterborne fluoro-silicon Polyurethane-Polyacrylate hybrid	Emulsion	2019	Zhou et al. (2019)
Polypropylene carbonate	Waterborne Polyurethane-Acrylate composite emulsion	Emulsion	2017	Ma et al. (2017)
Soybean oil	Soybean oil-based waterborne Polyurethane-Acrylate hybrid latex	Emulsion	2007	Lu and Larock (2007)
	Soybean oil-based waterborne Polyurethane-Acrylate hybrid latex	Emulsion	2019	Meng et al. (2019)
Silane	Silylated (Polyurethane-Acrylate hybrid/clay) nanocomposite	Emulsion	2007	Subramani et al. (2007)
	Fluorinated siliconated core–shell Polyurethane-Acrylate hybrid	Emulsion	2007	Dai et al. (2007)
	Waterborne Polyurethane-Acrylate hybrid nanocomposite	Emulsion	2012	Javaheriannaghash and Ghazavi (2012)
	Fluorinated UV-curable Polyurethane-Acrylate containing vinyltrimethoxysilane (VTMS)	Emulsion	2015	Park et al. (2015b)
SiO_2	Waterborne fluorinated Polyurethane-Acrylate/SiO_2 hybrid	Sol–Gel	2013	Yu et al. (2013)

listed in Table 2. Addition of these materials into the structure of WPUAHs is due to the increasing performance of WPUAHs. For example, fluorinated polymeric chains have hydrophobic behavior. Therefore, fluorinated WPUAHs have hydrophobicity and self-cleaning properties. The appropriate additives can improve the interactions and miscibility of the system. Also, the presence of the inorganic nanoparticles like SiO_2 leads to enhancement of the thermomechanical resistance properties of waterborne Polyurethane-Polyacrylate nanocomposites. However, the addition of the nanoparticles in WPUAHs must be lower than optimized content or equal to optimized content to prevent the inappropriate accumulation and agglomeration of the nanoparticle because of the inappropriate influence of the excessive content of nanoparticles on the performance of the WPUAHs.

The composition (Degrandi-Contraires et al., 2014; Pardini & Amalvy, 2008; Saeed & Shabir, 2013) and the chemical structure of the Polyurethane and Polyacrylate has a significant role in the applications of the WPUAHs. For example, the change in the hydroxyl-containing molecules, cyanate-containing molecules, and other additives have considerable influence on the properties of the WPUAHs (Lopez et al., 2013a; Šebenik & Krajnc, 2005; Zhu et al., 2008). The Polyurethane and Polyacrylate are the primary part of the WPUAHs. The effect of the chemical structure of the urethane and acrylate bonds has been investigated in several studies listed in Table 3.

Table 3 The effect of content and structure of urethane and acrylate on WPUAHs

Polymer	Subject	Method	Year	References
Acrylate	Fluorinated Polyurethane-Acrylate Hybrid	Copolymerization	2005	Li et al. (2005a)
	Polyurethane/Acrylate Hybrid	Emulsion	2010	Peruzzo et al. (2010)
	Waterborne Poly(urethane-urea)-Acrylate hybrid	Emulsion	2012	Yang et al. (2012)
	Waterborne Polyurethane-Acrylate hybrid emulsion	Emulsion	2013	Lee et al. (2013)
	Polyurethane-Acrylate hybrid	Emulsion	2015	Ma et al. (2015b)
	Crosslinkable waterborne Polyurethane-Acrylate emulsion	Emulsion	2016	Lee et al. (2016)
	Emulsifier-/solvent-free Polyurethane-Acrylate hybrid emulsion	Emulsion	2017	Ryu et al. (2017)
	Waterborne Polyurethane-Acrylate hybrid	Emulsion	2019	Honarkar (2019)
Urethane	Polyurethane-Acrylate hybrid emulsion	Emulsion	2005	Šebenik and Krajnc (2005)
	Waterborne Polyurethane-Polyacrylate pressure-sensitive adhesive	Miniemulsion	2011	Lopez et al. (2011a)
	Waterborne Polyurethane-Acrylate	Emulsion	2012	Peruzzo et al. (2012)

Table 4 Crosslinkable WPUAHs

Crosslinker	Subject	Method	Year	References
2,2 Bis(hydroxyl methyl) propionic acid (DMPA)	Waterborne Polyurethane-Acrylate hybrid	Emulsion	2012	Chiu et al. (2012)
Bisphenol A-glycidyl methacrylate (Bis-GMA)	Waterborne Polyurethane-Acrylate hybrid	Self-emulsification	2018	Alvarez et al. (2018)
Dimethylol propionic acid (DMPA)	Waterborne Polyurethane-Acrylate hybrid	Emulsion	2016	Cheon et al. (2016)
Pentaerythritol triacrylate (PETA)	Waterborne Polyurethane-Acrylate hybrid	Emulsion	2016	Cheon et al. (2016)
Trimethylolpropane (TMP)	Waterborne Polyurethane-Acrylate hybrid	Emulsion	2017	Che et al. (2017)
Ethylenediamine (EDA)	Waterborne Polyurethane-Acrylate hybrid	Emulsion	2017	Che et al. (2017)

The crosslink reactions have considerable potential to enhance the performance of the WPUAHs. Various crosslinkable monomers have been examined in the WPUAHs. The crosslink reactions influence the morphology and the properties of the WPUAHs. The crosslinkable monomers of WPUAHs are listed in Table 4.

Several methods have been investigated to prepare WPUAHs. The main route to prepare WPUAHs is emulsion polymerization. The methods of WPUAHs are listed in Table 5. According to Table 5, the miniemulsion has been used more than microemulsion and nanoemulsion methods to prepare WPUAHs. But the novel polymerization methods like RAFT polymerization needs more investigations to optimize their properties.

According to these publications, the most important parameters to characterize and categorize the WPUAHs are illustrated in Fig. 8.

The first important parameter to investigate WPUAHs is the probability of the reaction between urethane and acrylate segments. If it happens the copolymers include block copolymer or graft copolymer have been obtained. If the urethane and acrylate segments do not react together the physical interaction has been obtained. The physical interaction structures include composites and IPNs are the results of the physical interaction between urethane and acrylate segments and particles. The type of interactions between segments and particles of WPUAHs is considered as the most important parameter of WPUAHs. It determines the preparation of the copolymers, composites, or IPNs. The second important parameter is the miscibility of urethane and acrylate segments and particles into the structure of the hybrid. For example, increase the miscibility of segments leads to prepare IPN but the reduction of miscibility of the segments leads to prepare S-IPNs. The miscibility has a great influence on the properties and applications of WPUAHs but the reactivity is the most significant parameter. The third significant parameter of WPUAHs is the morphology of the particles. The morphology of the particles originates from the synthesis method, structure, and interactions between urethane and acrylate segments. The type of emulsifier and also, the preparation core–shell structure as the two most important parameters have a significant influence on the

Table 5 WPUAHs preparation methods

Method	Subject	Year	References
Microemulsion	Core–shell waterborne Polyacrylate-Polyurethane microemulsion	1999	Dong et al. (1999a)
	In situ Polyurethane-Polyacrylate microemulsion	2017	Wang et al. (2017)
Miniemulsion	Waterborne oil-modified Polyurethane-Acrylate coating	2000	Gooch et al. (2000)
	Waterborne Polyurethane-Acrylate hybrid	2005	Li et al. (2005b)
	Hybrid Polyurethane-Acrylate latex	2006	Wang et al. (2006)
	Waterborne Polyurethane-Polyacrylate pressure-sensitive adhesive	2011	Lopez et al. (2011a)
	Waterborne Polyurethane-Polyacrylate pressure-sensitive adhesive	2011	Lopez et al. (2011b)
	Waterborne Polyurethane-Polyacrylate pressure-sensitive adhesive	2011	Udagama et al. (2011)
	Waterborne Polyurethane-Acrylate pressure-sensitive adhesive	2011	Reyes et al. (2011)
	Waterborne Polyurethane-Acrylate hybrid	2012	Daniloska et al. (2012)
	Waterborne Polyurethane-Polyacrylate pressure-sensitive adhesive	2013	Lopez et al. (2013a)
	Waterborne Polyurethane-Polyacrylate pressure-sensitive adhesive	2013	Lopez et al. (2013b)
	Waterborne Polyurethane-Acrylate hybrid latex	2013	Degrandi-Contraires et al. (2013)
	Waterborne Polyurethane-Acrylate hybrid	2013	Ballard et al. (2013)
	Waterborne Polyurethane-Acrylate hybrid	2014	Hamzehlou et al. (2014)
	Waterborne Polyurethane-Acrylate hybrid adhesive	2014	Degrandi-Contraires et al. (2014)
	Waterborne Polyurethane-Acrylate hybrid	2015	Martínez-Rugerio et al. (2015)
Nanoemulsion	Nanograde Core–shell Polyurethane-Polyacrylate composite	2008	Chai and Tan (2008)
	Waterborne Polyurethane-Acrylate nanoemulsion	2014	Sultan et al. (2014)
RAFT	Waterborne Polyurethane-Polyacrylate copolymer	2015	Yu et al. (2015)

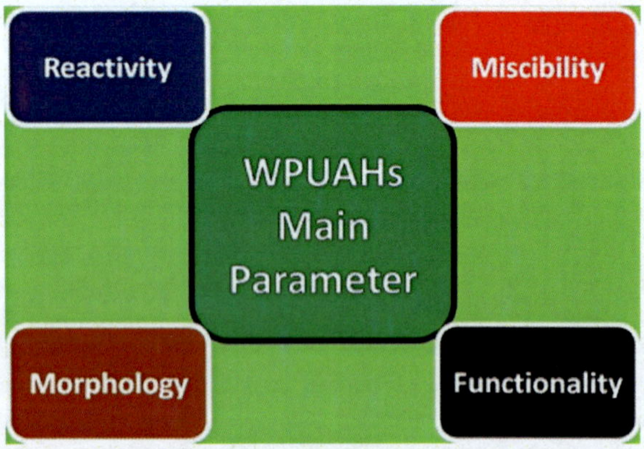

Fig. 8 The main parameters of the WPUAHs

morphology of WPUAHs. The morphology of the segments and particles of WPUAHs has a considerable influence on the stability and applications of WPUAHs. Also, the lateral bonds in WPUAHs have a considerable influence on the performance of the WPUAHs. The crosslinking is due to improving applications of the polymers. In this case, crosslinking reactions are due to enhance mechanical resistance of WPUAHs. The last important parameter of WPUAHs is the presence of functional groups in the urethane or acrylate particles. The modification of WPUAHs with fluorinated monomers is considered as the most important and most useful functional group into the structure of WPUAHs. The presence of the organofluorine compounds into the structure of WPUAHs is due to perform hydrophobicity and self-cleaning properties. Therefore, several studies have been investigated by fluorinated WPUAHs. Also, other substances like organosilanes are investigated in the structure of WPUAHs but most of the studies have been investigated about fluorinated WPUAHs because of the importance of the fluorinated polymers. The influence of these four parameters on WPUAHs classification is illustrated in Fig. 9.

4 Waterborne Polyurethane-Polyacrylate Hybrids (WPUAHs) Applications

The widespread applications of the polymers led to increased investigation and production of polymers. To the significant improvement of WPUAHs properties, some of the studies investigate through simultaneous use of multiple parameters like the fluorinated chain, crosslinking, and core–shell structure (Luo et al., 2014). The hybrids of Polyurethane-

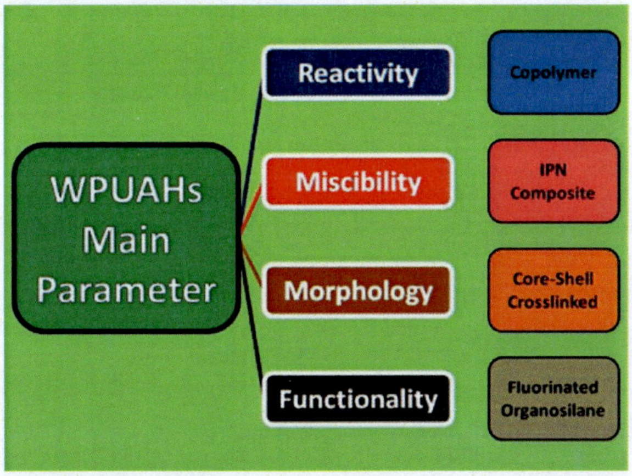

Fig. 9 The most important parameter of the WPUAHs with example

Polyacrylate have several applications because of their appropriate properties. Several studies examined the performance of WPUAHs in different applications include adhesive, floor coating, fluorescence coating, humidity-sensitive coating, mechanical durable coating, self-matting, self-segregation, topcoat exterior coating, UV-curable coating, water repellent coating, and wood coating. All applications of the WPUAHs are listed in Table 6. Chiefly, the WPUAHs have been used to prepare appropriate adhesives and coatings with considerable performances (Mehravar et al., 2019b). The applications of WPUAHs originate from the four important parameters of WPUAHs include chain interactions, particle miscibility, particle morphology, and chain functionality.

The main applications of the WPUAHs are pressure-sensitive adhesives and coatings that are examined in this chapter and listed in Table 6. According to Table 6, the contribution of adhesives and coatings in the applications of WPUAHs is illustrated in Fig. 10. Also, the investigation to enhance the performance of the WPUAHs is still ongoing. The efforts to improve WPUAHs as the pressure-sensitive adhesives and coatings are still ongoing.

Table 6 WPUAHs main applications

Feature	Subject	Method	Year	References
Adhesive	Waterborne Polyurethane-Polyacrylate pressure-sensitive adhesive	Miniemulsion	2011	Lopez et al. (2011a)
	Waterborne Polyurethane-Polyacrylate pressure-sensitive adhesive	Miniemulsion	2011	Lopez et al. (2011b)
	Waterborne Polyurethane-Polyacrylate pressure-sensitive adhesive	Miniemulsion	2011	Udagama et al. (2011)
	Waterborne Polyurethane-Acrylate pressure-sensitive adhesive	Miniemulsion	2011	Reyes et al. (2011)
	Waterborne Polyurethane-Polyacrylate pressure-sensitive adhesive	Miniemulsion	2013	Lopez et al. (2013a)
	Waterborne Polyurethane-Polyacrylate pressure-sensitive adhesive	Miniemulsion	2013	Lopez et al. (2013b)
	Waterborne Polyurethane-Acrylate hybrid latex	Miniemulsion	2013	Degrandi-Contraires et al. (2013)
	Waterborne Polyurethane-Acrylate hybrid polymer adhesive	Emulsion	2014	Liu et al. (2014)
	Waterborne Polyurethane-Acrylate hybrid adhesive	Miniemulsion	2014	Degrandi-Contraires et al. (2014)
	Emulsifier-free crosslinkable waterborne Polyurethane-Acrylate	Emulsion	2016	Lee et al. (2016)
	Waterborne Polyurethane-Polyacrylate pressure-sensitive adhesive	Miniemulsion	2017	Mehravar et al. (2017)
	Solvent-/emulsifier-free crosslinkable waterborne Polyurethane-Acrylate	Emulsion	2017	Che et al. (2017)
Antibacterial coating of textile	Silver/waterborne Polyurethane-Acrylate coating	Emulsion	2018	Zhang et al. (2018)
Chemical resistant coating	Waterborne Polyurethane-Acrylate hybrid emulsion	Semibatch emulsion	2014	Ou et al. (2014)
Damping coating	Waterborne Polyurethane-Acrylate hybrid latex	Emulsion	2018	Deng et al. (2018)
Floor coating	Grafted Polyurethane-Polyacrylate hybrid	Semibatch emulsion	2019	Mehravar et al. (2019a)

(continued)

Table 6 (continued)

Feature	Subject	Method	Year	References
Fluorescence dye	Waterborne Polyurethane-Acrylate hybrid	Emulsion	2017	Hu et al. (2017)
Humidity sensitive	Crosslinked core–shell waterborne Polyurethane-Acrylate copolymer	Emulsion	2013	Zhang et al. (2013)
Leather coating	Waterborne Polyurethane-Acrylate leather coating	Emulsion	2016	Bahadur et al. (2016)
Mechanical durable coating	Waterborne grafted Polyurethane-Acrylic hybrid	Miniemulsion	2011	Degrandi-Contraires et al. (2011)
	Crosslinked waterborne Polyurethane-Acrylate composite	Emulsion	2012	Xu et al. (2012)
	Waterborne Polyurethane-Acrylic hybrid latex	Miniemulsion	2013	Degrandi-Contraires et al. (2013)
Self-matting	Soybean oil-based waterborne Polyurethane-Acrylate hybrid latex	Emulsion	2019	Meng et al. (2019)
Self-segregation	Waterborne fluoro-silicon Polyurethane-Polyacrylate hybrid	Emulsion	2019	Zhou et al. (2019)
Topcoat exterior coating	Waterborne Polyurethane-Acrylate hybrid binder	Semibatch emulsion	2006	Aznar et al. (2006)
Textile finishing coating	Waterborne Polyurethane-Acrylate nanoemulsion	Nanoemulsion	2014	Sultan et al. (2014)
UV-curable	UV-Curable waterborne Polyurethane-Acrylate ionomer coating	Emulsion	1999	Wang et al. (1999)
	UV-curable Polyurethane-Acrylate/silica hybrid	Sol–Gel	2004	Zhang et al. (2004)
	Fluorinated UV-curable Polyurethane-Acrylate	Emulsion	2015	Park et al. (2015b)
Water repellent	Waterborne Polyurethane modified by stearyl Acrylate	Self-emulsification	2018	Sheng et al. (2018)
Wood coating	Polyurethane-Polyacrylate microemulsion as wood coating	Microemulsion	2017	Wang et al. (2017)

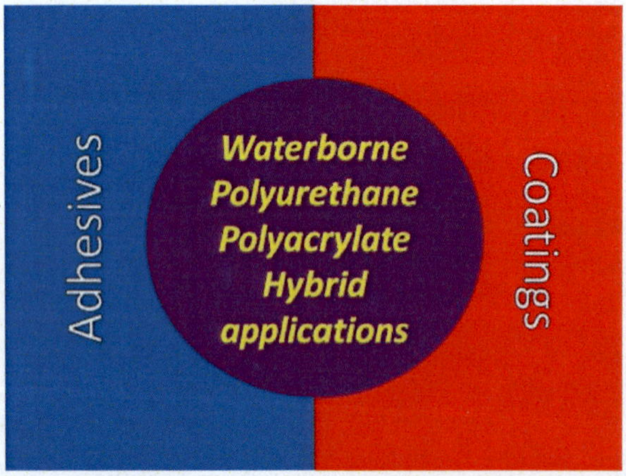

Fig. 10 The main applications of the WPUAHs

5 Conclusion

The waterborne Polyurethane-Polyacrylate hybrids (WPUAHs) have several appropriate applications that originate from the specific structure of these hybrids. The WPUAHs have been prepared via separated and in-situ methods. The waterborne Polyurethane-Polyacrylate copolymers were only obtained through in-situ preparation. The WPUAHs blends and IPNs have been prepared via both of the separated and in-situ methods. The chemical interaction between urethane and acrylate segments is due to a significant increase in the performance of the WPUAHs while physical interactions have lower enhancement. The most important parameters of WPUAHs are the reaction or interaction between urethane and acrylate segments, the miscibility of the particles, the morphology of the particles, and the functionality of the polymer chains. The WPUAHs have several applications, especially in pressure-sensitive adhesives and coatings.

References

Abdollahi, H., Ershad-langroudi, A., Salimi, A., & Rahimi, A. (2014). Anticorrosive coatings prepared using epoxy-silica hybrid nanocomposite materials. *Industrial and Engineering Chemistry Research, 53*, 10858–10869. https://doi.org/10.1021/ie501289g

Adler, H.-J., Jahny, K., & Vogt-Birnbrich, B. (2001). Polyurethane macromers—New building blocks for acrylic hybrid emulsions with outstanding performance. *Progress in Organic Coatings, 43*, 251–257. https://doi.org/10.1016/S0300-9440(01)00207-7

Alvarez, G. A., Fuensanta, M., Orozco, V. H., Giraldo, L. F., & Martín-Martínez, J. M. (2018). Hybrid waterborne polyurethane/acrylate dispersion synthesized with bisphenol A-glicidylmethacrylate

(Bis-GMA) grafting agent. *Progress in Organic Coatings, 118*, 30–39. https://doi.org/10.1016/j.porgcoat.2018.01.016

Athawale, V., & Kolekar, S. (1998). Interpenetrating polymer networks based on polyol modified castor oil polyurethane and polymethyl methacrylate. *European Polymer Journal, 34*, 1447–1451. https://doi.org/10.1016/S0014-3057(97)00282-6

Athawale, V. D., & Kulkarni, M. A. (2010). Core-shell and interpenetrating-network PU/AC hybrids: Synthesis and comparison—Waterborne polyurethane/polyacrylate hybrid dispersions. *Pigment & Resin Technology, 39*, 141–148. https://doi.org/10.1108/03699421011040767

Azadi, N., Parsimehr, H., & Ershad-Langroudi, A. (2020). Cultural heritage protection via hybrid nanocomposite coating. *Plastics, Rubber and Composites*. https://doi.org/10.1080/14658011.2020.1784589

Aznar, A. C., Pardini, O. R., & Amalvy, J. I. (2006). Glossy topcoat exterior paint formulations using water-based polyurethane/acrylic hybrid binders. *Progress in Organic Coatings, 55*, 43–49. https://doi.org/10.1016/j.porgcoat.2005.11.001

Bahadur, A., Shoaib, M., Saeed, A., & Iqbal, S. (2016). FT-IR spectroscopic and thermal study of waterborne polyurethane-acrylate leather coatings using tartaric acid as an ionomer. *E-Polymers, 16*, 463–474. https://doi.org/10.1515/epoly-2016-0154

Ballard, N., Carretero, P., & Asua, J. M. (2013). Effect of reaction temperature on adhesive properties of waterborne polyurethane/acrylic hybrids synthesized by semicontinuous miniemulsion polymerization. *Macromolecular Reaction Engineering, 7*, 504–514. https://doi.org/10.1002/mren.201300122

Barick, A. K., & Tripathy, D. K. (2010). Preparation and characterization of thermoplastic polyurethane/organoclay nanocomposites by melt intercalation technique: Effect of nanoclay on morphology, mechanical, thermal, and rheological properties. *Journal of Applied Polymer Science, 117*, 639–654. https://doi.org/10.1002/app.31303

Bayer, O. (1947). Das Di-Isocyanat-Polyadditionsverfahren (Polyurethane). *Angewandte Chemie, 59*, 257–272. https://doi.org/10.1002/ange.19470590901

Brown, R. A., Coogan, R. G., Fortier, D. G., Reeve, M. S., & Rega, J. D. (2005). Comparing and contrasting the properties of urethane/acrylic hybrids with those of corresponding blends of urethane dispersions and acrylic emulsions. *Progress in Organic Coatings, 52*, 73–84. https://doi.org/10.1016/j.porgcoat.2004.03.009

Chai, S. L., & Tan, H. M. (2008). Structure and property characterization of nanograde core-shell polyurethane/polyacrylate composite emulsion. *Journal of Applied Polymer Science, 107*, 3499–3504. https://doi.org/10.1002/app.27517

Che, J.-Y., Cheon, J.-M., Chun, J.-H., Park, C.-C., Lee, Y.-H., & Kim, H.-D. (2017). Preparation and properties of emulsifier-/solvent-free slightly crosslinked waterborne polyurethane-acrylic hybrid emulsions for footwear adhesives (III)–effect of trimethylol propane (TMP)/ethylene diamine (EDA) content. *Journal of Adhesion Science and Technology, 31*, 1872–1887. https://doi.org/10.1080/01694243.2017.1285744

Che, X. C., Jin, Y. Z., & Lee, Y. S. (2010). Preparation of nano-TiO2/polyurethane emulsions via in situ RAFT polymerization. *Progress in Organic Coatings, 69*, 534–538. https://doi.org/10.1016/j.porgcoat.2010.09.013

Chen, Q., van der Slot, S., Kreisig, S., Tian, M., Peters, R., & Loos, J. (2019). Revealing nanoscale chemical heterogeneity at the surface of water-based coatings prepared from urethane–Acrylic hybrids by photo-induced force microscopy. *Journal of Coatings Technology and Research, 16*, 1553–1558. https://doi.org/10.1007/s11998-019-00208-3

Chen, X., Zhang, C., Li, W., Chen, L., & Wang, W. (2018). Synthesis of waterborne polyurethane by the telechelic α, ω-di(hydroxy)poly (n-butyl acrylate). *Polymers, 10*, 219. https://doi.org/10.3390/polym10020219

Cheon, J.-M., Lee, S.-G., Chun, J.-H., Lee, D.-J., Lee, Y.-H., & Kim, H.-D. (2016). Preparation and properties of emulsifier-/NMP-free crosslinkable waterborne polyurethane-acrylic hybrid emulsions for footwear adhesives (II)—Effect of dimethylol propionic acid (DMPA)/pentaerylthritol triacrylate (PETA) content. *E-Polymers, 16*, 189–197. https://doi.org/10.1515/epoly-2016-0005

Chilaka, N., & Ghosh, S. (2014). Dielectric studies of poly (ethylene glycol)-polyurethane/poly (methylmethacrylate)/montmorillonite composite. *Electrochimica Acta, 134*, 232–241. https://doi.org/10.1016/j.electacta.2014.04.114

Chiu, H.-T., Yang, H.-M., Liu, C.-S., & Hsu, H.-Y. (2012). Synthesis, stability and properties of polyurethane/acrylic hybrids using m-TMXDI-based anionic poly(urethane-urea) dispersion. *Polymer—Plastics Technology and Engineering, 51*, 945–953. https://doi.org/10.1080/03602559.2012.680561

Dai, J. B., Zhang, X. Y., Chao, J., & Bai, C. Y. (2007). A new core-shell type fluorinated acrylic and siliconated polyurethane hybrid emulsion. *Journal of Coatings Technology and Research, 4*, 283–288. https://doi.org/10.1007/s11998-007-9042-z

Daniloska, V., Tomovska, R., & Asua, J. M. (2012). Hybrid miniemulsion photopolymerization in a continuous tubular reactor—A way to expand the characteristics of polyurethane/acrylics. *Chemical Engineering Journal, 184*, 308–314. https://doi.org/10.1016/j.cej.2012.01.040

Degrandi-Contraires, E., Lopez, A., Reyes, Y., Asua, J. M., & Creton, C. (2013). High-shear-strength waterborne polyurethane/acrylic soft adhesives. *Macromolecular Materials and Engineering, 298*, 612–623. https://doi.org/10.1002/mame.201200106

Degrandi-Contraires, E., Udagama, R., Bourgeat-Lami, E., McKenna, T., Ouzineb, K., & Creton, C. (2011). Mechanical properties of adhesive films obtained from PU−Acrylic hybrid particles. *Macromolecules, 44*, 2643–2652. https://doi.org/10.1021/ma2000702

Degrandi-Contraires, E., Udagama, R., McKenna, T., Bourgeat-Lami, E., Plummer, C. J. G., & Creton, C. (2014). Influence of composition on the morphology of polyurethane/acrylic latex particles and adhesive films. *International Journal of Adhesion and Adhesives, 50*, 176–182. https://doi.org/10.1016/j.ijadhadh.2014.01.025

Deng, R., Davies, P., & Bajaj, A. K. (2003). Flexible polyurethane foam modelling and identification of viscoelastic parameters for automotive seating applications. *Journal of Sound and Vibration, 262*, 391–417. https://doi.org/10.1016/S0022-460X(03)00104-4

Deng, Y., Zhou, C., Zhang, Q., Zhang, M., & Zhang, H. (2020). Structure and performance of waterborne polyurethane-acrylate composite emulsions for industrial coatings: Effect of preparation methods. *Colloid and Polymer Science, 298*, 139–149. https://doi.org/10.1007/s00396-019-04583-6

Deng, Y. J., Zhou, C., Zhang, M. Y., & Zhang, H. X. (2018). Effects of the reagent ratio on the properties of waterborne polyurethanes-acrylate for application in damping coating. *Progress in Organic Coatings, 122*, 239–247. https://doi.org/10.1016/j.porgcoat.2018.05.025

Dong, A., Wan, T., Feng, S., & Sun, D. (1999a). IR spectra studies of core-shell type waterborne polyacrylate-polyurethane microemulsions. *Journal of Polymer Science Part B: Polymer Physics, 37*, 2642–2650. https://doi.org/10.1002/(SICI)1099-0488(19990915)37:18%3c2642::AID-POLB8%3e3.0.CO;2-D

Dong, A., An, Y., Feng, S., & Sun, D. (1999b). Preparation and morphology studies of core-shell type waterborne polyacrylate-polyurethane microspheres. *Journal of Colloid and Interface Science, 214*, 118–122. https://doi.org/10.1006/jcis.1999.5847

Goharshenas Moghadam, S., Pazokifard, S., Parsimehr, H. (2020). Performance of room temperature cross-linkable acrylic latexes synthesized using DAAM and AAEMA functional monomers. In *Eco-Friendly and Smart Polymer System* (pp. 125–128). Springer International Publishing. https://doi.org/10.1007/978-3-030-45085-4_30

Gooch, J. W., Dong, H., & Schork, F. J. (2000). Waterborne oil-modified polyurethane coatings via hybrid miniemulsion polymerization. *Journal of Applied Polymer Science, 76*, 105–114. https://doi.org/10.1002/(SICI)1097-4628(20000404)76:1%3c105::AID-APP14%3e3.0.CO;2-8

Guo, A., Javni, I., & Petrovic, Z. (2000). Rigid polyurethane foams based on activated soybean meal. *Journal of Applied Polymer Science, 77*, 467–473. https://doi.org/10.1002/(SICI)1097-4628(20000711)77:2%3c467::AID-APP25%3e3.0.CO;2-F

Hamzehlou, S., Ballard, N., Carretero, P., Paulis, M., Asua, J. M., Reyes, Y., & Leiza, J. R. (2014). Mechanistic investigation of the simultaneous addition and free-radical polymerization in batch miniemulsion droplets: Monte Carlo simulation versus experimental data in polyurethane/acrylic systems. *Polymer, 55*, 4801–4811. https://doi.org/10.1016/j.polymer.2014.07.024

Hirose, M., Kadowaki, F., & Zhou, J. (1997). The structure and properties of core-shell type acrylic-polyurethane hybrid aqueous emulsions. *Progress in Organic Coatings, 31*, 157–169. https://doi.org/10.1016/S0300-9440(97)00032-5

Hirose, M., Zhou, J., & Nagai, K. (2000). The structure and properties of acrylic-polyurethane hybrid emulsions. *Progress in Organic Coatings, 38*, 27–34. https://doi.org/10.1016/S0300-9440(99)00081-8

Honarkar, H. (2019). The effect of different acrylic monomers contents on the properties of waterborne polyurethane/acrylate hybrids. *Polymer Science, Series B, 61*, 743–752. https://doi.org/10.1134/S1560090420010030

Hu, X., Ding, Y., Liu, J., Deng, Y., & Cheng, C. (2017). Synthesis and fluorescence properties of a waterborne polyurethane-acrylic hybrid polymeric dye. *Polymer Bulletin, 74*, 555–569. https://doi.org/10.1007/s00289-016-1729-9

HyeLin, K., YoungHee, L., JungSoo, K., ChaCheol, P., Hyun, P., HoHwan, C., & HanDo, K. (2016). Preparation and properties of crosslinkable waterborne polyurethane and polyurethane-acrylic hybrid emulsions and their crosslinked polymers. *Journal of Polymer Research, 23*, 240. https://doi.org/10.1007/s10965-016-1134-y

Javaheriannaghash, H., & Ghazavi, N. (2012). Preparation and characterization of water-based polyurethane-acrylic hybrid nanocomposite emulsion based on a new silane-containing acrylic macromonomer. *Journal of Coatings Technology and Research, 9*, 323–336. https://doi.org/10.1007/s11998-011-9373-7

Jiang, B., Tsavalas, J. G., & Sundberg, D. C. (2018). Morphology control in surfactant free polyurethane/acrylic hybrid latices—The special role of hydrogen bonding. *Polymer, 139*, 107–122. https://doi.org/10.1016/j.polymer.2018.01.054

Jiang, M., Zheng, Z., Ding, X., Cheng, X., & Peng, Y. (2007). Convenient synthesis of novel fluorinated polyurethane hybrid latexes and core-shell structures via emulsion polymerization process with self-emulsification of polyurethane. *Colloid and Polymer Science, 285*, 1049–1054. https://doi.org/10.1007/s00396-007-1658-0

Jiang, X., Kong, X.-Z., Zhu, X., Zhang, Z., & Zhang, Z.-G. (2010). Calculation of grafting and property characterization in polyurethane-acrylic hybrid materials prepared by emulsion process. *Macromolecular Chemistry and Physics, 211*, 2201–2210. https://doi.org/10.1002/macp.201000222

Jin, Z., Feng, W., Zhu, S., Sheardown, H., & Brash, J. L. (2010). Protein-resistant polyurethane by sequential grafting of poly (2-hydroxyethyl methacrylate) and poly(oligo(ethylene glycol) methacrylate) via surface-initiated ATRP. *Journal of Biomedical Materials Research, Part A, 95A*, 1223–1232. https://doi.org/10.1002/jbm.a.32935

Kast, H. (1985). Aspects of film formation with emulsion copolymers. *Die Makromolekulare Chemie, 10*, 447–461. https://doi.org/10.1002/macp.1985.020101985131

Kausar, A. (2019). Interpenetrating polymer network and nanocomposite IPN of polyurethane/epoxy: A review on fundamentals and advancements. *Polymer-Plastics Technology and Materials, 58*, 691–706. https://doi.org/10.1080/25740881.2018.1563114

Keddie, J. L. (1997). Film formation of latex. *Materials Science and Engineering R: Reports, 21*, 101–170. https://doi.org/10.1016/S0927-796X(97)00011-9

Kesselmeier, J., & Staudt, M. (1999). Biogenic volatile organic compounds (VOC): An overview on emission, physiology and ecology. *Journal of Atmospheric Chemistry, 33*, 23–88. https://doi.org/10.1023/A:1006127516791

Kim, B., & Shin, J. (2002). Modification of waterborne polyurethane by forming latex interpenetrating polymer networks with acrylate rubber. *Colloid and Polymer Science, 280*, 716–724. https://doi.org/10.1007/s00396-002-0672-5

Kim, B. K., Seo, J. W., & Jeong, H. M. (2003). Morphology and properties of waterborne polyurethane/clay nanocomposites. *European Polymer Journal, 39*, 85–91. https://doi.org/10.1016/S0014-3057(02)00173-8

Kim, I. H., Shin, J. S., Cheong, I. W., Kim, J. I., & Kim, J. H. (2002). Seeded emulsion polymerization of methyl methacrylate using aqueous polyurethane dispersion: Effect of hard segment on grafting efficiency. *Colloids and Surfaces A: Physicochemical and Engineering Aspects, 207*, 169–176. https://doi.org/10.1016/S0927-7757(02)00092-4

Kim, J.-W., Lee, D., Shum, H. C., & Weitz, D. A. (2008). Colloid surfactants for emulsion stabilization. *Advanced Materials, 20*, 3239–3243. https://doi.org/10.1002/adma.200800484

Kong, X., Li, S., Qu, J., & Chen, H. (2010). Self-emulsifying hydroxy acrylic polymer dispersions for two component waterborne polyurethane coatings. *Journal of Macromolecular Science, Part A, 47*, 368–374. https://doi.org/10.1080/10601320903539330

Krol, P., & Chmielarz, P. (2013). Synthesis of PMMA-b-PU-b-PMMA tri-block copolymers through ARGET ATRP in the presence of air. *Express Polymer Letters, 7*, 249–260. https://doi.org/10.3144/expresspolymlett.2013.23

Kukanja, D., Golob, J., Zupancic-Valant, A., & Krajnc, M. (2000). The structure and properties of acrylic-polyurethane hybrid emulsions and comparison with physical blends. *Journal of Applied Polymer Science, 78*, 67–80. https://doi.org/10.1002/1097-4628(20001003)78:1%3c67::AID-APP100%3e3.0.CO;2-4

Kukanja, D., Golob, J., & Krajnc, M. (2002). Kinetic investigations of acrylic-polyurethane composite latex. *Journal of Applied Polymer Science, 84*, 2639–2649. https://doi.org/10.1002/app.10440

Lee, S.-G., Cheon, J.-M., Chun, J.-H., Lee, Y.-H., Rahman, M.-M., Kim, H.-D. (2016). Preparation and properties of emulsifier/N-methylpyrrolidone-free crosslinkable waterborne polyurethane-acrylate emulsions for footwear adhesives. I. Effect of the acrylic monomer content. *Journal of Applied Polymer Science, 133*, 43758. https://doi.org/10.1002/app.43758

Lee, S. W., Lee, Y. H., Park, H., Do Kim, H. (2013). Effect of total acrylic/fluorinated acrylic monomer contents on the properties of waterborne polyurethane/acrylic hybrid emulsions. *Macromolecular Research, 21*, 709–718. https://doi.org/10.1007/s13233-013-1122-6

Li, G., Shen, Y., & Ren, Q. (2005a). Effect of fluorinated acrylate on the surface properties of cationic fluorinated polyurethane-acrylate hybrid dispersions. *Journal of Applied Polymer Science, 97*, 2192–2196. https://doi.org/10.1002/app.21852

Li, M., Daniels, E. S., Dimonie, V., Sudol, E. D., & El-Aasser, M. S. (2005b). Preparation of polyurethane/acrylic hybrid nanoparticles via a miniemulsion polymerization process. *Macromolecules, 38*, 4183–4192. https://doi.org/10.1021/ma048141z

Li, X.-G., Xie, Y.-B., Huang, M.-R., Umeyama, T., Ohara, T., & Imahori, H. (2019). Effective role of eco-friendly acetyl tributyl citrate in large-scale catalyst-free synthesis of waterborne polyurethanes without volatile organic compounds. *Journal of Cleaner Production, 237*, 117543. https://doi.org/10.1016/j.jclepro.2019.07.018

Liao, L., Li, X., Wang, Y., Fu, H., & Li, Y. (2016). Effects of surface structure and morphology of nanoclays on the properties of *Jatropha curcas* oil-based waterborne polyurethane/clay nanocomposites. *Industrial and Engineering Chemistry Research, 55*, 11689–11699. https://doi.org/10.1021/acs.iecr.6b02527

Lin, R., Zhang, Y., Li, H., Shi, Y., & Zhou, C. (2019). Tailoring the morphology and properties of waterborne polyurethanes by incorporation of acrylic monomers. *Progress in Organic Coatings, 135*, 65–73. https://doi.org/10.1016/j.porgcoat.2019.05.019

Lipatov, Y., Kosyanchuk, L., & Nesterov, A. (2002). Phase separation in blends of linear polymers formedin situ according to different mechanisms. *Polymer International, 51*, 772–780. https://doi.org/10.1002/pi.925

Liu, Z., Jones, F. R., Zhai, Z., & Feng, L. (2014). Interfacial adhesion of polyamide 66 fibres to an aqueous polyurethane-acrylic hybrid polymer adhesive. *Journal of Adhesion, 90*, 310–326. https://doi.org/10.1080/00218464.2013.795115

Lopez, A., Degrandi, E., Canetta, E., Keddie, J. L., Creton, C., & Asua, J. M. (2011a). Simultaneous free radical and addition miniemulsion polymerization: Effect of the diol on the microstructure of polyurethane-acrylic pressure-sensitive adhesives. *Polymer, 52*, 3021–3030. https://doi.org/10.1016/j.polymer.2011.04.053

Lopez, A., Degrandi-Contraires, E., Canetta, E., Creton, C., Keddie, J. L., & Asua, J. M. (2011b). Waterborne polyurethane−Acrylic hybrid nanoparticles by miniemulsion polymerization: applications in pressure-sensitive adhesives. *Langmuir, 27*, 3878–3888. https://doi.org/10.1021/la104830u

Lopez, A., Reyes, Y., Degrandi-Contraires, E., Canetta, E., Creton, C., Keddie, J. L., & Asua, J. M. (2013a). Simultaneous free-radical and addition miniemulsion polymerization: Effect of the chain transfer agent on the microstructure of polyurethane-acrylic pressure-sensitive adhesives. *Macromolecular Materials and Engineering, 298*, 53–66. https://doi.org/10.1002/mame.201100390

Lopez, A., Reyes, Y., Degrandi-Contraires, E., Canetta, E., Creton, C., & Asua, J. M. (2013b). Waterborne hybrid polymer particles: Tuning of the adhesive performance by controlling the hybrid microstructure. *European Polymer Journal, 49*, 1541–1552. https://doi.org/10.1016/j.eurpolymj.2013.01.030

Lu, Y., & Larock, R. C. (2007). New hybrid latexes from a soybean oil-based waterborne polyurethane and acrylics via emulsion polymerization. *Biomacromolecules, 8*, 3108–3114. https://doi.org/10.1021/bm700522z

Lu, Y., Xia, Y., & Larock, R. C. (2011). Surfactant-free core-shell hybrid latexes from soybean oil-based waterborne polyurethanes and poly(styrene-butyl acrylate). *Progress in Organic Coatings, 71*, 336–342. https://doi.org/10.1016/j.porgcoat.2011.03.027

Luo, Q., Shen, Y., Li, P., Wang, C., & Zhao, Z. (2014). Synthesis and characterization of crosslinking waterborne fluorinated polyurethane-acrylate with core-shell structure. *Journal of Applied Polymer Science, 131*, 40970. https://doi.org/10.1002/app.40970

Ma, C. C. M., Du, Y. C., & Wang, F. Y. (2002). Thermal degradation-behaviors of poly (dimethylsiloxane)-urethane-graft-poly (methyl methacrylate) copolymers based on various diisocyanates. *Journal of Materials Science, 37*, 1247–1252. https://doi.org/10.1023/A:1014388010394

Ma, G., Guan, T., Hou, C., Wu, J., Wang, G., Ji, X., & Wang, B. (2015a). Preparation, properties and application of waterborne hydroxyl-functional polyurethane/acrylic emulsions in two-component coatings. *Journal of Coatings Technology and Research, 12*, 505–512. https://doi.org/10.1007/s11998-014-9647-y

Ma, G., Guan, T., Wu, J., Hou, C., Wang, G., Qin, G., & Wang, B. (2015b). Effect of 2-ethylhexyl acrylate and N-acryloylmorpholine on the properties of polyurethane/acrylic hybrid materials. *Journal of Applied Polymer Science, 132*, 41463. https://doi.org/10.1002/app.41463

Ma, L., Song, L., Li, F., Wang, H., & Liu, B. (2017). Preparation and properties of poly (propylene carbonate)-based waterborne polyurethane-acrylate composite emulsion. *Colloid and Polymer Science, 295*, 2299–2307. https://doi.org/10.1007/s00396-017-4198-2

Martínez-Rugerio, G., Alegría, Á., Daniloska, V., Tomovska, R., Paulis, M., & Colmenero, J. (2015). Dielectric relaxations of acrylic-polyurethane hybrid materials. *Polymer, 74*, 21–29. https://doi.org/10.1016/j.polymer.2015.07.055

McCook, N. L., Boesl, B., Burris, D. L., & Sawyer, W. G. (2006). Epoxy, ZnO, and PTFE nanocomposite: Friction and wear optimization. *Tribology Letters, 22*, 253–257. https://doi.org/10.1007/s11249-006-9089-5

Mehravar, S., Roschmann, K. J., Arocha, P. U., Reck, B., Agirre, A., Tomovska, R., Asua, J. M., & Ballard, N. (2019a). Correlating microstructure and performance of PU/(meth)acrylic hybrids as hardwood floor coating. *Progress in Organic Coatings, 131*, 417–426. https://doi.org/10.1016/j.porgcoat.2019.03.002

Mehravar, S., Ballard, N., Agirre, A., Tomovska, R., & Asua, J. M. (2017). Relating polymer microstructure to adhesive performance in blends of hybrid polyurethane/acrylic latexes. *European Polymer Journal, 87*, 300–307. https://doi.org/10.1016/j.eurpolymj.2016.12.031

Mehravar, S., Ballard, N., Tomovska, R., & Asua, J. M. (2019b). Polyurethane/acrylic hybrid waterborne dispersions: Synthesis, properties and applications. *Industrial and Engineering Chemistry Research, 58*, 20902–20922. https://doi.org/10.1021/acs.iecr.9b02324

Meng, Y., Lv, P., Liu, Q., Liao, B., Pang, H., & Liu, W. (2019). Preparation and characterization of soybean oil-based waterborne polyurethane/acrylate hybrid emulsions for self-matting coatings. *New Journal of Chemistry, 43*, 19193–19199. https://doi.org/10.1039/C9NJ04538D

Mishra, V., Thomas, D. A., & Sperling, L. H. (1996). Partition coefficient for small amounts of monomer, solvent, or plasticizer in a two-phased polymer blend or IPN. *Journal of Polymer Science Part b: Polymer Physics, 34*, 2105–2108. https://doi.org/10.1002/(SICI)1099-0488(19960915)34:12%3c2105::AID-POLB16%3e3.0.CO;2-5

Mo Jeong, H., & Hun Lee, S. (2003). Properties of waterborne polyurethane/PMMA/clay hybrid materials. *Journal of Macromolecular Science, Part B, 42*, 1153–1167. https://doi.org/10.1081/MB-120024811

Moradi, G., Monazzam, M., Ershad-Langroudi, A., Parsimehr, H., & Keshavarz, S. T. (2020). Organoclay nanoparticles interaction in PU:PMMA IPN foams: Relationship between the cellular structure and damping-acoustical properties. *Applied Acoustics, 164*, 107295. https://doi.org/10.1016/j.apacoust.2020.107295

Moradi, G., Nassiri, P., Ershad-Langroudi, A., & Monazzam, M. R. (2018). Acoustical, damping and thermal properties of polyurethane/poly(methyl methacrylate)-based semi-interpenetrating polymer network foams. *Plastics, Rubber and Composites, 47*, 221–231. https://doi.org/10.1080/14658011.2018.1468146

Ou, J., Yang, Y., Gan, J., Ha, C., & Zhang, M. (2014). Preparation, properties, and applications of acrylic-polyurethane hybrid emulsions in extinction electrophoresis. *Journal of Applied Polymer Science, 131*, 40078. https://doi.org/10.1002/app.40078

Pardini, O. R., & Amalvy, J. I. (2008). FTIR,1H-NMR spectra, and thermal characterization of water-based polyurethane/acrylic hybrids. *Journal of Applied Polymer Science, 107*, 1207–1214. https://doi.org/10.1002/app.27188

Park, J. M., Jeon, J. H., Lee, Y. H., Lee, D. J., Park, H., Chun, H. H., Do Kim, H. (2015b). Synthesis and properties of UV-curable polyurethane acrylates containing fluorinated acrylic monomer/vinyltrimethoxysilane. *Polymer Bulletin, 72*, 1921–1936. https://doi.org/10.1007/s00289-015-1380-x

Park, Y. G., Lee, Y. H., Rahman, M. M., Park, C. C., Do Kim, H. (2015). Preparation and properties of waterborne polyurethane/self-cross-linkable fluorinated acrylic copolymer hybrid emulsions using a solvent/emulsifier-free method. *Colloid and Polymer Science, 293*, 1369–1382. https://doi.org/10.1007/s00396-015-3504-0

Patrício, P. S. O., de Sales, J. A., Silva, G. G., Windmöller, D., & Machado, J. C. (2006). Effect of blend composition on microstructure, morphology, and gas permeability in PU/PMMA blends. *Journal of Membrane Science, 271*, 177–185. https://doi.org/10.1016/j.memsci.2005.06.064

Pedraza, E. P., & Soucek, M. D. (2005). Effect of functional monomer on the stability and film properties of thermosetting core-shell latexes. *Polymer, 46*, 11174–11185. https://doi.org/10.1016/j.polymer.2005.08.069

Peruzzo, P. J., Anbinder, P. S., Pardini, O. R., Costa, C. A., Leite, C. A., Galembeck, F., & Amalvy, J. I. (2010). Polyurethane/acrylate hybrids: effects of the acrylic content and thermal treatment on the polymer properties. *Journal of Applied Polymer Science, 116*, 2694–2705. https://doi.org/10.1002/app31795

Peruzzo, P. J., Anbinder, P. S., Pardini, O. R., Vega, J., Costa, C. A., Galembeck, F., & Amalvy, J. I. (2011). Waterborne polyurethane/acrylate: Comparison of hybrid and blend systems. *Progress in Organic Coatings, 72*, 429–437. https://doi.org/10.1016/j.porgcoat.2011.05.016

Peruzzo, P. J., Anbinder, P. S., Pardini, O. R., Vega, J. R., & Amalvy, J. I. (2012). Influence of diisocyanate structure on the morphology and properties of waterborne polyurethane-acrylates. *Polymer Journal, 44*, 232–239. https://doi.org/10.1038/pj.2011.111

Poomalai, P. (2005). Siddaramaiah, studies on poly(methyl methacrylate) (PMMA) and thermoplastic polyurethane (TPU) blends. *Journal of Macromolecular Science, Part A, 42*, 1399–1407. https://doi.org/10.1080/10601320500205764

Ravey, M., Pearce, E. M. (1997). Flexible polyurethane foam. I. Thermal decomposition of a polyether-based, water-blown commercial type of flexible polyurethane foam. *Journal of Applied Polymer Science, 63*, 47–74. https://doi.org/10.1002/(SICI)1097-4628(19970103)63:1<47::AID-APP7>3.0.CO;2-S

Reyes, Y., Lopez, A., & Asua, J. M. (2011). Modeling the microstructure of acrylic-polyurethane hybrid polymers synthesized by miniemulsion polymerization. *Macromolecular Reaction Engineering, 5*, 352–360. https://doi.org/10.1002/mren.201100010

Rosthauser, J. W., & Nachtkamp, K. (1986). Waterborne polyurethanes. *Journal of Coated Fabrics, 16*, 39–79. https://doi.org/10.1177/152808378601600103

Ruckenstein, E. (1996). Microemulsions, macroemulsions, and the Bancroft rule. *Langmuir, 12*, 6351–6353. https://doi.org/10.1021/la960849m

Ryu, Y.-S., Lee, Y.-H., Kim, J.-S., Park, C.-C., & Kim, H.-D. (2017). Preparation and properties of emulsifier-/solvent-free polyurethane-acrylic hybrid emulsions for binder materials: Effect of the glycidyl methacrylate/acrylonitrile content. *Journal of Applied Polymer Science, 134*, 44497. https://doi.org/10.1002/app.44497

Saeed, A., & Shabir, G. (2013). Synthesis of thermally stable high gloss water dispersible polyurethane/polyacrylate resins. *Progress in Organic Coatings, 76*, 1135–1143. https://doi.org/10.1016/j.porgcoat.2013.03.009

Santamaria-Echart, A., Arbelaiz, A., Saralegi, A., Fernández-d'Arlas, B., Eceiza, A., Corcuera, M. (2015). Relationship between reagents molar ratio and dispersion stability and film properties of waterborne polyurethanes. *Colloids and Surfaces A: Physicochemical and Engineering Aspects, 482*, 554–561. https://doi.org/10.1016/j.colsurfa.2015.07.012

Sebenik, U., Golob, J., & Krajnc, M. (2003). Comparison of properties of acrylic-polyurethane hybrid emulsions prepared by batch and semibatch processes with monomer emulsion feed. *Polymer International, 52*, 740–748. https://doi.org/10.1002/pi.1127

Šebenik, U., & Krajnc, M. (2005). Properties of acrylic-polyurethane hybrid emulsions synthesized by the semibatch emulsion copolymerization of acrylates using different polyurethane particles. *Journal of Polymer Science Part A: Polymer Chemistry, 43*, 4050–4069. https://doi.org/10.1002/pola.20896

Sheng, L., Zhang, X., Ge, Z., Liang, Z., Liu, X., Chai, C., & Luo, Y. (2018). Preparation and properties of waterborne polyurethane modified by stearyl acrylate for water repellents. *Journal of Coatings Technology and Research, 15*, 1283–1292. https://doi.org/10.1007/s11998-018-0096-x

Shi, R., Zhang, X., & Dai, J. (2013). Synthesis of TDI-polyurethane/polyacrylate composite emulsion by solvent-free method and performances of the latex film. *Journal of Macromolecular Science, Part A, 50*, 350–357. https://doi.org/10.1080/10601325.2013.755893

Shin, M., Lee, Y., Rahman, M., & Kim, H. (2013). Synthesis and properties of waterborne fluorinated polyurethane-acrylate using a solvent-/emulsifier-free method. *Polymer, 54*, 4873–4882. https://doi.org/10.1016/j.polymer.2013.07.005

Sperling, L. H. (1994). Interpenetrating polymer networks: An overview. In: D. Klempner, L. H. Sperling, & L. A. Utracki (Eds.), *Interpenetrating Polymer Networks* (pp. 3–38). American Chemical Society. https://doi.org/10.1021/ba-1994-0239.ch001

Subramani, S., Choi, S.-W., Lee, J.-Y., & Kim, J. H. (2007). Aqueous dispersion of novel silylated (polyurethane-acrylic hybrid/clay) nanocomposite. *Polymer, 48*, 4691–4703. https://doi.org/10.1016/j.polymer.2007.06.023

Sultan, M., Islam, A., Gull, N., Bhatti, H. N., & Safa, Y. (2014). Structural variation in soft segment of waterborne polyurethane acrylate nanoemulsions. *Journal of Applied Polymer Science, 132*, 41706. https://doi.org/10.1002/app.41706

Sun, J., Fang, H., Wang, H., Yang, S., Xiao, S., & Ding, Y. (2017). Waterborne epoxy-modified polyurethane-acrylate dispersions with nano-sized core-shell structure particles: Synthesis, characterization, and their coating film properties. *Journal of Polymer Engineering, 37*, 113–123. https://doi.org/10.1515/polyeng-2016-0003

Taylor, J. W., & Winnik, M. A. (2004). Functional latex and thermoset latex films. *Journal of Coatings Technology and Research, 1*, 163–190. https://doi.org/10.1007/s11998-004-0011-5

Udagama, R., Degrandi-Contraires, E., Creton, C., Graillat, C., McKenna, T. F. L., & Bourgeat-Lami, E. (2011). Synthesis of acrylic−Polyurethane hybrid latexes by miniemulsion polymerization and their pressure-sensitive adhesive applications. *Macromolecules, 44*, 2632–2642. https://doi.org/10.1021/ma200073d

Wang, C., Chu, F., Guyot, A., Gauthier, C., & Boisson, F. (2006). Hybrid acrylic–polyurethane latexes: Emulsion versus miniemulsion polymerization. *Journal of Applied Polymer Science, 101*, 3927–3941. https://doi.org/10.1002/app.22889

Wang, C., Li, X., Du, B., Li, P., Lai, X., & Niu, Y. (2014). Preparation and properties of a novel waterborne fluorinated polyurethane–Acrylate hybrid emulsion. *Colloid and Polymer Science, 292*, 579–587. https://doi.org/10.1007/s00396-013-3107-6

Wang, C., Li, X., Wang, H., Fei, G., & Wen, H. (2017). In-situ polyurethane/polyacrylate microemulsion formation: The effects of acrylic content in wood coating application. *Iranian Polymer Journal, 26*, 753–763. https://doi.org/10.1007/s13726-017-0560-y

Wang, W., Guo, Y., & Otaigbe, J. U. (2008). Synthesis and characterization of novel biodegradable and biocompatible poly (ester-urethane) thin films prepared by homogeneous solution polymerization. *Polymer, 49*, 4393–4398. https://doi.org/10.1016/j.polymer.2008.07.057

Wang, Y., Qiu, F., Xu, B., Xu, J., Jiang, Y., Yang, D., & Li, P. (2013). Preparation, mechanical properties and surface morphologies of waterborne fluorinated polyurethane-acrylate. *Progress in Organic Coatings, 76*, 876–883. https://doi.org/10.1016/j.porgcoat.2013.02.003

Wang, Z., Gao, D., Yang, J., & Chen, Y. (1999). Synthesis and characterization of UV-curable waterborne polyurethane-acrylate ionomers for coatings. *Journal of Applied Polymer Science, 73*, 2869–2876. https://doi.org/10.1002/(SICI)1097-4628(19990929)73:14%3c2869::AID-APP10%3e3.0.CO;2-S

Widmaier, J.-M., & Bonilla, G. (2006). In situ synthesis of optically transparent interpenetrating organic/inorganic networks. *Polymers for Advanced Technologies, 17*, 634–640. https://doi.org/10.1002/pat.774

Wu, G., Kong, Z., Chen, J., Huo, S., & Liu, G. (2014). Preparation and properties of waterborne polyurethane/epoxy resin composite coating from anionic terpene-based polyol dispersion. *Progress in Organic Coatings, 77*, 315–321. https://doi.org/10.1016/j.porgcoat.2013.10.005

Wu, Y., Feng, Y., & Guo, P. (2019). The polyurethane/polyacrylate soap-free emulsion formation: Effects of hydrophilic polyurethane. *Journal of Macromolecular Science, Part A, 56*, 741–749. https://doi.org/10.1080/10601325.2018.1561151

Xin, H., Shen, Y., & Li, X. (2011b). Novel cationic polyurethane-fluorinated acrylic hybrid latexes: Synthesis, characterization and properties. *Colloids Surfaces A Physicochem. Eng. Asp., 384*, 205–211. https://doi.org/10.1016/j.colsurfa.2011.03.056

Xin, H., Shen, Y.-D., & Li, X.-R. (2011a). Synthesis and properties of cationic polyurethane-fluorinated acrylic hybrid latexes by emulsifier-free emulsion polymerization and the solvent-free method. *Polymer Bulletin, 67*, 1849–1863. https://doi.org/10.1007/s00289-011-0523-y

Xu, H., Qiu, F., Wang, Y., Yang, D., Wu, W., Chen, Z., & Zhu, J. (2012). Preparation, mechanical properties of waterborne polyurethane and crosslinked polyurethane-acrylate composite. *Journal of Applied Polymer Science, 124*, 958–968. https://doi.org/10.1002/app.35127

Yang, Z., Peng, H., Wang, W., & Liu, T. (2012). Effect of acrylic monomer content on the properties of waterborne poly (urethane-urea)/acrylic hybrid materials. *Journal of Applied Polymer Science, 124*, 5113–5121. https://doi.org/10.1002/app35662

Yi, T., Ma, G., Hou, C., Li, S., Zhang, R., Wu, J., Hao, X., & Zhang, H. (2017). Polyurethane-acrylic hybrid emulsions with high acrylic/polyurethane ratios: Synthesis, characterization, and properties. *Journal of Applied Polymer Science, 134*, 44488. https://doi.org/10.1002/app.44488

Yu, L. Z., Wang, Y. Y., Qiu, F. X., Xu, J. C., Xu, B. B., Jiang, Y., Yang, D. Y., & Li, P. L. (2013). Preparation, characterisation and properties of waterborne fluorinated polyurethane acrylate/SiO_2 hybrid. *Plastics, Rubber and Composites, 42*, 385–392. https://doi.org/10.1179/1743289813Y.0000000057

Yu, W., Du, M., Ye, W., Lv, W., & Zheng, Q. (2014). Relaxation behavior of layered double hydroxides filled dangling chain-based polyurethane/polymethyl methacrylate nanocomposites. *Polymer, 55*, 2455–2463. https://doi.org/10.1016/j.polymer.2014.03.039

Yu, Y., Wang, J., & Zong, J. (2015). Synthesis of α, ω-vinyl telechelic polyurethane macromonomer and preparation of polyurethane-poly (methyl methacrylate) copolymer emulsions via RAFT emulsion polymerization). *Designed Monomers and Polymers, 18*, 242–250. https://doi.org/10.1080/15685551.2014.999464

Zare, Y. (2016). Study of nanoparticles aggregation/agglomeration in polymer particulate nanocomposites by mechanical properties. *Composites. Part A: Applied Science and Manufacturing, 84*, 158–164. https://doi.org/10.1016/j.compositesa.2016.01.020

Zhang, C., Zhang, X., Dai, J., & Bai, C. (2008). Synthesis and properties of PDMS modified waterborne polyurethane-acrylic hybrid emulsion by solvent-free method. *Progress in Organic Coatings, 63*, 238–244. https://doi.org/10.1016/j.porgcoat.2008.05.011

Zhang, L., Zeng, Z., Yang, J., & Chen, Y. (2004). Characterization and properties of UV-curable polyurethane-acrylate/silica hybrid materials prepared by the sol-gel process. *Polymer International, 53*, 1431–1435. https://doi.org/10.1002/pi.1536

Zhang, S.-F., Wang, R.-M., He, Y.-F., Song, P.-F., & Wu, Z.-M. (2013). Waterborne polyurethane-acrylic copolymers crosslinked core-shell nanoparticles for humidity-sensitive coatings. *Progress in Organic Coatings, 76*, 729–735. https://doi.org/10.1016/j.porgcoat.2013.01.003

Zhang, T., Wu, W., Wang, X., & Mu, Y. (2010). Effect of average functionality on properties of UV-curable waterborne polyurethane-acrylate. *Progress in Organic Coatings, 68*, 201–207. https://doi.org/10.1016/j.porgcoat.2010.02.004

Zhang, X., Zhu, M., Wang, W., & Yu, D. (2018). Silver/waterborne polyurethane-acrylate's antibacterial coating on cotton fabric based on click reaction via ultraviolet radiation. *Progress in Organic Coatings, 120*, 10–18. https://doi.org/10.1016/j.porgcoat.2018.03.004

Zhao, J., Zhou, T., Zhang, J., Chen, H., Yuan, C., Zhang, W., & Zhang, A. (2014b). Synthesis of a waterborne polyurethane-fluorinated emulsion and its hydrophobic properties of coating films. *Industrial and Engineering Chemistry Research, 53*, 19257–19264. https://doi.org/10.1021/ie5040732

Zhao, Z., Li, X., Li, P., Wang, C., & Luo, Q. (2014a). Study on properties of waterborne fluorinated polyurethane/acrylate hybrid emulsion and films. *Journal of Polymer Research, 21*, 460. https://doi.org/10.1007/s10965-014-0460-1

Zhou, X., Tu, W., & Hu, J. (2006). Preparation and characterization of two-component waterborne polyurethane comprised of water-soluble acrylic resin and HDI Biuret. *Chinese Journal of Chemical Engineering, 14*, 99–104. https://doi.org/10.1016/S1004-9541(06)60044-X

Zhou, X., Li, Y., Fang, C., Li, S., Cheng, Y., Lei, W., & Meng, X. (2015). Recent advances in synthesis of waterborne polyurethane and their application in water-based Ink: A review. *Journal of Materials Science and Technology, 31*, 708–722. https://doi.org/10.1016/j.jmst.2015.03.002

Zhou, Y., Liu, C., Gao, J., Chen, Y., Yu, F., Chen, M., & Zhang, H. (2019). A novel hydrophobic coating film of water-borne fluoro-silicon polyacrylate polyurethane with properties governed by surface self-segregation. *Progress in Organic Coatings, 134*, 134–144. https://doi.org/10.1016/j.porgcoat.2019.04.078

Zhu, X., Jiang, X., Zhang, Z., & Kong, X. Z. (2008). Influence of ingredients in water-based polyurethane-acrylic hybrid latexes on latex properties. *Progress in Organic Coatings, 62*, 251–257. https://doi.org/10.1016/j.porgcoat.2007.12.006

Zhu, Z., Li, R., Zhang, C., & Gong, S. (2018). Preparation and properties of high solid content and low viscosity waterborne polyurethane—Acrylate emulsion with a reactive emulsifier. *Polymers, 10*, 154. https://doi.org/10.3390/polym10020154

Applications of Cationic Waterborne Polyurethanes

Anil M. Palve, Kwadwo Mensah-Darkwa, Felipe M. de Souza, and Ram K. Gupta

Abstract

The acknowledgment of the hazards caused by using toxic chemicals and volatile organic solvents and components to both environment and human health drove governments and international organizations to consider some restrictions on the chemical industries. These actions forced both companies and the research community to pave a new path for science that proved to be challenging as well as much rewarding. Through this perspective of implementing sustainable routes to the industries, many renewable sources were started to be used both in research and development and industries. The materials synthesized using renewable resources and their properties turned out to be similar or even better than some petrochemical-based materials. Waterborne polyurethanes are one of the sustainable polymers which can be synthesized using renewable resources. These materials provide ease and green approaches by replacing volatile organic solvents with water in many procedures. Most importantly, a new set of materials with enhanced properties were developed for the usage in biomedical, coatings, adhesives, electromagnetic shielding, water purification, gas analysis, self-healing, and many more. This chapter discusses the green approaches and applications to get a clear and profound picture of the chemistry of cationic waterborne polyurethanes.

Keywords

Waterborne polyurethanes • Polyols • Vegetable oils • Biomedical • Coatings • Flame-retardant

1 Introduction

Polymers and polymer-based compounds find usage in our daily life and industrial sectors like packaging, electronics, pharmaceutical, health, construction, defense, aerospace, automotive, etc. (Campos et al., 2013; Francis & Kumar, 2016; Mohammad, 2007). This is mostly because of lightweight, low price, excellent corrosion resistance, and facile synthesis of these compounds, just to mention a few (Signoret et al., 2019; Sundriyal et al., 2020; Yang et al., 2018). With the ever-increasing demand for polymer-based materials coupled with environmental issues associated with fossil feedstocks (which are the main source for the raw materials) (Al-Jahwari & Pervez, 2020; Biron, 2020; Muthuraj & Mekonnen, 2018; Rydz et al., 2020), there has been a re-invigorated drive to research into robust green technologies and environmentally friendly materials.

Polyurethane (PU) is undoubtedly the most versatile class of polymeric materials (Akindoyo et al., 2016; Kozlowski et al., 2011; Mello et al., 2009). In 2018, the global polymer production was 359 million tons, and China reached 30% of the world's plastic production. Europe consumed about 61.8 million tons with an estimated 7.9% been polyurethanes, listed as the fifth most used plastic. The USA was the first trade partner of the European Plastics Industry in the same reporting year (PlasticsEurope, 2019). Polyurethane was first reported in the literature in 1947 by Otto Bayer and his research group in Germany (Bayer, 1947). It was not until

A. M. Palve
Department of Chemistry, Mahatma Phule ASC College, Panvel, Navi-Mumbai, Maharashtra 410206, India

K. Mensah-Darkwa
Department of Materials Engineering, College of Engineering, Kwame Nkrumah University of Science and Technology, Kumasi, Ghana

F. M. de Souza · R. K. Gupta (✉)
Kansas Polymer Research Center, Pittsburg State University, Pittsburg, KS 66762, USA
e-mail: rgupta@pittstate.edu

F. M. de Souza · R. K. Gupta
Department of Chemistry, Pittsburg State University, Pittsburg, KS 66762, USA

Inamuddin et al. (eds.), *Sustainable Production and Applications of Waterborne Polyurethanes*, Advances in Science, Technology & Innovation, https://doi.org/10.1007/978-3-030-72869-4_3

1954 when commercial production of polyurethane foams was realized due to the commercialization of polyisocyanates (Seymour & Kauffman, 1992). Over the past six-plus decades, it has become a material of interest in various industries such as the coating, paints, ink, foams, sealant, and adhesive industries (Gayki et al., 2015; Ji et al., 2015; Zhu et al., 2016) because of their strong bonding to the surface and stability under harsh conditions (Akindoyo et al., 2016; Rahman et al., 2009; Sin et al., 2013; Stanzione et al., 2020; Zhang et al., 2020a).

At present, though many advances have been made in the researches and industrial applications of polyurethanes, however, there is over-reliance on raw materials such as methane, ethylene, and benzene, and these volatile organic compounds (VOCs) represent a substantial drawback since they are derived from petroleum sources (Irfan et al., 2018; Liang et al., 2018a; Lu & Larock, 2010a). Polyurethanes produced using these materials are not environmentally friendly and also raise the question of sustainability (Andjelkovic et al., 2005; Damant, 1986). Coupled with the ever-increasing strict environmental policy on the promotion of sustainable production and consumption, several developing countries have restricted their use (Paul, 1990; Rasshofer & Schomer, 2003). Furthermore, potential human exposure to these dangerous chemicals poses a health risk, and these health effects can range from relatively mild conditions to serious diseases such as cancer (Dales & Raizenne, 2004; Huang & Tsuang, 2014; Pappas et al., 2000; Ramírez et al., 2012; Rumchev et al., 2007). Recent developments in government regulations on health, safety, and environmental and the risk of using conventional polyurethane motivated scientists to synthesize eco-friendly polymeric materials (Chaudhari et al., 2013; Dieterich et al., 1970; Gaikwad et al., 2015; Gogoi & Karak, 2014; Noreen et al., 2016).

Waterborne polyurethanes (WPU) have exhibited many advantages over conventional VOCs-based polyurethanes and have emerged as one of the environmentally friendly alternatives (Noreen et al., 2016). With the growing environmental awareness, waterborne polyurethanes offer an opportunity to use sustainable raw materials to produce environmentally sustainable polymers (Fang et al., 2018; Liang et al., 2020; Mehravar et al., 2019; Zafar et al., 2019), particularly, polyols derived from vegetable oils (Liang et al., 2018b, 2020; Lu & Larock, 2010a, 2010b; Man et al., 2019). Waterborne polyurethanes' manufacturing process is known to be similar to the existing manufacturing process used for conventional polyurethanes; hence, there can be a seamless transfer of the technology (Liang et al., 2018b, 2020; Lu & Larock, 2010a, 2010b; Man et al., 2019). Waterborne polyurethanes are derived from the same reaction used to synthesize polyurethanes; however, eco-friendly solvents are used than VOCs. Waterborne polyurethanes are conventional types of polyurethanes that have gone through polyaddition reaction to add urethane (–NH–CO–O–) and urea (–NH–CO–NH–) groups in the molecular chain (Kreye et al., 2013). Notwithstanding, the change is a vital component, and waterborne polyurethanes can preserve the original properties of VOCs-based polyurethanes, such as resistance to low-temperature drying, high abrasion resistance, good mechanical properties, UV resistance, low viscosity, and typical polyurethane performance (Szycher & Szycher, 2012). Waterborne polyurethanes can tolerate acid and alkali environments and are not easily affected by harsh environments such as temperature, weather, and UV light. Waterborne polyurethanes have found wide applications in the automotive, electrical and electronic, building and construction, agricultural, packaging, household leisure, and sports industries (PlasticsEurope, 2019). They are also extensively used in papermaking, paint, ink, leather, adhesive, and textile industries (Lei et al., 2014; Li et al., 2014; Qu et al., 2000, 2002; Wu & Hu, 2016; Yan, 2007). Due to their versatility and environment-friendly nature (both in synthesis and application), waterborne polyurethanes research, development, and technologies are well reported in the literature (Das et al., 2018; Gurunathan et al., 2016).

Waterborne polyurethanes are classified based on the dispersion state and the ionic-type side group (Fig. 1). The dispersion state of waterborne polyurethanes is divided into an aqueous solution, colloidal, and emulsion based on the particle sizes ranging from 0.001 to 0.1 μm. One advantage of using water-based dispersion methods is the flexibility that offers the introduction of ionic and/or non-ionic hydrophilic backbone in the structure. Cationic waterborne polyurethanes exhibit excellent adhesion to many surfaces like fabrics and leather due to opposite charges on these surfaces (Ahmadi & Ahmad, 2020; Tenorio-Alfonso et al., 2020). Over the decades, researchers have been improvising to design and fabricate waterborne polyurethanes with varied properties for applications in various fields. This write-up is seeking to expand on the applications of cationic waterborne polyurethanes, and this falls under the hydrophilic side groups which are divided into cationic, anionic, and non-ionic (Shin & Choi, 2018).

2 Materials and Methods Used for Cationic Waterborne Polyurethanes

Polyurethane is a polymeric material synthesized from two main components: one being isocyanates and the other being polyols. It has become a material of interest, because it can be used for foaming, adhesives, coatings, paints, etc. As a polymer, it can be anionic or cationic. An anionic polymer is one in which the active center to which successive monomers are added during the chain-growth process is a negative

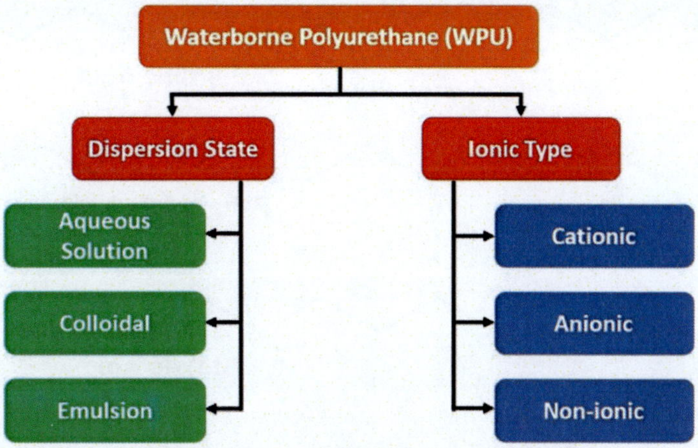

Fig. 1 Classification of waterborne polyurethanes

ion. A cationic polymer on the other hand, during the propagation reaction, the active center is a positive ion, and hence, the added monomers behave as electron donors. A waterborne or more appropriately, water-based polyurethane is one in which a discontinuous phase of polyurethane is dispersed in a stable continuous phase of water. Cationic waterborne polyurethanes usually have application as coating, adhesives, etc., and therefore are of great interest. Waterborne polyurethanes consist of hard (owing to the isocyanates) and soft (owing to the polyols) segments, and the materials that are employed in synthesizing cationic waterborne polyurethanes are crucial because their functionalities dictate their properties. The ability to produce polyols of varying structures using various raw materials and processing methods is vital in the preparation of polyurethanes. Like, primary hydroxyl groups have a high affinity for isocyanates compared to secondary hydroxyls. Eco-friendly materials used for the synthesis of polyols for polyurethanes are described below.

Vegetable oils are widely used as polyol blends for bio-based waterborne polyurethanes, due to their high abundance, economical, and non-toxicity (Liang et al., 2020). Recently, considerable research and publications are concentrated on bio-based chemicals to develop eco-friendly cationic waterborne polyurethanes. The syntheses and applications have been very much reported. Typically, polyols derived from vegetable oils are synthesized taking advantage of unsaturation in their chemical structure (triglycerides). Triglycerides in vegetable oils consist of three carbon chains formed between glycerol and various fatty acids which can be functionalized using several green processes to synthesize polyols (Fig. 2) (Xia & Larock, 2010). Depending on the environment and growing conditions, the fatty acid content (Fig. 3) of different vegetable oils may differ. Polyols derived from a variety of seeds/nuts have gained significant interest and are used extensively for the synthesis of eco-friendly polymers. Waterborne polyurethanes produced from vegetable-based polyols have potential applications in several fields.

Castor oil-based polyols have been applied to disperse cationic waterborne polyurethanes where *N*-methyl diethanolamine (MDEA) acted as an ion center (Liang et al., 2018a). Various waterborne polyurethane films were prepared with varying amounts of MDEA. The synthesized cationic waterborne polyurethanes exhibited enhanced hydrophilicity, due to the reduction in the size of cationic waterborne polyurethanes with an increase in MDEA concentration in the dispersion. The superior thermomechanical and antibacterial properties of the synthesized cationic waterborne polyurethanes suggest their applications in coatings for surgical devices and the food industry. There was also a recent study that used polyol blends of citric acid (octahydro-2,5-pentalenediol, OPD) with castor oil to produce waterborne polyurethane (Liang et al., 2020).

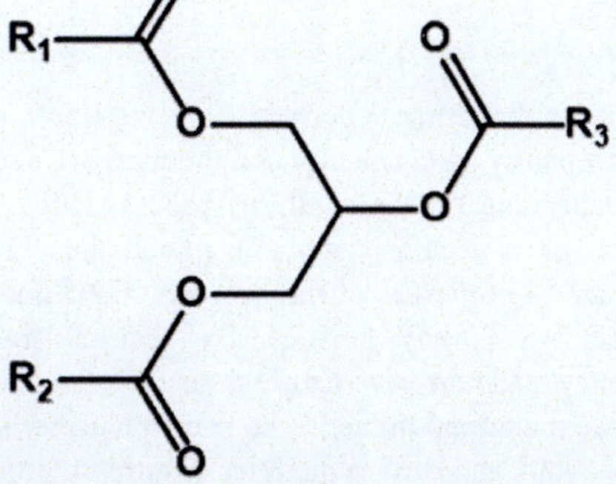

Fig. 2 General structure of plant oil (R_1, R_2, and R_3 represent fatty acids). "Adapted with permission (Xia & Larock, 2010). Copyright (2010), Royal Society of Chemistry."

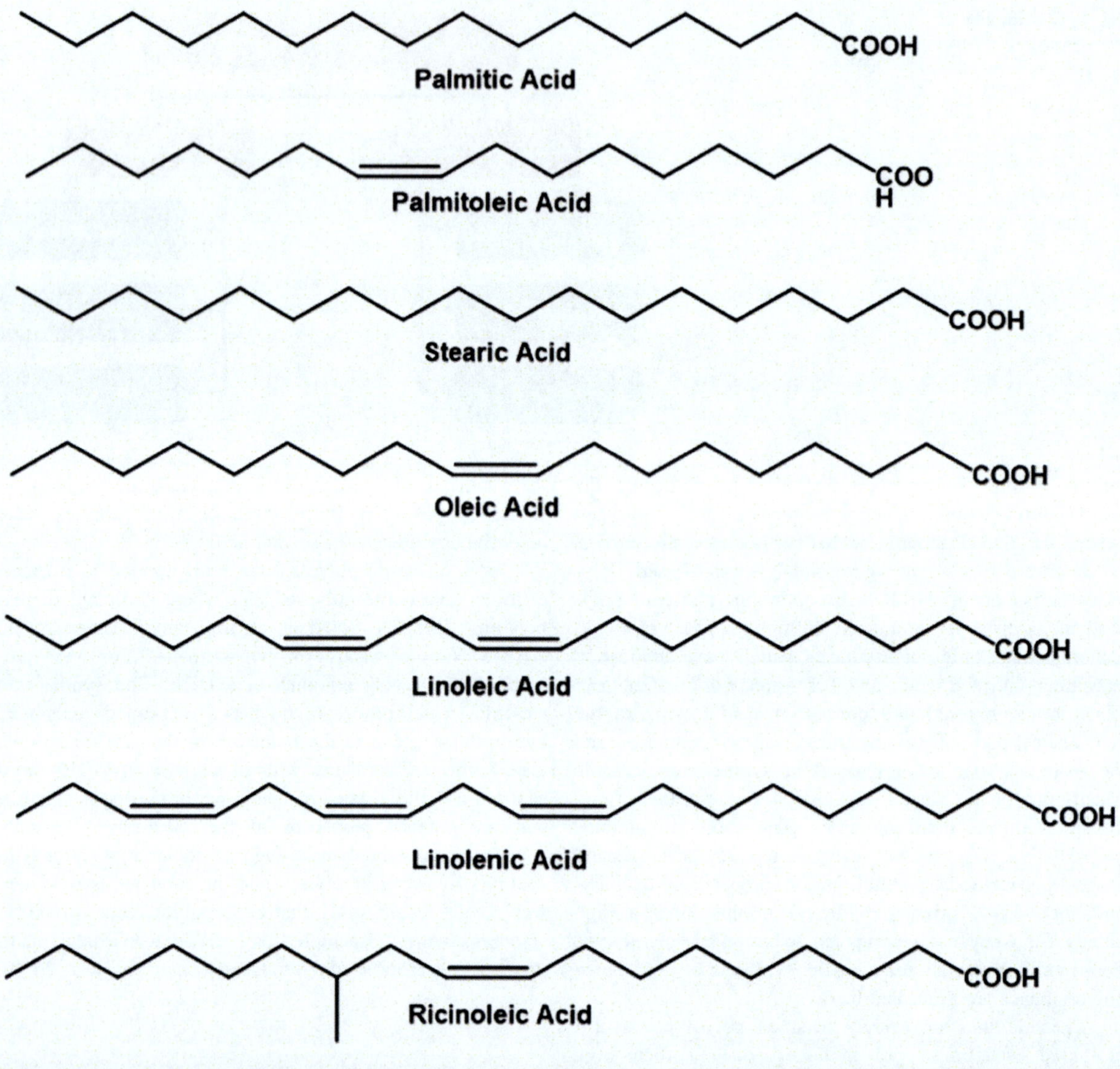

Fig. 3 General structures of selected fatty acids in plant oils. "Adapted with permission (Xia & Larock, 2010). Copyright (2010), Royal Society of Chemistry."

Waterborne polyurethanes derived from vegetable oils tend to perform poorly mechanically and thermally (Cheng et al., 2019). Depending on the synthesis process (Fig. 4), OPD can affect the molecular chain movement, thus improving the mechanical properties of the polymer. OPD content can also affect some characteristics like thermal transitions, transparency, and corrosive nature to name a few. However, they reported minimal influence on some characteristics like thermal stability and surface qualities. Figure 4b portrays the network of partial replacements castor oil with OPD blends showing rigid cyclic structures that are responsible for the improved mechanical performance.

Thermomechanical characteristics of environment-friendly waterborne polyurethane dispersions prepared using jatropha oil-derived polyols indicated its promising applications in coatings (Saalah et al., 2015). The samples showed good thermal stability up to 286 °C. In another report, tung oil-based polyol was applied for the synthesis of cationic waterborne polyurethane dispersions (Man et al., 2019). The cured films showed enhanced thermomechanical behavior for the films having a higher amount of the tung oil-derived polyol. These samples also exhibited antibacterial properties, making them suitable for high-performance coating in pathogen control in paints (Man et al., 2019).

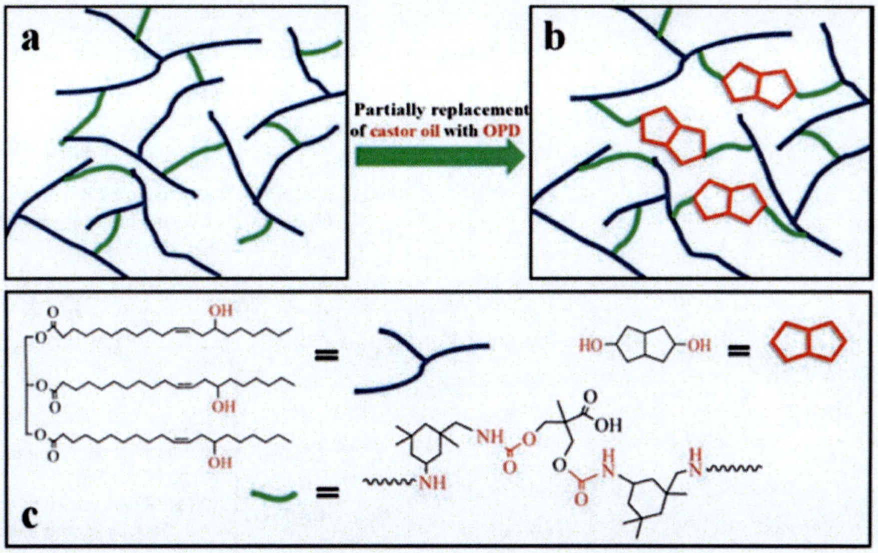

Fig. 4 Network of the polyurethane films prepared using **a** castor oil, **b** castor oil and OPD, and **c** schematics of waterborne polyurethanes. "Reproduced from Liang et al. (2020). Copyright (2020), American Chemical Society"

Lu and his research groups used methoxylated soybean oil-based polyols (MSOLs) to produce aqueous cationic polyurethane (Lu & Larock, 2010b). The variation in thermomechanical properties of the synthesized samples based on the functionality of the prepared polyol was studied. The sample exhibited temperature steadiness up to 200 °C, but started decreasing when the MSOLs content increased from 135 to 176 mg KOH/g. This behavior was explained based on the increased concentration of reactive functional groups (Petrović et al., 2008). The mechanical properties were also dependent on the hydroxyl functional groups.

The idea of using waste forms of oil to produce renewable polymer has always been a fascinating proposition. In a work, waste frying oils were used to synthesize cationic waterborne polyurethanes in a two-step process (alcoholysis reaction and esterification). The route of the chemical process is shown in Fig. 5 (Phunphoem et al., 2019). They examined the effects of emulsifiers on the physical, chemical, mechanical, as well as the antimicrobial behavior of the dispersed polymer. The polymer showed resistance to water and dilute acid (5% v/v H_2SO_4) along with enhanced microbial inhibition for *S. aureus*. Sukhawipat and his team developed a cationic waterborne polyurethane as a substitute for natural rubber in the adhesive industry (Sukhawipat et al., 2020a). By using 2,4-diisocyanatotoluene, *N,N*-Bis (2-hydroxyethyl)methylamine, ethylene glycol, plus natural rubber, the synthesized sample exhibited excellent shear properties. Waterborne polyurethanes developed using renewable resources are environment-friendly and have the potential for commercialization. There is also the possibility of utilizing surfactant-free chemicals in the polymerization process. In this study, Fei and his research group used dimethylaminoethyl methacrylate-b-alkyl methacrylates block copolymers via catalytic chain transfer polymerization (Fei et al., 2019). This work explained the mechanism of the different hydrophobic side chains (terminal double bonds) and how they affect the morphology and rheology of cation waterborne polyurethane dispersions. Quaternary ammonium groups, which are also known to have antimicrobial properties (Gao & Kyratzis, 2012), were used to stabilize waterborne polyurethane with carboxylate groups (Li et al., 2015). Waterborne polyurethanes synthesized using polyester diol along with isophorone diisocyanate exhibited excellent thermal stability. The antimicrobial properties of the quaternary ammonium groups can be investigated in tandem with waterborne polyurethanes for coating applications.

Various methods are investigated for efficient synthesis of waterborne polyurethane dispersions. There is a considerable research effort directed toward the use of novel raw materials for improved performance, as well as an equal effort in developing cheap and facile preparation techniques. There are various synthesis approaches that have been developed for the preparation of waterborne polyurethane, and the more traditional process includes pre-polymer emulsification, acetone method, melt dispersion method, and ketamine/ketazine processes. Recently, innovative techniques such as homogeneous solution polymerization (HSP), mini-emulsion polymerization process (MEPP), additional fragmentation chain transfer method, as well as atom transfer radical polymerization are developed (Madbouly & Otaigbe, 2009).

Pre-polymer emulsification along with the acetone process has gained wide use in industries. They are widely used

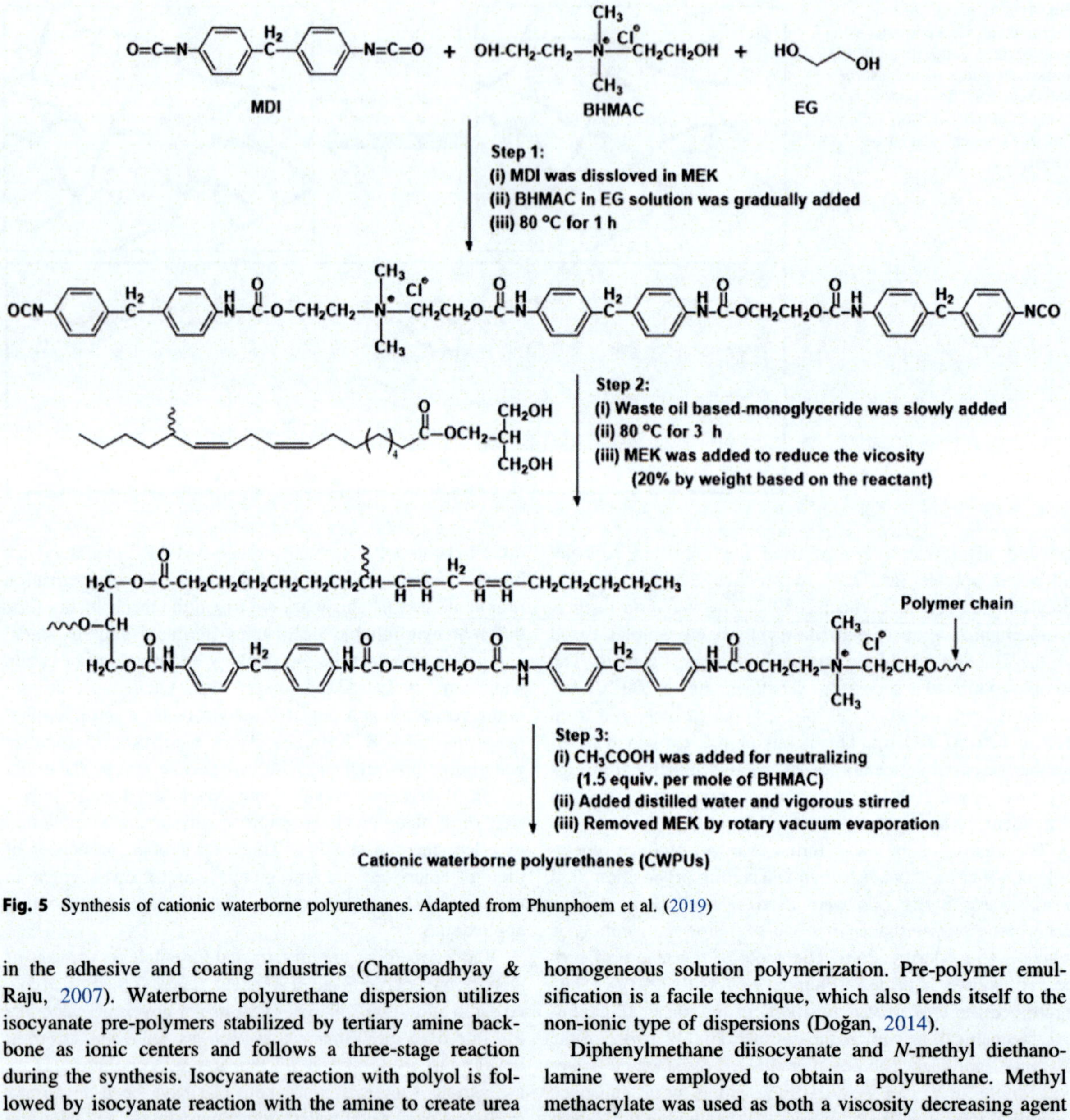

Fig. 5 Synthesis of cationic waterborne polyurethanes. Adapted from Phunphoem et al. (2019)

in the adhesive and coating industries (Chattopadhyay & Raju, 2007). Waterborne polyurethane dispersion utilizes isocyanate pre-polymers stabilized by tertiary amine backbone as ionic centers and follows a three-stage reaction during the synthesis. Isocyanate reaction with polyol is followed by isocyanate reaction with the amine to create urea and reaction of remaining isocyanate with water to reduce the toxicity of waterborne polyurethane. Pre-polymer emulsification is a versatile process commonly used to obtain a large variety of polyurethanes. This technique effectively reduces the use of solvent-borne synthesis. In this process, isocyanate groups with terminal pre-polymers are reacted with ionic groups that form a backbone, and water is mixed for the dispersion of chemicals. A homogeneous mixture of water and surfactant modified with hexamethylenediamine (HMDA) can be also used as the dispersion agent (Madbouly et al., 2005). This process is called the homogeneous solution polymerization. Pre-polymer emulsification is a facile technique, which also lends itself to the non-ionic type of dispersions (Doğan, 2014).

Diphenylmethane diisocyanate and *N*-methyl diethanolamine were employed to obtain a polyurethane. Methyl methacrylate was used as both a viscosity decreasing agent and a monomer. By using different initiators, they obtained different particle morphologies (Zhang et al., 2006). To synthesize novel waterborne polyurethanes, Wei et al. synthesized a surfactant pre-polymer and dispersed it in polytetrahydrofuran. The films made from these samples exhibited good mechanical properties, yet no excess isocyanate and external emulsifiers were used (Wei et al., 1998). In more recent work, a pre-polymer process was used to synthesize nanocomposites of polyurethanes with ZnO for their possible applications in coatings (Zhang et al., 2011). A polyurethane having amphiphilic nature was used to treat

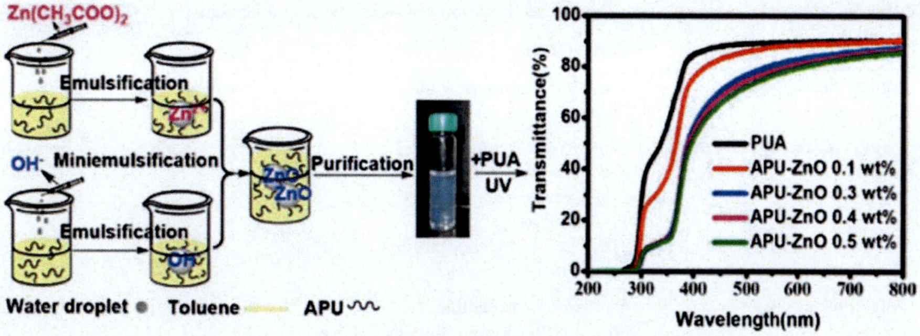

Fig. 6 Schematics for the preparation of nanocomposites and their optical behavior. "Reproduced from Zhang et al., (2020b). Copyright (2020), American Chemical Society"

ZnO nanoparticles to prepare the nanocomposites which displayed improved properties like resistance to sunlight and strong adhesion to the surface. The amphiphilic polyurethane and ZnO nanoparticles were dispersed in toluene using the inverse emulsion method, and the composition of 0.3 wt% ZnO nanoparticles in the polyurethane showed enhanced mechanical and UV-shielding properties as shown in Fig. 6 (Zhang et al., 2020b).

Another approach to disperse waterborne polyurethanes is the acetone route, where the hydrophilic isocyanate pre-polymer with diamines is mixed in acetone to get poly (urethane)-poly(urea) compounds. Water is used to remove the acetone from the dispersion. Due to the low-boiling point of acetone, generally vacuum distillation process is used to remove acetone. Nanoparticles of waterborne polyurethanes were synthesized using poly (1,4-butylene adipate) along with dihydroxy acid using the acetone process. The results showed that the narrow distribution of dispersed polyurethanes improved the thermomechanical properties of the resulting polymer (Shendi et al., 2017). Nanda and his research group studied the effects of chemical adaptability to waterborne polyurethanes prepared via the acetone process and observed that the resulting properties such as the size of dispersed particles and viscosity of the dispersion depend on the concentration of ionic compound and amount of pre-neutralization (Nanda & Wicks, 2006). Using a modified acetone process, Fang et al. produced a self-healing waterborne polyurethane by using Diels–Alder diol in the polymer chains (Fang et al., 2018). The samples containing Diels–Alder diol bonds were synthesized by polyaddition of isocyanate and hydroxyl groups. The samples exhibited excellent thermal-driven healing offering a new paradigm in smart materials.

Melt dispersion process was used to consume the pre-polymer via urea reaction, which forms biuret at the end of the polymeric chain. The reaction takes place around 130 °C in a solvent-free environment (Rosthauser & Nachtkamp, 1986). To promote the chain extension, a reaction with formaldehyde can be performed. The reaction introduces a $-CH_2-$ links between chain end to raise the degree of polymerization. After getting the desired degree of polymerization, the temperature can drop, and water can be mixed to create a dispersed polyurethane. Water can be directly added before or after the methylation reaction with a role to disperse the polyurethane and decrease viscosity. Another advantage of this process is that other compounds such as primary amines, ammonia or hydrazine can be used as an end-cap of isocyanate for the formation of urea end groups (Rosthauser & Nachtkamp, 1986).

The ketamine/ketazine process is based on a diamine compound with a protecting group of a ketone (ketamine) or dihydrazine protected by a ketone group (ketazine). These compounds can be added straight to the pre-polymer mix, and in sequential step, water can be added. Water has two roles in this process, besides promoting the dispersion of the polyurethane, it also causes the hydrolysis of the ketamine and ketazine yielding a diamine or a dihydrazine, respectively. These reagents can then combine to the end of isocyanates to promote the chain extension. The aromatic isocyanates are known to be more reactive toward the water, but by using this method this reaction can be avoided because both the diamine and dihydrazines are more reactive with isocyanate than water.

3 Applications of Cationic Waterborne Polyurethanes

Waterborne polyurethanes are environment-friendly and being used for various applications because of low VOCs. This section of the chapter will enlighten on the applications of cationic waterborne polyurethanes in biomedical, biodegradable, shape memory, electrical and electronic appliances, thermal and corrosion coating, sensors, different protective coating, etc. (Table 1).

3.1 Biomedical Applications

Cationic waterborne polyurethanes are often used in biomedical applications since they provide better biocompatibility, high tear and abrasion resistance, ability to bound

Table 1 Materials, process, and applications of cationic waterborne polyurethanes

Polyols	Components	Process	Application	References
Polyethylene glycol	Dimethyl propionic acid Chitosan from crab shell	Emulsification	Biomedical	Atef El-Sayed et al. (2010)
Methoxylated soybean oil polyol	*N*-methyldiethanolamine *N*-ethyldiethanolamine 1,4-piperazinediethanol Triethanolamine 2,2′,2″,2‴-(ethylenedinitrilo) tetra-ethanol	Emulsification	Biomedical	Xia et al. (2012)
Polyethylene glycol	*N*-ethyldiethanolamine Propyl chloride-functionalized silica Benzyl chloride-functionalized silica	Emulsification	Biomedical	Xu et al. (2015)
Castor oil	*N*-methyldiethanolamine 2-Hydroxy methacrylate	Emulsification	Antibacterial	Liu et al. (2016)
Polyethylene glycol	*N*-alkyl-diethanolamine 2,2,3,3-tetrafluoro-1,4-butanediol	Emulsification	Hydrophobic coating	Król et al. (2013)
Polyethylene glycol	*N*-methyldiethanolamine *N*,*N*-diethylethyl-enediamine	Emulsification	Biodegradable Gene carrier	Yang et al. (2004)
Polyethylene glycol Poly-(Ɛ-caprolactone)	L-Lysine *N*,*N*,*N*′,*N*′-tetramethyl-L-lysine	Solvent exchange	Biodegradable Biocompatible Antibacterial	Zhang et al. (2017)
Thiolene vegetable oils-based polyols	*N*-methyldiethanolamine	Emulsification	Biodegradable	Liang et al. (2018b)
Poly(1,4-butylene adipate)	Dimethyl propionic acid	Solvent exchange	Hydrophobic coating	Zhao et al. (2014)
Graphene oxide	*N*,*N*-dimethylethanolamine	Acetone	Anti-corrosion	Wen et al. (2019)
Polypropylene glycol-400 Polypropylene glycol-1000 Polypropylene glycol-2000	*N*-methyldiethanolamine	Self-emulsification	Anti-corrosion	Gurunathan et al. (2013)
Polycaprolactone polyol	Dimethylol butanoic acid Hydroxyethyl methacrylate 3-methyl propyltrimethoxy silane Graphene oxide	Emulsification	Anti-corrosion Hydrophobic coating	Wang et al. (2018a)
Diethanolamine	Potassium peroxodisulfate GN grafted w/AEMA	Emulsification	Interference shielding	Hsiao et al. (2015)
Polyester polyol	Dimethyl propionic acid Acrylic acid Acrylamide	Solvent exchange	Solid-state electrolyte	Wang et al. (2018b)
Polyether polyol	Dimethyl propionic acid Organophosphonate oligomer Phosphorus–nitrogen with biphosphonyl	Solvent exchange	Flame-retardant coating	Gu and Luo (2015)
N-210 commercial polyol	Dimethyl propionic acid *N*,*N*-bis(2-hydroxyethyl) aminomethyl-phosphonate Polyphosphate	Acetone	Flame-retardant coating	Yin et al. (2017)
Polypropylene glycol-2000	Dimethyl propionic acid DOPO-DAM	Acetone	Flame-retardant coating	Wang et al. (2019b)
Cellulose nanocrystal		Emulsification	Hydrophobic coating	Chowdhury et al. (2020)
N-(2-hydroxyethyl)-maleimide Polypropylene glycol	1,4-butanediol	Acetone	Self-healing	Fang et al. (2018)
Polypropylene glycol	Dimethyl propionic acid 2-hydroxy ethyldisulfide	Emulsification	Self-healing	Zhang et al. (2019)

(continued)

Table 1 (continued)

Polyols	Components	Process	Application	References
Polytetramethylene ether glycol	Dimethyl propionic acid 2-hydroxymethyl methacrylate 2-amino-4-hydroxy-6-methylpyrimidine Ethylenediamine	Emulsification	Self-healing Conductor	Yang et al. (2020)
Polycaprolactone polyol	Dimethyl propionic acid Ethylenediamine	Self-emulsification	Shape memory Biomedical	Wang et al. (2018c)
Poly(neopentylglycoladipate) Poly(oxyethylene alkyl amine)	*N*-methyldiethanolamine	Emulsification	Thermal protection Mechanical protection	Li et al. (2017)
Polyester diol 1,4-butanediol	Dimethyl propionic acid *N*-methyldiethanolamine	Acetone	Bulletproof	Li et al. (2015)
Citric acid-PEG Polycaprolactone polyol	Butanediol	Emulsification	Antibacterial	Bramhecha and Sheikh (2019)
Hydroxyl telechelic natural rubber Hydroxylated rubber seed oil	Dimethyl propionic acid Methyl methacrylate Butyl acrylate	Emulsification	Thermal protection Mechanical protection	Sukhawipat et al. (2020b)
Polytetramethylene ether glycol	Titanium dioxide/reduced graphene oxide 2,2-dimethylolbutyric acid Hydroxyethyl methacrylate	Acetone	Self-cleaning	Chen et al. (2019)

bacteria and other microbes. Cationic waterborne polyurethanes are among the best biocompatible polymers and superior to many commercially available polymers such as polyethylene, silicone rubber, and polyvinyl chloride. In the biomedical equipment, it has been observed that most infections are induced by the bacteria sticking together and duplicating. To overcome this problem, it is important to kill and repel microbes at the same time. In the last decade, different reports on waterborne polyurethane extended with chitosan and tested for antibacterial properties (Atef El-Sayed et al., 2010; Helander et al., 2001). Natural resources such as soybean-oil-based cationic polyurethane coatings displayed enhanced inhibition against gram-positive Listeria Monocytogenes (LM). Zone inhibition (diameter) for the film and dispersion for LM were 12.52 ± 0.14, 12.66 ± 0.22, respectively. The main reason for the excellent activity was due to the ammonium species involved in destructive interaction with the cell wall (Xia et al., 2012). In the other case, where cationic waterborne polyurethanes synthesized from soybean oil were also found inhibitory activity against above mentioned foodborne pathogens as well as *S. aureus* and *E. coli* (Bakhshi et al., 2013; Garrison et al., 2014). Garrison et al. reported that higher concentrations of cations in the polymer enhanced the antibacterial activity toward pathogen (Garrison et al., 2014). 'Grafting to' method was applied in the preparation of polyurethane as well as polyurethane-polyethylene glycol-coated silica particles, and antibacterial activities were studied for *S. aureus* and *E. coli* (Xu et al., 2015). The antibacterial activities were due to the presence of silica, cations in the polymer, and functionalization.

Lysozyme-based polyurethane coatings have shown enhanced antibacterial activities in many surgical devices and food industries (Liu et al., 2016). Cationic waterborne polyurethane synthesized from castor oil (Liang et al., 2018a), waste frying oil (Phunphoem et al., 2019), tung oil (Man et al., 2019), etc., were also reported as antibacterial agents. The properties of the coatings were significantly improved by using two layers where the top layer was antibacterials with an underneath layer of antifouling properties using quaternary ammonium salt waterborne polyurethanes (He et al., 2016). Wu et al. prepared cationic waterborne polyurethanes which showed 100% antibacterial activities toward *E. coli* and *S. aureus*. Furthermore, it displayed the capability to create compounds with plasmid DNA and low cytotoxicity (Wu et al., 2016). Networked waterborne polyurethane (CPTMGPU) having Gemini quaternary ammonium salt showed repeatable contact-active antibacterial efficiency (Zhang et al., 2018) (Fig. 7). Also, CPTMGPU-based coating showed promising characteristics for surgical instruments as well as on the implants to fight against contagious diseases due to its biocompatibility and non-toxic nature (Zhang et al., 2018).

Injury therapeutic assessments with *P. aeruginosa* microbes and their lipopolysaccharide (LPS) used in mice showed that imidazolium cationic polyurethane foams fit as

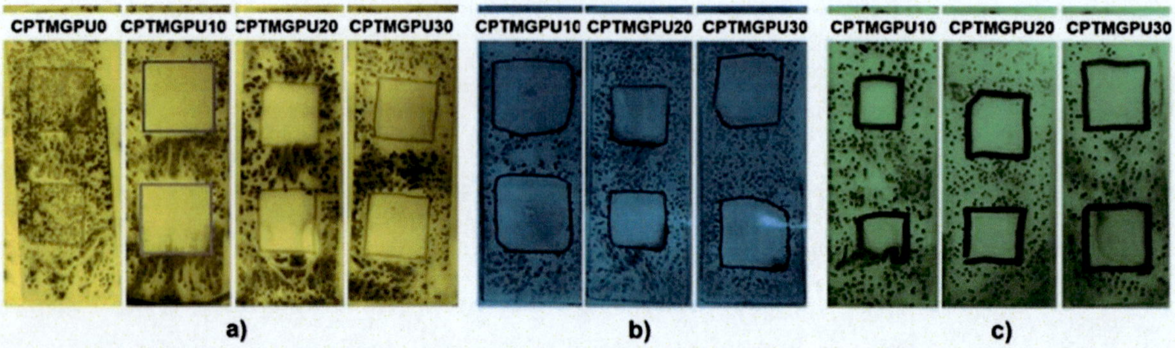

Fig. 7 Photographs of CPTMGPU films after the contact-killing test repeated for **a** 1, **b** 6, and **c** 10 times. "Adapted with permission from Zhang et al. (2018). Copyright (2018) American Chemical Society."

antibacterial polymers able to adsorb endotoxin, showing potential for clinic applications (Fig. 8) (Ding et al., 2019). To decrease the immune response or allergy of macrophage, studies were performed in poly(l-lactic acid)-based waterborne polyurethane that consisted of introducing calcium ions (Huang et al., 2018). The scheme that shows the waterborne polyurethane nanoparticles (PU-NPs) attune the polarization of macrophage is represented in Fig. 9. This study suggested possible uses in the therapeutic field by helping to prevent inflammation reactions and regulate unstable scenarios related to macrophage (Huang et al., 2018).

3.2 Biodegradable Coatings

To date, there are several cationic waterborne polyurethanes for biodegradable applications. Normally, the polyurethane coatings present polar segments that aid their dispersion in water, and to enhance this property, segments with ions can be implemented. On the other hand, other types of coatings based on polyurethane require a polar group to impart a hydrophobic character. Some examples for this case are ceramic, textiles, paper, wood, etc. Few attempts have been carried out to incorporate F, Si, etc., elements into cationic waterborne polyurethanes. Krol et al. synthesized fluorine incorporated waterborne cationomer polyurethane for coatings (Król et al., 2013). The incorporation of fluorine (about 2–5%) in cationic waterborne polyurethanes decreases the surface free energy and increases hydrophobicity. The incorporation of Si into cationic waterborne polyurethanes also lowers the surface free energy of polyurethane. Besides, the higher the amount of Si, the higher the absorption of water. The X-ray photoelectron spectroscopy analysis of this cationic waterborne polyurethane reveals that Si tends to migrate to the surface with ease causing an increase of the hydrophobic nature (Yu et al., 2019). Non-toxic and easily degradable cationic polyurethanes were reported for gene delivery applications (Yang et al., 2004). Transfect with high efficiency and low harm to HEK 293 cells were observed reaching results similar to established gene carriers like poly (20dimethylaminoethyl methacrylate) (Yang et al., 2004). A biodegradable coating for equipment used in surgery and prosthesis is another important area for the sustainable production of cationic waterborne polyurethanes. Waterborne polyurethane that is biodegradable and presents a salt of Gemini quaternary ammonium in its structure was used both as coatings for surface implants as well as tissue scaffolds to avoid infections that can occur in bio-based materials. In this research article, authors have performed an in vitro study of cationic waterborne polyurethanes to analyze the degradation of enzymes and interactions with cells (Zhang et al., 2017). In addition to this, other types of waterborne polyurethane dispersions that were based on linseed, olive, grape, castor, rice bran, canola, and corn oil were obtained and found application as coatings used for either protection or aesthetics (Liang et al., 2018b).

3.3 Weather Protective Coating

Due to high surface energy, common waterborne polyurethanes show an adverse effect on water and soil repellency. To overcome such problems, it is necessary to incorporate strong hydrophobic components such as fluorinated monomers in waterborne polyurethanes. Due to the highest electronegativity and low polarizable nature of fluorine atom, fluorinated waterborne polyurethanes show good water resistance. The following example gives evidence of the use of waterborne polyurethane-fluorinated acrylate (WFPU) emulsions for water-repellent property. The high hydrophobicity along with the capacity to form films and wonderful mechanical properties allows cross-linked WFPU (2.93 wt% of cross-link agent with

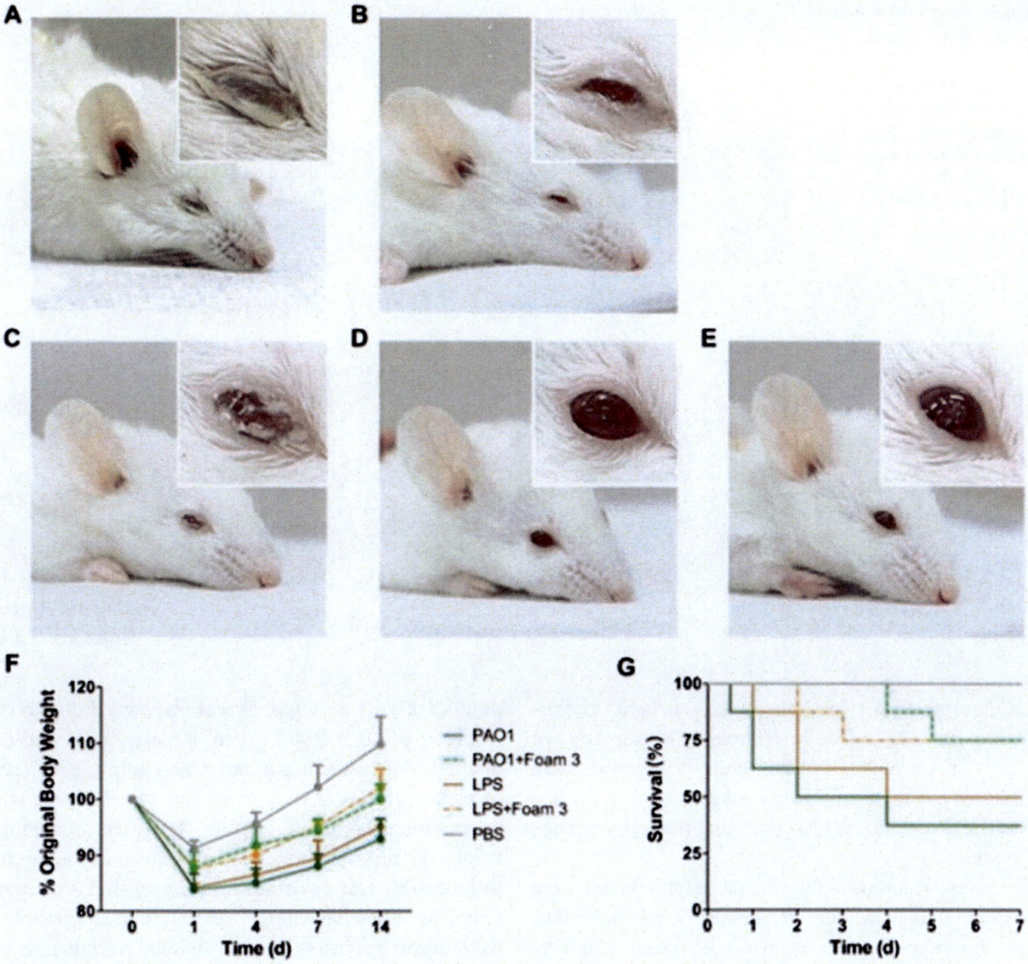

Fig. 8 Assessment of mice models. **a–e** Images systemic status for the rodents at 12 h after the operation from PAO1 (**a**) and LPS (**c**). **b** and **d** show the controls that were positive and mice that were treated individually with Foam 3, respectively. **e** shows the PBS negative control. Both body weight and survival rates were demonstrated in (**f**) and (**g**), respectively, showing mice divided into five groups to analyze the healing test. "Adapted with permission from Ding et al. (2019). Copyright (2019) American Chemical Society."

Fig. 9 Projected mechanism of macrophage polarization. "Adapted with permission from Huang et al. (2018). Copyright (2018) American Chemical Society."

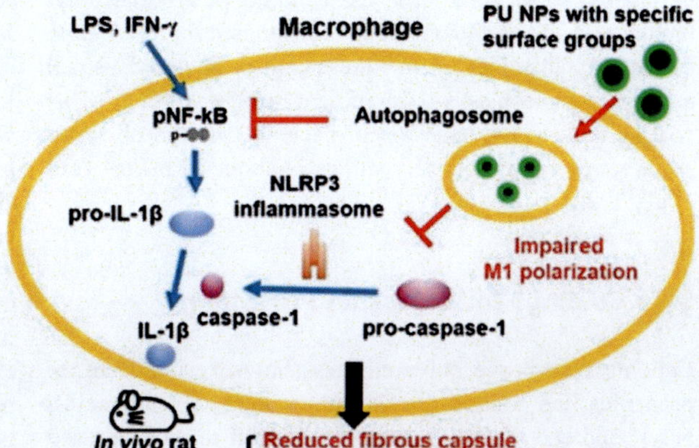

Fig. 10 Digital images of cotton interacting with water. **a** neat material, **b** coated with the waterborne polyurethane, and **c** coated with the cross-linked coating. "Adapted with permission from Zhao et al. (2014). Copyright (2014) American Chemical Society."

30 wt% of fluorinated acrylate) for water-repellent applications (Zhao et al., 2014). The water-repellent properties of cotton that was uncoated, coated, and the coating with cross-link agent for the absorption of water were described in Fig. 10. The droplet on WFPU does not show any change after ten minutes.

In other cases, waterborne fluorinated polyurethanes were used to modify the surface of cotton-based materials. Waterborne fluorinated polyurethanes with linear segments of $CF_3CF_2CH_2$– were more hydrophobic than waterborne fluorinated polyurethanes containing $(CF_3)_2CH$– group. The contact angle for materials treated with the polyurethane for water was 146° when 8 wt% $CF_3CF_2CH_2$– containing groups were used, which increased the hydrophobicity of the polymer (Zhu et al., 2008). The cationic waterborne polyurethanes made of silicone have been studied for antifouling coating applications. After the coating is cured, small emulsifier molecules would present on the surface. This coating reduces the adhesion of bacteria, and the observed surface free energy of the coatings is about 20 mJ/m^2 (Liu et al., 2020).

3.4 Coatings for Corrosion Protection

Cationic waterborne polyurethanes show excellent surface properties like adhesion, resistance against solvents, acids, bases, and eco-friendly aspects enabling it to then be used for corrosion protection. Around 3% of gross domestic product (GDP) on a global scale is wasted due to corrosion (Renner et al., 2006). One largely used anti-corrosion method is by inserting a chromate coating over the surface of metals. It is necessary to replace the chromate coating by environmental-friendly coating. The uses of chromate conversion coating are now banned in most of the courtiers due to its health and environmental impacts. To overcome this issue, an opposite charge coating that contained cationic waterborne polyurethane and anionic vermiculite clay was used as an anti-corrosion agent for aluminum metal. Thirty bilayers of this material with a thickness of only 300 nm can maintain the corrosion protection for at least 5 days. This effective anti-corrosion is observed because of the compact barrier effect by the nano-brick wall arrangement and the lower degree of ionization of cationic polyurethane as shown in Fig. 11 (Qin et al., 2018).

In addition to this, few novel materials were also tested for anti-corrosion property such as covalently bonded waterborne polyurethanes with graphene oxide (GO), which provided efficient corrosion protection after the mixing of GO and modified GO (IP-GO). But with an increase in the concentration of modified IP-GO, anti-corrosion behavior was displayed better. The features of this novel coating only displayed a distinct difference in a digital image and minor corrosion for the same environments of the salt spray experiment. The outcomes show that modified GO can produce a network structure that is efficient to prevent the permeation of corrosive agents (Wen et al., 2019). Likewise, few WPU were also employed for different substrates like

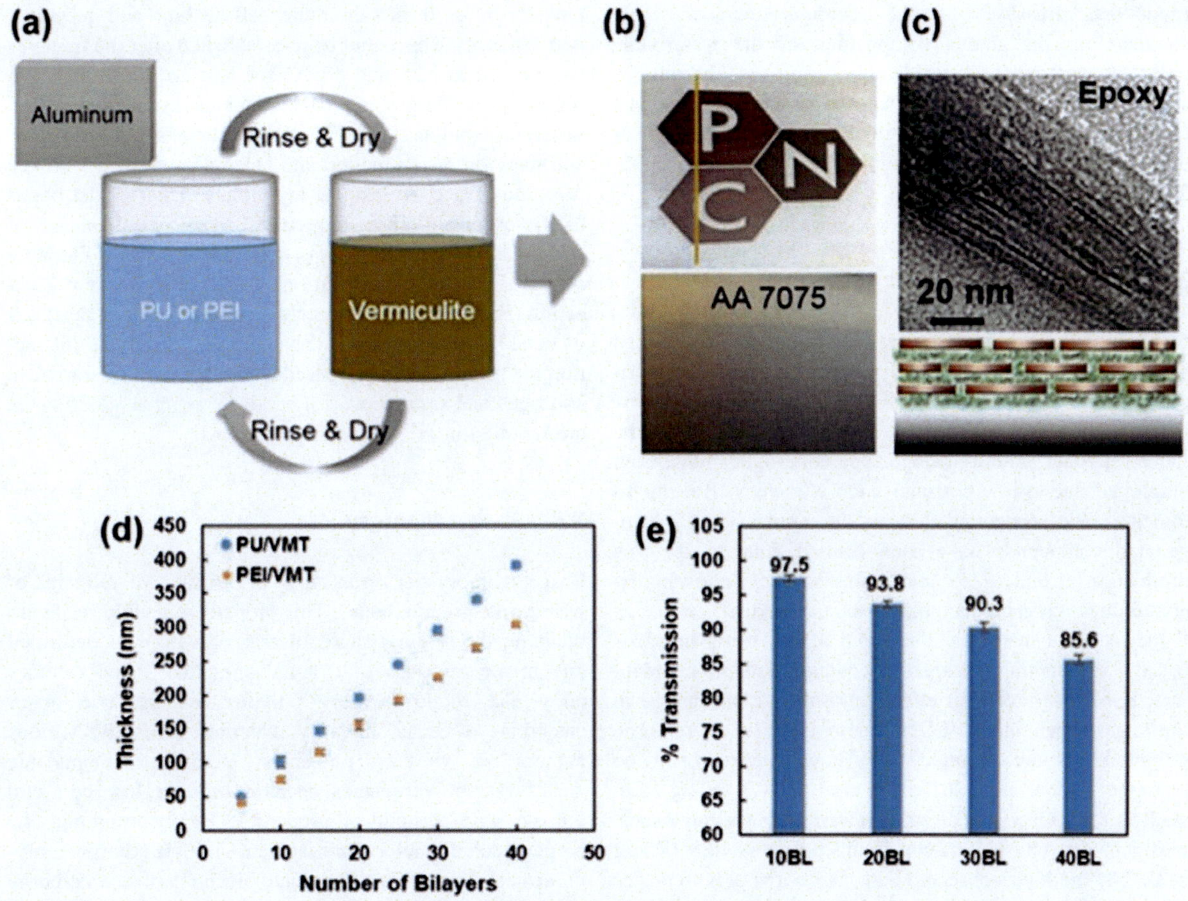

Fig. 11 a Graphic of the layer by the layer assembly process, **b** photo multilayer coatings on PET, uncoated PET, at the right side and left side of the yellow line, respectively, and aluminum, **c** TEM showing the cross section a (PU/VMT)10 and graphic of the thin-film structure, **d** thickness versus the number of bilayers, and **e** % transmittance versus the number of bilayers. "Adapted with permission from Qin et al. (2018). Copyright (2018) American Chemical Society."

mild steel and tinplate for anti-corrosion coating purposes (Gurunathan et al., 2013; Wang et al., 2018a).

3.5 Electrical and Electronic Applications

Frequencies within the range of 8.0–12.5 GHz (X band) can be used for military and telecom applications in cellphones, radar, and TVs. The sensitive optoelectronic devices and electromagnetic interference (EMI) have become a significant difficulty due to the rising utilization of high operating frequencies and bandwidths. For commercial applications, the EMI defending value of 20 dB is measured as the benchmark value for marketable claim purposes. Hsiao et al. reported the use of r-GO/WPU combination for satisfactory electrical conductivity of around 16.9 S/m and EMI shielding effectiveness value approximately up to 34 dB above the frequency interval of 8.0–12.5 GHz (Hsiao et al., 2014). Afterward, they have prepared newer composites using graphene nanosheet (loading of 5 vol%) with aminoethyl methacrylate and WPU and investigated for high electrical conduction of around 43.6 S/m and EMI shielding effect of 38 dB inside the X band frequency range (Hsiao et al., 2015).

The waterborne polyurethane derivative with carbon is reported for outstanding adhesive and electrolyte-uptake applications. The unique waterborne polyurethane with poly (acrylic acid) and acid-treated carbon paper exhibited an interesting all-solid-state, stretchy, no crammed supercapacitor applications (Wang et al., 2018b). Similarly, potassium poly(acrylate) containing waterborne polyurethane (WPU-PAAK) can be used as a solid electrolyte as well as an adhesive in producing carbon-supported flexible electrodes and quasi solid-state electrical double-layer capacitors. Because of the configuration of ionic channels/net, the unique composition of the waterborne polyurethanes binder considerably improves the charge storage capacity of activated carbon or carbon nanotube complexes compared to poly(vinylidene fluoride), which is one of the most used commercial binder s(Wang et al., 2019a).

3.6 Flame Inhibition Coatings

Flame inhibition coatings are fire-resistant chemicals applied in residential, commercial, and various industries. The primary goals of the flame-retardant coatings are to stop, slow the spreading of a fire, or reduce the fire intensity. In the flame retardant, initially halogenated compounds have been employed due to low cost and good efficiency. But due to toxicity and bioaccumulation in the human body, halogenated compounds have been banned. Due to the low production of toxic gases and high efficiency, phosphorus-containing compounds play an important role as flame-retardant materials. The trend of not using halogen-based and adopting phosphorus-containing flame-retardant waterborne polyurethanes effectively showed an increase in limiting oxygen index (LOI). These types of waterborne polyurethanes are reported and showed effective flame resistance (Gu & Luo, 2015; Yin et al., 2017). Wang et al. studied the LOI value of plain waterborne polyurethanes, nitrogen, and phosphorus-containing polyurethanes (Wang et al., 2019b). It was observed that waterborne polyurethanes showed only 18.4%, which indicated that waterborne polyurethanes were flammable; however, after the addition of flame retardant by 9 wt% to waterborne, polyurethanes enhanced the LOI value up to 31.0%.

3.7 Sensors

Sensors are another area where waterborne polyurethane showed promising applications such as humidity sensor and self-healing flexible sensor. Chowdhury et al. showed that nanocellulose crystal-based waterborne polyurethane nanocomposite coatings have a humidity sensitive application. The reversible change in the color of coating has been observed as humidity changes. In addition to this, the material showed good mechanical strength which leads to stability after the alter of water attack (Chowdhury et al., 2020). Healable sensors are another prime area because it has potential application in wearable electronic devices, soft robots, and strain sensors. Mostly, self-healing property is based on reversible interactions and which is completed during a bond-destroying and rejoining reaction. Waterborne polyurethanes based on Diels–Alder reaction (WPU-DA-x) have been used in admirable self-healing and recycling performances. This could be re-established after the fractures were cured at 125 and 65 °C for half an hour and 24 h, respectively (Fang et al., 2018). This environmental benign waterborne polyurethane-DA-x firstly breaks into little parts, and then the dissemination and entanglement of the broken parts took place. At last, the rejoining of Diels–Alder bonds to re-establish the properties. The mechanism of self-repairing WPU-DA-x is shown in Fig. 12. A literature survey showed that in addition to reversible covalent interaction (Fang et al., 2018), mechanisms of exchange reactions of tunable disulfide bonds (Zhang et al., 2019), the interaction of more species consequence between hierarchical hydrogen and strong covalent bonds (Yang et al., 2020), etc., are reported in self-healing applications.

3.8 Shape Memory

Shape memory performance is the distinctive property of waterborne polyurethanes. This property is mainly related to retaining the original shape of materials from its deformed one in the response of external stimulus. Shape memory alloy and shape memory polymer are the two major categories of shape memory materials. Recently, various authors are working on shape memory biodegradable waterborne polyurethanes because of their low cost and tunable physicochemical properties. By incorporating the biodegradable molecular segment in waterborne polyurethanes, it is possible to prepare biocompatible waterborne polyurethanes for shape memory applications. Different authors reported the use of biodegradable shapes elastomers based on waterborne polyurethanes for different applications such as bone tissue engineering (Wang et al., 2018c), shape memory biodegradable stents, or scaffold (Chien et al., 2017).

3.9 Coatings for Other Applications

Because of the soft and hard segment of polyurethanes, the use of hybrid materials diversifies properties and applications of waterborne polyurethanes. The majority of applications of waterborne polyurethanes are covered in the above subsections. Herein, other applications such as thermal coating, mechanical, filtration, and textile have been explained. The thermal stability is an important property of the cationic waterborne polyurethanes and applicable in various areas like bulletproof composite material, thermal coating, etc. A waterborne polyurethane containing cationic and nonionic groups with poly(ethylene glycol methyl ether) (PEGME) was studied for the improved thermal properties. The thermal degradation studies demonstrated a slight

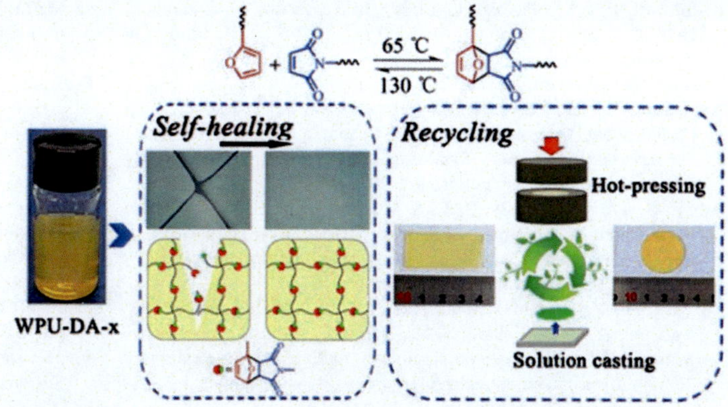

Fig. 12 Graphic diagram of the self-healing mechanism. "Adapted with permission from Fang et al. (2018). Copyright (2018) American Chemical Society."

increment with the increase of the length of the PEGME (Li et al., 2017). Comparative studies of thermal stability were performed for cationic waterborne polyurethane that contained quaternary ammonium groups along with anionic waterborne polyurethanes that contained carboxylate groups that found application as composites for bullet resistance vestment (Li et al., 2015). The results demonstrated higher thermal stability for quaternary ammonium groups rather than carboxylated groups.

Waterborne polyurethanes are also utilized in the water–oil separation phenomenon. Vasquez et al. synthesized hydrophilic and underwater oleophobic foams of waterborne polyurethane using graphite (PUEGr) and calcium carbonate (PUEGr_t) (Vásquez et al., 2019). These materials find applications as filters that use gravity as the gradient. The PUEGr and PUEGr_t foams have a satisfactory rejection of oil that reached 96.85% and 99.99% with a flow rate of 9988 L/m^2h and 8547 L/m^2h, respectively. The maximum water absorption capacity for these materials was 11.5 ± 1.0 g/g and 8.1 ± 0.3 g/g, respectively. In addition to this, the foams function as 3D filters that do not require surfactants to form oil-in-water mixtures (Vásquez et al., 2019).

Cotton tissue displayed a removal of 6303 g water/day while the fabric coated with the polyurethane, which was synthesized by Bramhecha et al. showed removal of 50 g of water/day under the same situation (Bramhecha & Sheikh, 2019). Because of the nonexistence of hydrophilic groups as well as the porous nature of polyurethane, over 99% of water vapors were obstructed so this material could be used in waterproof breathable coating (Bramhecha & Sheikh, 2019). The waterborne polyurethanes are also used in green coating (Sukhawipat et al., 2020b) and degradation of dyes. The waterborne polyurethanes composites with 0.5% C-TiO_2 demonstrated effective self-cleaning property with a high conversion rate of over 88% of methyl orange under 6 h visible light irradiation (Chen et al., 2019). Due to the smaller amount amounts of organic solvents, no or less free isocyanate waterborne polyurethanes are replaced by conventional polyurethanes. This also enhances the health and environmental safety.

4 Concluding Remarks and Future Perspective

In conclusion, this chapter covers the main applications of cationic waterborne polyurethanes synthesized using sustainable sources and environmentally friendly methods. The possible combinations of green methodologies and renewable resources can provide a range of waterborne polyurethanes for various applications. Some of the strongest factors that attracted the attention of waterborne polyurethane were the competitive price of starting materials and the green conversion to value-added products that find applications in specific areas such as biodegradable scaffolds for cell regeneration, anti-bacterial coatings, composites with electromagnetic shielding effect, nanoparticles dispersions for photocatalysis, water purification, and self-healing engineering materials. All these aspects of low cost, versatile properties, and renewable sources fit on the concept of sustainability, which is the effective usage of renewable sources for future generations showing that the efforts of the scientific community, as well as industry, are capable of leading the technologic development into a more sustainable direction.

References

Ahmadi, Y., & Ahmad, S. (2020). Recent progress in the synthesis and property enhancement of waterborne polyurethane nanocomposites: Promising and versatile macromolecules for advanced applications. *Polymer Reviews, 60*, 226–266.

Akindoyo, J. O., Beg, M., Ghazali, S., Islam, M. R., Jeyaratnam, N., & Yuvaraj, A. R. (2016). Polyurethane types, synthesis and applications—A review. *RSC Advances, 6*, 114453–114482.

Al-Jahwari, F. S., & Pervez, T. (2020). In S. Hashmi & Choudhury IABT-E of R and SM (Eds.), *The potential of environmental-friendly biopolymers as an alternative to conventional petroleum-based polymers* (pp. 200–206). Elsevier.

Andjelkovic, D. D., Valverde, M., Henna, P., Li, F., & Larock, R. C. (2005). Novel thermosets prepared by cationic copolymerization of various vegetable oils—Synthesis and their structure-property relationships. *Polymer (guildf), 46*, 9674–9685.

Atef El-Sayed, A., El Gabry, L. K., & Allam, O. G. (2010). Application of prepared waterborne polyurethane extended with chitosan to impart antibacterial properties to acrylic fabrics. *Journal of Materials Science: Materials in Medicine, 21*, 507–514.

Bakhshi, H., Yeganeh, H., Mehdipour-Ataei, S., Shokrgozar, M. A., Yari, A., & Saeedi-Eslami, S. N. (2013). Synthesis and characterization of antibacterial polyurethane coatings from quaternary ammonium salts functionalized soybean oil based polyols. *Materials Science and Engineering C, 33*, 153–164.

Bayer, O. (1947). Das di-isocyanat-polyadditionsverfahren (polyurethane). *Angew Chemie, 59*, 257–272.

Biron, M. (2020). 12—Plastics sustainability: Prospective. In Biron MBT-APG to PS (Ed.), *Plastics design library* (pp. 595–627). William Andrew Publishing.

Bramhecha, I., & Sheikh, J. (2019). Development of sustainable citric acid-based polyol to synthesize waterborne polyurethane for antibacterial and breathable waterproof coating of cotton fabric. *Industrial and Engineering Chemistry Research, 58*, 21252–21261.

Campos, E., Branquinho, J., Carreira, A. S., Carvalho, A., Coimbra, P., Ferreira, P., & Gil, M. H. (2013). Designing polymeric microparticles for biomedical and industrial applications. *European Polymer Journal, 49*, 2005–2021.

Chattopadhyay, D. K., & Raju, K. (2007). Structural engineering of polyurethane coatings for high performance applications. *Progress in Polymer Science, 32*, 352–418.

Chaudhari, A., Gite, V., Rajput, S., Mahulikar, P., & Kulkarni, R. (2013). Development of eco-friendly polyurethane coatings based on neem oil polyetheramide. *Industrial Crops and Products, 50*, 550–556.

Chen, G., Ouyang, S., Deng, Y., Chen, M., Zhao, Y., Zou, W., & Zhao, Q. (2019). Improvement of self-cleaning waterborne polyurethane-acrylate with cationic TiO_2/reduced graphene oxide. *RSC Advances, 9*, 18652–18662.

Cheng, Z., Li, Q., Yan, Z., Liao, G., Zhang, B., Yu, Y., Yi, C., & Xu, Z. (2019). Design and synthesis of novel aminosiloxane crosslinked linseed oil-based waterborne polyurethane composites and its physicochemical properties. *Progress in Organic Coatings, 127*, 194–201.

Chien, Y. C., Chuang, W. T., Jeng, U. S., & Hsu, S. H. (2017). Preparation, characterization, and mechanism for biodegradable and biocompatible polyurethane shape memory elastomers. *ACS Applied Materials & Interfaces, 9*, 5419–5429.

Chowdhury, R. A., Clarkson, C. M., Shrestha, S., El Awad Azrak, S. M., Mavlan, M., & Youngblood, J. P. (2020). High-performance waterborne polyurethane coating based on a blocked isocyanate with cellulose nanocrystals (CNC) as the polyol. *ACS Applied Polymer Materials, 2*, 385–393.

Dales, R., & Raizenne, M. (2004). Residential exposure to volatile organic compounds and asthma. *Journal of Asthma, 41*, 259–270.

Damant, G. H. (1986). The flexible polyurethane foam industry's response to tough fire safety regulations. *Journal of Fire Sciences, 4*, 174–191.

Das, S., Kumar, S., Samal, S. K., Mohanty, S., & Nayak, S. K. (2018). A review on superhydrophobic polymer nanocoatings: Recent development and applications. *Industrial and Engineering Chemistry Research, 57*, 2727–2745.

de Mello, D., Pezzin, S. H., & Amico, S. C. (2009). The effect of post-consumer PET particles on the performance of flexible polyurethane foams. *Polymer Testing, 28*, 702–708.

Dieterich, D., Keberle, W., & Witt, H. (1970). Polyurethane ionomers, a new class of block polymers. *Angewandte Chemie International Edition in English, 9*, 40–50.

Ding, Y., Sun, Z., Shi, R., Cui, H., Liu, Y., Mao, H., Wang, B., Zhu, D., & Yan, F. (2019). Integrated endotoxin adsorption and antibacterial properties of cationic polyurethane foams for wound healing. *ACS Applied Materials & Interfaces, 11*, 2860–2869.

Doğan, N. A. (2014). Synthesis and characterization of novel waterborne polyurethane dispersions.

Fang, Y., Du, X., Jiang, Y., Du, Z., Pan, P., Cheng, X., & Wang, H. (2018). Thermal-driven self-healing and recyclable waterborne polyurethane films based on reversible covalent interaction. *ACS Sustainable Chemistry and Engineering, 6*, 14490–14500.

Fei, G., Geng, H., Wang, H., Liu, X., Liao, Y., Shao, Y., & Wang, M. (2019). Optimization of waterborne poly(urethane-acrylate) nanoemulsions based on cationic polymerizable macrosurfactants with different hydrophobic side chain length. *Polymers (Basel), 11*.

Francis, R., & Kumar, D. S. (2016). *Biomedical applications of polymeric materials and composites*. Wiley.

Gaikwad, M. S., Gite, V. V., Mahulikar, P. P., Hundiwale, D. G., & Yemul, O. S. (2015). Eco-friendly polyurethane coatings from cottonseed and karanja oil. *Progress in Organic Coatings, 86*, 164–172.

Gao, Y., & Kyratzis, I. L. (2012). Antimicrobial finishing of wool using an oxidative pretreatment to enhance the exhaustion of quaternary ammonium compounds. *Journal of Applied Polymer Science, 125*, E71–E78.

Garrison, T. F., Zhang, Z., Kim, H. J., Mitra, D., Xia, Y., Pfister, D. P., Brehm-Stecher, B. F., Larock, R. C., & Kessler, M. R. (2014). Thermo-mechanical and antibacterial properties of soybean oil-based cationic polyurethane coatings: Effects of amine ratio and degree of crosslinking. *Macromolecular Materials and Engineering, 299*, 1042–1051.

Gayki, B. K., Thorat, P. V., & Tayde, S. S. (2015). A review on synthesis and characterization of castor oil based polyurethane adhesive. *International Journal of Emerging Technology and Advanced Engineering, 5*, 95–97.

Gogoi, S., & Karak, N. (2014). Biobased biodegradable waterborne hyperbranched polyurethane as an ecofriendly sustainable material. *ACS Sustainable Chemistry and Engineering, 2*, 2730–2738.

Gu, L., & Luo, Y. (2015). Flame retardancy and thermal decomposition of phosphorus-containing waterborne polyurethanes modified by halogen-free flame retardants. *Industrial and Engineering Chemistry Research, 54*, 2431–2438.

Gurunathan, T., Arukula, R., Suk Chung, J., & Rao, C. R. K. (2016). Development of environmental friendly castor oil-based waterborne polyurethane dispersions with polyaniline. *Polymers for Advanced Technologies, 27*, 1535–1540.

Gurunathan, T., Rao, C. R. K., Narayan, R., & Raju, K. V. S. N. (2013). Synthesis, characterization and corrosion evaluation on new cationomeric polyurethane water dispersions and their polyaniline composites. *Progress in Organic Coatings, 76*, 639–647.

He, W., Zhang, Y., Li, J., Gao, Y., Luo, F., Tan, H., Wang, K., & Fu, Q. (2016). A novel surface structure consisting of contact-active antibacterial upper-layer and antifouling sub-layer derived from gemini quaternary ammonium salt polyurethanes. *Science and Reports, 6*, 1–9.

Helander, I. M., Nurmiaho-Lassila, E. L., Ahvenainen, R., Rhoades, J., & Roller, S. (2001). Chitosan disrupts the barrier properties of the outer membrane of Gram-negative bacteria. *International Journal of Food Microbiology, 71*, 235–244.

Hsiao, S. T., Ma, C. C. M., Liao, W. H., Wang, Y. S., Li, S. M., Huang, Y. C., Bin, Y. R., & Liang, W. F. (2014). Lightweight and flexible reduced graphene oxide/water-borne polyurethane composites with high electrical conductivity and excellent electromagnetic interference shielding performance. *ACS Applied Materials & Interfaces, 6*, 10667–10678.

Hsiao, S. T., Ma, C. C. M., Tien, H. W., Liao, W. H., Wang, Y. S., Li, S. M., Yang, C. Y., Lin, S. C., & Bin, Y. R. (2015). Effect of covalent modification of graphene nanosheets on the electrical property and electromagnetic interference shielding performance of a water-borne polyurethane composite. *ACS Applied Materials & Interfaces, 7*, 2817–2826.

Huang, Y.-C.T., & Tsuang, W. (2014). Health effects associated with faulty application of spray polyurethane foam in residential homes. *Environmental Research, 134*, 295–300.

Huang, Y. J., Hung, K. C., Hung, H. S., & Hsu, S. H. (2018). Modulation of macrophage phenotype by biodegradable polyurethane nanoparticles: Possible relation between macrophage polarization and immune response of nanoparticles. *ACS Applied Materials & Interfaces, 10*, 19436–19448.

Irfan, M., Bhat, S. I., & Ahmad, S. (2018). Reduced graphene oxide reinforced waterborne soy alkyd nanocomposites: Formulation, characterization, and corrosion inhibition analysis. *ACS Sustainable Chemistry and Engineering, 6*, 14820–14830.

Ji, S., Cao, W., Yu, Y., & Xu, H. (2015). Visible-light-induced self-healing diselenide-containing polyurethane elastomer. *Advanced Materials, 27*, 7740–7745.

Kozlowski, R., Malgorzata, M., & Bozena, M. (2011). Comfortable, flexible upholstery fire barriers on base of bast, wool and thermostable fibres. *Polymer Degradation and Stability, 96*, 396–398.

Kreye, O., Mutlu, H., & Meier, M. A. R. (2013). Sustainable routes to polyurethane precursors. *Green Chemistry, 15*, 1431–1455.

Król, P., Król, B., Stagraczyński, R., & Skrzypiec, K. (2013). Waterborne cationomer polyurethane coatings with improved hydrophobic properties. *Journal of Applied Polymer Science, 127*, 2508–2519.

Lei, L., Zhong, L., Lin, X., Li, Y., & Xia, Z. (2014). Synthesis and characterization of waterborne polyurethane dispersions with different chain extenders for potential application in waterborne ink. *Chemical Engineering Journal, 253*, 518–525.

Li, J., Zheng, W., Zeng, W., Zhang, D., & Peng, X. (2014). Structure, properties and application of a novel low-glossed waterborne polyurethane. *Applied Surface Science, 307*, 255–262.

Li, M., Liu, F., Li, Y., & Qiang, X. (2017). Synthesis of stable cationic waterborne polyurethane with a high solid content: Insight from simulation to experiment. *RSC Advances, 7*, 13312–13324.

Li, Y., Zheng, J., Luo, Y., Zhou, H., & Wang, Q. (2015). Effects of different ionizable groups on the thermal properties of waterborne polyurethanes used in bulletproof composites. *Journal of Applied Polymer Science, 132*, 1–9.

Liang, H., Feng, Y., Lu, J., Liu, L., Yang, Z., Luo, Y., Zhang, Y., & Zhang, C. (2018b). Bio-based cationic waterborne polyurethanes dispersions prepared from different vegetable oils. *Industrial Crops and Products, 122*, 448–455.

Liang, H., Li, Y., Huang, S., Huang, K., Zeng, X., Dong, Q., Liu, C., Feng, P., & Zhang, C. (2020). Tailoring the performance of vegetable oil-based waterborne polyurethanes through incorporation of rigid cyclic rings into soft polymer networks. *ACS Sustainable Chemistry and Engineering, 8*, 914–925.

Liang, H., Liu, L., Lu, J., Chen, M., & Zhang, C. (2018a). Castor oil-based cationic waterborne polyurethane dispersions: Storage stability, thermo-physical properties and antibacterial properties. *Industrial Crops and Products, 117*, 169–178.

Liu, K., Su, Z., Miao, S., Ma, G., & Zhang, S. (2016). UV-curable enzymatic antibacterial waterborne polyurethane coating. *Biochemical Engineering Journal, 113*, 107–113.

Liu, S., Zhang, Z., & Qi, Y. (2020) Effect of emulsifier on the structure and properties of waterborne silicone antifouling coating. *Coatings, 10*.

Lu, Y., & Larock, R. C. (2010a). Aqueous cationic polyurethane dispersions from vegetable oils. *Chemsuschem, 3*, 329–333.

Lu, Y., & Larock, R. C. (2010b). Soybean oil-based, aqueous cationic polyurethane dispersions: Synthesis and properties. *Progress in Organic Coatings, 69*, 31–37.

Madbouly, S. A., & Otaigbe, J. U. (2009). Recent advances in synthesis, characterization and rheological properties of polyurethanes and POSS/polyurethane nanocomposites dispersions and films. *Progress in Polymer Science, 34*, 1283–1332.

Madbouly, S. A., Otaigbe, J. U., Nanda, A. K., & Wicks, D. A. (2005). Rheological behavior of aqueous polyurethane dispersions: Effects of solid content, degree of neutralization, chain extension, and temperature. *Macromolecules, 38*, 4014–4023.

Man, L., Feng, Y., Hu, Y., Yuan, T., & Yang, Z. (2019). A renewable and multifunctional eco-friendly coating from novel tung oil-based cationic waterborne polyurethane dispersions. *Journal of Cleaner Production, 241*, 118341.

Mehravar, S., Ballard, N., Tomovska, R., & Asua, J. M. (2019). Polyurethane/acrylic hybrid waterborne dispersions: Synthesis, properties and applications. *Industrial and Engineering Chemistry Research, 58*, 20902–20922.

Mohammad, F. (2007). *Specialty polymers: Materials and applications*. International Publishing House.

Muthuraj, R., & Mekonnen, T. (2018). Recent progress in carbon dioxide (CO_2) as feedstock for sustainable materials development: Co-polymers and polymer blends. *Polymer (guildf), 145*, 348–373.

Nanda, A. K., & Wicks, D. A. (2006). The influence of the ionic concentration, concentration of the polymer, degree of neutralization and chain extension on aqueous polyurethane dispersions prepared by the acetone process. *Polymer (guildf), 47*, 1805–1811.

Noreen, A., Zia, K. M., Zuber, M., Tabasum, S., & Saif, M. J. (2016). Recent trends in environmentally friendly water-borne polyurethane coatings: A review. *Korean Journal of Chemical Engineering, 33*, 388–400.

Pappas, G. P., Herbert, R. J., Henderson, W., Koenig, J., Stover, B., & Barnhart, S. (2000). The respiratory effects of volatile organic compounds. *International Journal of Occupational and Environmental Health, 6*, 1–8.

Paul, K. T. (1990). British regulations for upholstered furniture. In *Fire and polymers* (pp. 30–498). American Chemical Society.

Petrović, Z. S., Guo, A., Javni, I., Cvetković, I., & Hong, D. P. (2008). Polyurethane networks from polyols obtained by hydroformylation of soybean oil. *Polymer International, 57*, 275–281.

Phunphoem, S., Saravari, O., & Supaphol, P. (2019). Synthesis of cationic waterborne polyurethanes from waste frying oil as antibacterial film coatings. *International Journal of Polymer Science, 2019*.

PlasticsEurope, M. R. (2019). Plastics—The facts 2018/2019: An analysis of European plastics production, demand and waste data.

Qin, S., Cubides, Y., Lazar, S., Ly, R., Song, Y., Gerringer, J., Castaneda, H., & Grunlan, J. C. (2018). Ultrathin transparent nanobrick wall anticorrosion coatings. *ACS Applied Nano Materials, 1*, 5516–5523.

Qu, J., Chen, W., Tu, W., & Chen, H. (2000). Development of leather finishes. *FINE Chemicals, 4*.

Qu, J., Tu, W., & Chen, H. (2002). Synthesis and characterization of two component waterborne polyurethane paints. *Journal of Chemical Engineering of Chinese Universities, 16*, 212–216.

Rahman, M. M., Kim, H.-D., & Lee, W.-K. (2009). Properties of waterborne polyurethane adhesives: Effect of chain extender and polyol content. *Journal of Adhesion Science and Technology, 23*, 177–193.

Ramírez, N., Cuadras, A., Rovira, E., Borrull, F., & Marcé, R. M. (2012). Chronic risk assessment of exposure to volatile organic compounds in the atmosphere near the largest Mediterranean industrial site. *Environment International, 39*, 200–209.

Rasshofer, W., & Schomer, D. (2003) European and national ELV regulations on car recovery, their impact on polyurethane applications in the automotive industry and a proposal by the German plastics industry to solve the plastics recovery issue.

Renner, F. U., Stierle, A., Dosch, H., Kolb, D. M., Lee, T. L., & Zegenhagen, J. (2006). Initial corrosion observed on the atomic scale. *Nature, 439*, 707–710.

Rosthauser, J. W., & Nachtkamp, K. (1986). Waterborne polyurethanes. *Journal of Coated Fabrics, 16*, 39–79.

Rumchev, K., Brown, H., & Spickett, J. (2007). Volatile organic compounds: Do they present a risk to our health? *Reviews on Environmental Health, 22*, 39.

Rydz, J., Sikorska, W., Musioł, M., Zawidlak-Węgrzyńska, B., & Duale, K. (2020). In S. Hashmi & Choudhury IABT-E of R and SM (Eds.), *Sustainable future alternative: (Bio)degradable polymers for the environment* (pp. 274–284). Elsevier.

Saalah, S., Abdullah, L. C., Aung, M. M., Salleh, M. Z., Awang Biak, D. R., Basri, M., & Jusoh, E. R. (2015). Waterborne polyurethane dispersions synthesized from jatropha oil. *Industrial Crops and Products, 64*, 194–200.

Seymour, R. B., & Kauffman, G. B. (1992). Polyurethanes: A class of modern versatile materials. *Journal of Chemical Education, 69*, 909.

Shendi, H. K., Omrani, I., Ahmadi, A., Farhadian, A., Babanejad, N., & Nabid, M. R. (2017). Synthesis and characterization of a novel internal emulsifier derived from sunflower oil for the preparation of waterborne polyurethane and their application in coatings. *Progress in Organic Coatings, 105*, 303–309.

Shin, E. J., & Choi, S. M. (2018) Advances in waterborne polyurethane-based biomaterials for biomedical applications.

Signoret, C., Caro-Bretelle, A.-S., Lopez-Cuesta, J.-M., Ienny, P., & Perrin, D. (2019). MIR spectral characterization of plastic to enable discrimination in an industrial recycling context: I. Specific case of styrenic polymers. *Waste Management, 95*, 513–525.

Sin, S. L., Kumar, J. N., Tan, H. R., He, C., Liu, Y., & Xu, J. (2013). The effect of different clay dispersion methods on the properties of polyurethane/clay nanocomposites. *Australian Journal of Chemistry, 66*, 1039–1047.

Stanzione, M., Oliviero, M., Cocca, M., Errico, M. E., Gentile, G., Avella, M., Lavorgna, M., Buonocore, G. G., & Verdolotti, L. (2020). Tuning of polyurethane foam mechanical and thermal properties using ball-milled cellulose. *Carbohydrate Polymers, 231*, 115772.

Sukhawipat, N., Raksanak, W., Kalkornsurapranee, E., Saetung, A., & Saetung, N. (2020). A new hybrid waterborne polyurethane coating synthesized from natural rubber and rubber seed oil with grafted acrylate. *Progress in Organic Coatings, 141*, 105554.

Sukhawipat, N., Saetung, N., Pilard, J. F., Bistac, S., & Saetung, A. (2020). Effects of molecular weight of hydroxyl telechelic natural rubber on novel cationic waterborne polyurethane: A new approach to water-based adhesives for leather applications. *International Journal of Adhesion and Adhesives, 99*.

Sundriyal, P., Pandey, M., & Bhattacharya, S. (2020). Plasma-assisted surface alteration of industrial polymers for improved adhesive bonding. *International Journal of Adhesion and Adhesives, 101*, 102626.

Szycher, M. (2012). Waterborne polyurethanes. In M. Szycher (Ed.), *Szychers handbook of polyurethanes* (2nd ed., pp. 449–494). CRC Press.

Tenorio-Alfonso, A., Sánchez, M. C., & Franco, J. M. (2020). A review of the sustainable approaches in the production of bio-based polyurethanes and their applications in the adhesive field. *Journal of Polymers and the Environment*, 1–26.

Vásquez, L., Campagnolo, L., Athanassiou, A., & Fragouli, D. (2019). Expanded graphite-polyurethane foams for water-oil filtration. *ACS Applied Materials & Interfaces, 11*, 30207–30217.

Wang, H., Qin, S., Yang, X., Fei, G., Tian, M., Shao, Y., & Zhu, K. (2018a). A waterborne uniform graphene-poly(urethane-acrylate) complex with enhanced anticorrosive properties enabled by ionic interaction. *Chemical Engineering Journal, 351*, 939–951.

Wang, H., Wang, S., Du, X., Wang, H., Cheng, X., & Du, Z. (2019b). Synthesis of a novel flame retardant based on DOPO derivatives and its application in waterborne polyurethane. *RSC Advances, 9*, 7411–7419.

Wang, J. A., Lin, S. C., Wang, Y. S., Ma, C. C. M., & Hu, C. C. (2019a). Bi-functional water-born polyurethane-potassium poly (acrylate) designed for carbon-based electrodes of quasi solid-state supercapacitors: Establishing ionic tunnels and acting as a binder. *Journal of Power Sources, 413*, 77–85.

Wang, J. A., Lu, Y. T., Lin, S. C., Wang, Y. S., Ma, C. C. M., & Hu, C. C. (2018b). Designing a novel polymer electrolyte for improving the electrode/electrolyte interface in flexible all-solid-state electrical double-layer capacitors. *ACS Applied Materials & Interfaces, 10*, 17871–17882.

Wang, Y. J., Jeng, U. S., & Hsu, S. H. (2018c). Biodegradable water-based polyurethane shape memory elastomers for bone tissue engineering. *ACS Biomaterials Science & Engineering, 4*, 1397–1406.

Wei, X., Ying, Y., & Yu, X. (1998). A novel synthetic strategy to aromatic-diisocyanate-based waterborne polyurethanes. *Journal of Applied Polymer Science, 70*, 1621–1626.

Wen, J. G., Geng, W., Geng, H. Z., Zhao, H., Jing, L. C., Yuan, X. T., Tian, Y., Wang, T., Ning, Y. J., & Wu, L. (2019). Improvement of corrosion resistance of waterborne polyurethane coatings by covalent and noncovalent grafted graphene oxide nanosheets. *ACS Omega, 4*, 20265–20274.

Wu, G. H., & Hsu, S. H. (2016) Synthesis of water-based cationic polyurethane for antibacterial and gene delivery applications. *Colloids Surfaces B Biointerfaces, 146*, 825–832.

Wu, Q., & Hu, J. (2016). Waterborne polyurethane based thermoelectric composites and their application potential in wearable thermoelectric textiles. *Composites. Part b, Engineering, 107*, 59–66.

Xia, Y., & Larock, R. C. (2010). Vegetable oil-based polymeric materials: Synthesis, properties, and applications. *Green Chemistry, 12*, 1893–1909.

Xia, Y., Zhang, Z., Kessler, M. R., Brehm-Stecher, B., & Larock, R. C. (2012). Antibacterial soybean-oil-based cationic polyurethane coatings prepared from different amino polyols. *Chemsuschem, 5*, 2221–2227.

Xu, Q., Sardon, H., Chan, J. M. W., Hedrick, J. L., & Yang, Y. Y. (2015). Polyurethane-coated silica particles with broad-spectrum antibacterial properties. *Polymer Chemistry, 6*, 2011–2022.

Yan, Y. Z. X. Y. Z. (2007). Application of polyurethane in paper making. *Shanghai Paper Making, 2*.

Yang, T., Chin, W., Cherng, J., & Shau, M. (2004). Synthesis of novel biodegradable cationic polymer: N, N-diethylethylenediamine polyurethane as a gene carrier. *Biomacromolecules, 5*, 1926–1932.

Yang, X., Guo, Y., Luo, X., Zheng, N., Ma, T., Tan, J., Li, C., Zhang, Q., & Gu, J. (2018). Self-healing, recoverable epoxy elastomers and their composites with desirable thermal conductivities by incorporating BN fillers via in-situ polymerization. *Composites Science and Technology, 164*, 59–64.

Yang, Y., Ye, Z., Liu, X., & Su, J. (2020). A healable waterborne polyurethane synergistically cross-linked by hydrogen bonds and covalent bonds for composite conductors. *Journal of Materials Chemistry C, 8*, 5280–5292.

Yin, X., Luo, Y., & Zhang, J. (2017). Synthesis and characterization of halogen-free flame retardant two-component waterborne polyurethane by different modification. *Industrial and Engineering Chemistry Research, 56*, 1791–1802.

Yu, Q., Pan, P., Du, Z., Du, X., Wang, H., & Cheng, X. (2019). The study of cationic waterborne polyurethanes modified by two different forms of polydimethylsiloxane. *RSC Advances, 9*, 7795–7802.

Zafar, F., Ghosal, A., Sharmin, E., Chaturvedi, R., & Nishat, N. (2019). A review on cleaner production of polymeric and nanocomposite coatings based on waterborne polyurethane dispersions from seed oils. *Progress in Organic Coatings, 131*, 259–275.

Zhang, H., Pan, H., Zhang, Q., & Chen, D. (2006). Studies on the particle morphology of waterborne cationic polyurethane/polyacrylate microemulsions. *Journal of Macromolecular Science, Part A Pure and Applied Chemistry, 43*, 1793–1800.

Zhang, M., Zhao, F., & Luo, Y. (2019). Self-healing mechanism of microcracks on waterborne polyurethane with tunable disulfide bond contents. *ACS Omega, 4*, 1703–1714.

Zhang, S., Jiang, J., Yang, C., Chen, M., & Liu, X. (2011). Facile synthesis of waterborne UV-curable polyurethane/silica nanocomposites and morphology, physical properties of its nanostructured films. *Progress in Organic Coatings, 70*, 1–8.

Zhang, S., Zhang, D., Bai, H., & Ming, W. (2020b). ZnO Nanoparticles coated with amphiphilic polyurethane for transparent polyurethane nanocomposites with enhanced mechanical and UV-shielding performance. *ACS Applied Nano Materials, 3*, 59–67.

Zhang, W., Xue, W., Zeng, Z., & Li, J. (2020a). Synthesis and properties of flame-retardant reactive hot melt polyurethane adhesive. *Journal of Adhesion Science and Technology, 34*, 178–191.

Zhang, Y., He, W., Li, J., Wang, K., Li, J., Tan, H., & Fu, Q. (2017). Gemini quaternary ammonium salt waterborne biodegradable polyurethanes with antibacterial and biocompatible properties. *Materials Chemistry Frontiers, 1*, 361–368.

Zhang, Y., He, X., Ding, M., He, W., Li, J., Li, J., & Tan, H. (2018). Antibacterial and biocompatible cross-linked waterborne polyurethanes containing Gemini quaternary ammonium salts. *Biomacromolecules, 19*, 279–287.

Zhao, J., Zhou, T., Zhang, J., Chen, H., Yuan, C., Zhang, W., & Zhang, A. (2014). Synthesis of a waterborne polyurethane-fluorinated emulsion and its hydrophobic properties of coating films. *Industrial and Engineering Chemistry Research, 53*, 19257–19264.

Zhu, M.-J., Qing, F.-L., & Meng, W.-D. (2008). Novel waterborne polyurethanes containing short fluoroalkyl chains: Synthesis and applications on cotton fabrics. *Journal of Applied Polymer Science, 109*, 1911–1915.

Zhu, Y., Romain, C., & Williams, C. K. (2016). Sustainable polymers from renewable resources. *Nature, 540*, 354–362.

Waterborne Polyurethanes Additive Technologies

Pallavi Jain and Sapna Raghav

Abstract

Waterborne polyurathenes (WBPUs) possess flexible properties and are environmentally friendly, these properties of WBPUs have attracted growing interest in an extensive range of industrial applications. While wide work on WBPU synthesis has been reported, still some primary component roles in synthesis remain uncertain. WBPU can be formulated in minimal to no co-solvent coatings and adhesives forming films that form films. These are highly adherent to numerous materials that include polymeric fibres and glass. These eco-friendly WBPUs do not contaminate the soil or generate wastewater as well as are non-flammable and safe. Generally, WBPUs are water insoluble and hydrophobic and for dispersion of WBPUs in water, these need to be modified. This chapter summarizes various modified WBPUs composites along with their properties and applications.

Keywords

Waterborne polyurathenes • Modified WBPUs • Environmental friendly • Hydrophobic • Water insoluble

1 Introduction

Polyurathenes (PUs) are polymers that show high performance and are produced by changing the flexible and rigid parts. The preparation of the ultimate polymer includes the following two consecutive steps. In the first step, an intermediate polymer is formed by reacting the polyol and isocyanate together which has an average molecular weight and this polymer is known as prepolymer. The prepolymer is in general a thick viscous liquid. In the second step, this prepolymer is reacted with a diamine or diol chain extender (CE) to form a high molecular weight (HMW) polymer (Hepburn, 1982).

Different end-polymers can be produced by altering the nature of the catalysts, diisocyanates (DIs) and polyols or other additives. In general, the PUshave fantastic properties including toughness, flexibility, chemical resistance, scratch resistance and abrasion resistance. In most cases, a larger number of PU have a great amount of volatile organic compounds (VOCs) and few of them are isocynate free. The organic solvents are hazardous and cause environmental pollution. For this reason, the aqueous PU dispersion mechanism came to light in the late 1960s (Arnoldus, 1990). The WBPU processing theory is intended to build polymers with numerous hydrophilic groups for achieving water solubility. As the general nature of PU is quite conflicting with water, that is why this alteration in the core nature was made. This amendment in their nature is made by inculcating monomers of hydrophilic nature which have ionic functionality together with sulfonate, carboxylate, or quaternary ammonium groups.

PU ionomers and internal emulsifier is the name being given to these ionic moieties (Dieterich, Keberle, & Witth, 1970). The application of WBPU holds three foremost motives. Firstly, the emphasis has been laid on decrementing the emission of solvent into the environment, by the administrative organization. On the second hand, these solvents are costly and lastly, the most important reason is the characteristics of these WBPU give the ease to use them in wide areas of applications (Garcia-Pacios et al., 2013; Rosthauser & Nachtkamp, 1987). These polymers are environment friendly as, they are safe, non-flammable and in addition, the air is not polluted or waste water is not generated by them. The systems are safe for the environment because only water is being evaporated. Various industrial

P. Jain (✉)
Department of Chemistry, Delhi-NCR Campus, SRM-IST, Modinagar, 210204, India
e-mail: pallavij@srmist.edu.in

S. Raghav
Department of Chemistry, NBGSM College, Gurugram University, Haryana, 122103, India

Inamuddin et al. (eds.), *Sustainable Production and Applications of Waterborne Polyurethanes*, Advances in Science, Technology & Innovation, https://doi.org/10.1007/978-3-030-72869-4_4

applications listed as ink-binder, glass-fibres, coatings, synthetic leathers, membranes, adhesives, biomaterials, paper-sizing, textiles for water-proofing, packaging films are held by these WBPUs (Bai et al., 2007; Honarkar, 2017; Liu, Li, & Sun, 2017; Meng et al., 2009; Poussard et al., 2016; Zhou et al., 2015).

A widespread area of industrial applications such as textiles, beauty-care, constructions, automotive and bio-pharmaceuticals make use of PUs and therefore, they are multipurpose in nature (Akindoyo et al., 2016; Behl et al., 2010). The property of WBPUs of being safe and non-flammable has led to a new advancement of producing eco-friendly coatings as well as adhesives (Chattopadhyay & Raju, 2007). The characteristics of WBPU are mostly dependent on the polymer's chemical configuration since the dispersion of WBPU includes a dispersion of polymeric chain of PUs in water. The arrangement and characteristics of WBPUs are largely determined by its main components (diols, polyols, DIs, additives, neutralizers and CEs) used in its production. However, very different and conflicting results are stated in the literature, as a result of the deep research made to determine the function of these building components. The inculcation of various synthesis procedures and functional groups results in these differences further, making the comparison study challenging. For Instance, a higher Young's modulus (YM) was resulted by polycarbonate (PC) based WBPUs in comparison to polyether (PE) based WBPUs, as reported by Tanaka and Kunimura, (Tanaka & Kunimura, 2002) on the contrary, PC-based PUs results in lower YM compared to polyester/PE-based PUs as reported by Garcia-Pacios et al. (2013).

Systematic synthesis of WBPUs is a way to keenly study the detailed roles of these key parameters as the single variables are secluded. Around 300 WBPU materials generated in the present date are reported to have a systematic tuning in terms of several components concentrations which in turn, has made understanding the individual role played by each component in directing WBPU physicochemical properties quite efficient. As in the field of hair styling, WBPUs having optimum hydrophobicity and mechanical properties can provide high performance in styling and the sensory advantages of WBPU polymer styling are improved by the inculcation of selected additives (fatty acid based). The WBPUs are potential materials for sturdy and durable products for styling providing a natural feel, as stated in both *in-vitro* and *in-vivo* testing.

Many kinds of syntheses are performed to analyze the important constituents. Sodium carbonate was considered as an appropriate neutralizer in the syntheses owing to the safety that it offers (Akesson et al., 1986). A total of three distinct polyols have been observed and they are hexanediol-derived PC, PE and PC derived from hexanediol/pentanediol. They all were taken in amounts of 30–65 wt%. The initial conclusions displayed that all three polyols had the almost same effect on WBPU mechanical properties. Although, it was also observed that the PC derived polyols produced PUs with more YM in contrast to those yielded by PE-based polyols or hexanediol/pentanediol-based PC polyols. The large amount of YM yield due to PC-based WBPUs was because of extra H-bonding present between the carbonyl groups of the softer part (Zhu et al., 2016a). Another study with similar results concluded that the extra H-bonding from PC-based polyols proceeded via improved mixing of phases in correspondence to enhanced mechanical properties (Tanaka & Kunimura, 2002).

Two extenders of chain, i.e. 1-lysine and butanediol were deployed to enhance the molecular weight of WBPU. The first and second one reacts chemically with isocyanate (NCO) groups to develop urea and urethane bonds, respectively (Delebecq et al., 2013). The 1:1 molar ratio of sum the of amine (NH_2) and hydroxyl (OH) groups and the NCO groups were taken in all types of syntheses. For instance, when the content of polyol was made constant and the amount of 1-lysine was enhanced, the amount of 1,4-butanediol was reduced appropriately. WBPUs mechanical properties were tuned within a wide range by customizing the 1-lysine amounts (0–12 wt%). A gradual enhancement in the YM was noticed with increasing the concentration of 1-lysine. This impact was due to the strong cohesion forces between the bonds of urea in contrast to the urethane bonds leading to stronger materials. Similarly, an effect for specific diamine CEs was observed (Santamaria-Echart et al., 2016). 4–6 wt% of 1-lysine was that to be used to make a PU material which is strong as well as flexible. Functional groups that can be ionized in 1-lysine, have positively impacted the stability of the compound and therefore WBPU dispersions (WBPUDs) shelf life (Chen et al., 2002). DI monomers played a vital role in finding physicochemical attributes of WBPU. The study included two distinct isocyanates, i.e. 4,4′-dicyclohexylmethane diisocynate (H12MDI) and isophorone diisocynate (IPDI), which were analyzed in depth so as to optimize PU properties that will be applied in hair styling implementations. H12MDI was found to result in PUs with more crystalline domains and hence formed products having higher toughness and tensile strength (TS) (Akindoyo et al., 2016). In an impartial comparison, WBPUs were synthesized taking similar molar concentrations of H12MDI and IPDI and characterized by different techniques (Zhou et al., 2016). The characterization concluded a greater degree orientation within the structure more domains of crystal were present within the material. No major distinction was seen in the mechanical properties and this surprising result might have occurred due to the availability of many isomers within H12MDI which were employed. The isomeric composition of H12MD1 hugely impacted the resulting WBPUs mechanical properties (Saralegi et al., 2013). In a study, in contrast to IPDI derived WBPUs, the tensile strength was reported to be a little less for H12MDI derived WBPUs (Barikani, Ebrahimi, & Mohaghegh, 2007), while other studies suggested that H12MDI derived PUs showed lower TS as well as particle size (Hourston et al., 1999). Therefore, the same kind of mechanical attributes of IPDI and H12MDI derived WBPUs seen in the research likely rose due to the utilization of mixed H12MDI isomers (Kang et al., 2018).

2 Structure of WBPUs

WBPUs are created from two major segments namely, the emulsifier and the backbone. The emulsifiers are materials with hydrophilic nature and amphiphilic polymers similar to graft polymers. The WBPUDs get stabilized by these emulsifiers. The hard portions having ionic groups are hydrophilic and the soft portions are vital hydrophobic (water-hating) parts. The emulsifiers are categorized as external and internal emulsifiers. The later emulsifiers have ionic centres like cationic, anionic and zwitterions as well as non-ionic centres and hence, the non-ionic WBPUDs consist of a hydrophilic group like polyethylene oxide (PEO). Such dispersions in a variety of pH ranges are seen as colloidally stable. The synthesis scheme for the non-ionic dispersions is demonstrated in Fig. 1 (Szycher, 2013).

3 Properties

3.1 Non-adhesive Behaviour of WBPUDs

The two key components of WBPUDs affect the coating industries positively. The dispersed PU phase is produced by the macrodiol monomers in addition to DI that bonded together by urethane linkages. The dispersion of these polymerized substances is achieved after its suspension in water (Guo, Huang, & Qu, 2018a; Si et al., 2016). In the present years, WBPUDs have attracted researchers worldwide. The readily dispersible polymer and the potential in the thin films to condense the polymer content make it a safer alternative over dispersions carried by solvent (Xu et al., 2013). The other benefits of using PUs in such dispersions include customizing hard and soft sections of PUs, the appealing quality of texture and films, as well as the wetting capacity of polymeric films on surfaces with high energy. The vital concern is, however, in the lower contents of solid and slower solvent evaporation. The organic solvent dispersal has long dominated the polymer industry. It was down-casted only when we reached the final peak of the global environment disruption. After that, the scientists started looking for greener substitute solvents (Mishra, Desai, & Patel, 2018; Pardoa et al., 2018). Ultimately, WBPUs have emerged as natural alternatives for solvent-borne dispersions. These dispersions that are waterborne managed the solvent derived dispersion in practical, environmental and technical concerns (Akram et al., 2018; Alvarez et al., 2018). Since the WBPUDs are employed as emulsions, coating and adhesives the huge usage of the solvent has been substituted (Lewandowski et al., 2002). The restrictions presently have been related to the usage of solvent dependent systems. Various environmental concerns have resulted in the formation of rules like usage of aqueous dispersions of PUs to prevent environmental depletion (Lei Wa et al., 2017).

PU is a major well-known polymer because of its distinct structure. It is composed of DI-bonded macrodiols via urethane linkages. The polymer building block consists of alternating hard and soft portions supported by its DIs, macrodiol, in addition to a polymer chain terminator (Mathew et al., 2018; Shah et al., 2018). The variety in the structure of PU components of DIs, macrodiols and CE resulted suitable transformation in the product's main structure (Gao, Zhang, & Dong, 2013; Kwak, Kim, & Lee, 2005).

The present working concentrates on the thermal as well as mechanical drawbacks that are accountable for WBPUDs non-adhesive behaviour. A sequential composition of WBPUDs has been created by means of appropriate mixing of macrodiols of polypropylene glycol (PPG) and hydroxyl-terminated polybutadiene (HTPB). Use of 1,6-hexane diisocyanate (HDI) has created the urethane linkages. The polymeric components are formed using pre-polymer methods to create environmentally safe dispersions in de-ionized water. The dispersions were tackled to assess the sum of adhesivity.

The PU's composition plays an important role in the dispersion's adhesiveness and it has been studied. The dispersions debonding were investigated by deploying probe tack test. The studies showed that increasing the HTPB molecular weight (upto 3000 g/mol) caused the reduction of WBPUDs adhesive properties. The adhesion energy for adhesive films was much lower (10–42 J/m^2). According to Probe track analysis, the HMW composition of HTPB with a variety of compositions of PPG was not aptly suited for the coating formation of films. The hard and soft segment values were seen as 66% and 76% respectively. Dynamic mechanical analysis showed a rise in the storage module with an increase in HTPB content along with an enhancement in PPG (2700 g/mol) molecular weight up to 1.42 MPa.

This has caused material rigidity and low adhesion energies. The observations of Differential scanning calorimetry (DSC) findings reflected an out-of-range glass transition temperature (Tg) that did not support the tack activity from − 68 °C to − 57 °C. The irrelevant enhancement in the Tg value eventually produced a major loss in the

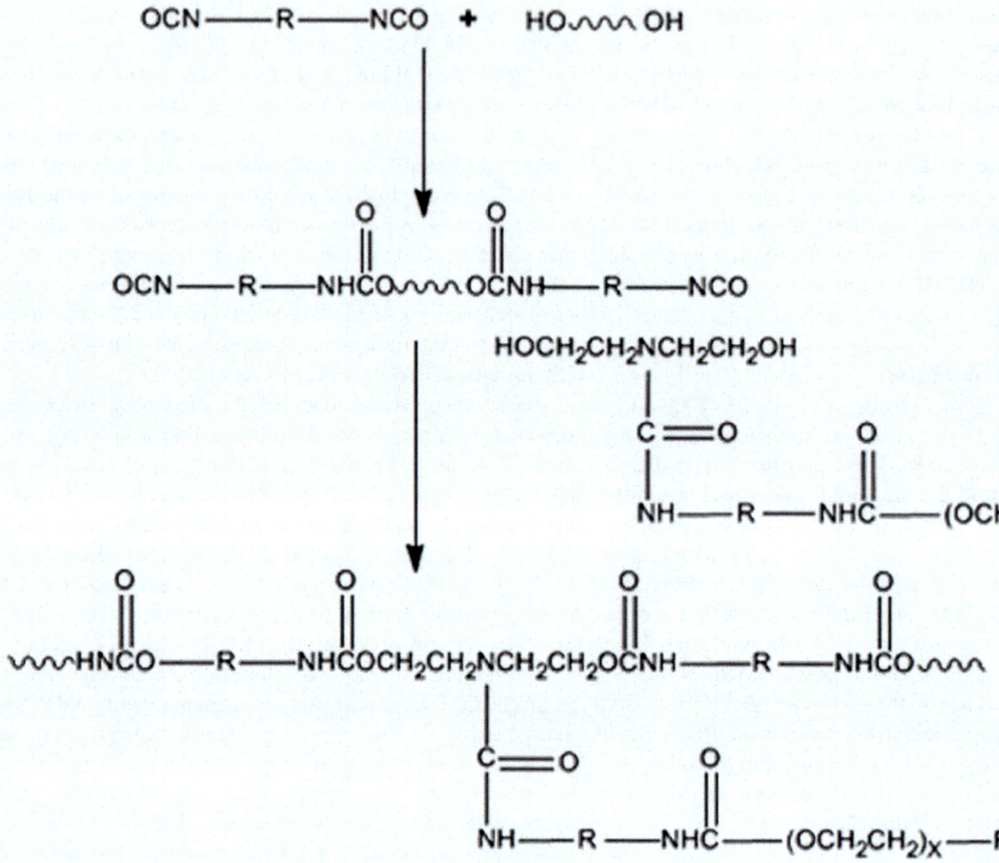

Fig. 1 Scheme of synthesis for the non-ionic dispersions (Szycher, 2013)

behaviour of the tack. Selecting the correct monomer molecular weight is very significant in order to manage the final properties of the polymer dispersions. Though, the dispersion strength with HMW HTPB macrodiols was not an ideal condition for adhesion production (Akram et al., 2019).

3.2 Rheological Properties of WBPU Paints

WBPU paint should be a viscous liquid as well as the storage modulus (G′) must be smaller in comparison to the loss modulus (G″) and the viscosity is towards the lower side, which leads to the paint flow (Deka & Dey, 2019). The viscoelastic properties of WBPU paints are the rulers of their application. Though, obtaining the precise viscoelastic attributes of PU paints through the conventional rotary rheometer (RR) is complex due to the rapid evaporation of solvent throughout the measurements. Therefore, other methods for characterizing the rheological properties of WBPU paints are appropriate and essential (Bhavsar, Raj, & Parmar, 2013; Kästner, 2001; Kuan et al., 2005; Madbouly & Otaigbe, 2009; Osterhold & Wagner, 2002).

The two separate WBPU paints were examined for its rheological properties by utilizing macro-rheological and micro-rheological methods. Solvent evaporation during macro-rheological measurement on a RR cannot be completely excluded, which has an impact on the proven results of the rheological experiments. Thus, the results of linear oscillatory frequency sweep and steady shear obtained from RR measurement were used merely for qualitative analysis. Since solvent evaporation can be ignored during micro-rheological measurements on a Diffusing Wave Spectroscopy (DWS), the results of the G′ and G″ versus frequency are accurate in comparison to the results received from the RR measurement. Consequently, for quantitative analysis, the effects of G′ and G″ versus frequency from DWS measurements were utilized. The G′ for both low-frequency WBPUs is greater than G″ and the opposite is true for the experimental angular frequency range. The modulus values at the similar frequency and the low shear rate viscosity for the two PU paints showed a difference, describing their application difference (Zhanga et al., 2015).

3.3 Effect on the Performance of Two-Component WBPU Additive Coatings

To reason the demand of augmenting environmental concern and higher performance prerequisites, a lot of efforts are being made for the production of water-based coatings that have low content are free of formaldehyde. In the process of this development, two-component WBPU coating systems have played a vital role. On the other hand, this increase in the use of WBPU coatings is becoming an ongoing challenge for formulators. As a matter of fact, the majority of waterborne solvents has a surface tension (ST) of 72.5 dynes/cm whereas, the organic solvents have less than 30 dynes/cm, therefore the formulators face many difficulties in dealing with the waterborne systems. In order to make the task of getting the desired flow/levelling and the rest of surface performance requirements, the formulators usually make use of additives to enhance the surface properties of WBPU coatings. Moreover, to attain the rheological properties for various applications, waterborne systems often require numerous thickeners. Formulators have been able to achieve much consistency in terms of both aesthetics and performance in the waterborne coating systems due to various new waterborne additives available in the market at the current date. The properties of coatings have been intensely affected by the existence of these additives and their interactions.

Statistical experimental design method was used to analyze how a coating's performance in a two-component WBPU coating system can be affected by numerous additives and solvents. The different coating performance parameters, for example, chemical resistivity, viscosity, gloss and hardness can get affected by different solvents, catalysts as wellas solvents at diverse levels, were evaluated.

The additives affected the coating performances in different ways, has been divulged through statistical analysis. In addition to this, the performance properties of the coating have been manipulated due to various two-factor interactions between additives. The one-at-a-time experimental methods can fail to discover these interactions. Coating systems' performance can be made grander by using the information attained from this study (Feng, Lunney, & Wargo, 1999).

3.4 WBPU with High Solid Content and Elasticity

WBPUs were developed utilizing mainly water as a dispersal medium owing to environmentally sustainable and versatile high-performance products to be applied to water based adhesives, coatings and inks (Noreen et al., 2016; Rahman & Lee, 2010; Rahman, Kim, & Lee, 2009; Tao et al., 2013). With regard to the addition of hydrophilic chain extenders, PU can spread firmly to aqueous media, chiefly having carboxylic acid (COOH) (Liu et al., 2018; Yen & Kuo, 2015) and sulfonic acid (SO_3H) and their salts (Honarkar, Barmar, & Barikani, 2016). Moreover, hydrophilic properties for PU can also be acquired by hydrophilic soft segments such as polyethylene glycol (PEG) (Yen & Kuo, 1997).

A WBPU with a solid content of about 45% has been prepared by utilizing polytetrahydrofuran and 2,2-dimethylopropionic acid (DMPA) as a soft segment and hydrophilic chain extenders, respectively, In addition, IDPI and HDP were employed as mixed DIs. WBPU capped by anion-terminated in hard segments was less stable in comparison to anion-terminated in soft segments (Lee & Kim, 2009). Another WBPU with 50% solid content, with an EB to 2000% has been developed by employing DMPA and poly (1,4-butylene adipate) (PBA) glycol as CE. Nevertheless, the crystallinity of PBA got better but it resulted in that the PU could not be water dispersable and the augmentation of solid content was regulated (Li & Sun, 2007). Heat resistance (HR), poor gloss, water resistance (WR) and solvent resistance (SR) are some of the disadvantages caused by the limitations of the WBPU molecular structure (Al-Deyab, Al-Hazmi, & El-Newehy, 2010; Peng & Joy, 2015). To overcome these advantages, less viscous WBPU-acrylate emulsions with 46% solid content were prepared. The emulsion produced had the advantages of better film stability, smaller particle size ($\approx$ 160 nm diameter), better performance and narrow distribution (Zhu et al., 2018).

However, owing to organic emulsifier utilization, the WBPU-acrylate emulsions was not possible in the true sense of aqueous dispersion. By pre-polymerization of IPDI with polycaprolacetone (PCA) and polytetrahydrofuran (PTHM) and subsequent reaction of chain extension by hydrophilic extenders of DMPA and trimethylolpropane (TMP), high elasticity WBPU with 60% solid content was prepared. In aqueous dispersion and dry-state, the size of the PUs can be illustrated by scanning electron microscope (SEM) and dynamic light scattering (DLS). The atomic force microscope (AFM) and micro mechanical stretching machine were employed to evaluate the mechanical properties of WBPU film (Liu et al., 2019).

4 Application

4.1 Graphene-Modified WBPU Materials

Graphene oxide (GO) is obtained via graphite oxidation and is a functionalized grapheme (FG). The functional group substitutes only a few carbon atoms, while not demolishing the entire graphene's crystalline cells. So, GO also preserves graphene's crystalline characteristics. Meanwhile, GO sheets hold a large percentage of oxygen (glycol, epoxy, ketone, carboxyl and OH functional groups) capable of adjusting the strengths of their interaction of Vander Waals forces, resulting in improved compatibility with organic polymer. There is also certain carboxyl and carbonyl groups at the boundary of the sheet which makes more hydrophilic GO sheets, making them more suitable to disperse into certain solvents (Cunha & Paiva, 2019; Zheng et al., 2019).

Using the FG, a novel form of WBPU was developed in order to prepare materials capable of increasing the HR and WR of WBPU. The customized PU materials were evaluated for their wear and HR along with mechanical properties. An increase in the initial temperature of thermal decomposition (from 217 to 244 °C) was observed. The mechanical property of customized PU films first declined and then increased as the FG enhanced. Additionally, the wear resistance of customized PU films decreased as a result of increased FG. Besides at 0.02 wt% FG content, the HR was found to be finest whereas when the content was > 0.04 wt%, the resulted products were stable. The study is considered to be very helpful in the development of composite materials which are environmentally friendly and high performance (Guo et al., 2018b).

4.2 Antiglare (AG)WBPU

WBPU is now considered to be eco-friendly, cheap, washable, safe and apyrous (Chiu, Yu, & Li, 2010; Guo et al., 2009; Jana et al., 2016; Jin et al., 2015; Mahdi & Tan, 2016; Wei et al., 2013). By adjusting the formula, the particle size of WBPU can be easily modified. The particle size of WBPU emulsions were influenced by the contents of poly (n-butyl acrylate-styrene) (PBS), dimethylol-propionic acid (DMPA) and styrene (Guo et al., 2012). Low-glossed coatings have been reported and its influence on size and morphology was analyzed. This allowed for the manufacture of nanostructures with light diffusivity and high transmittance by adjusting the formula (Li et al., 2014). Until recently, work on WBPU coating with AG effect was still lacking. This issue has been resolved by the addition of nano-microspheres into the emulsion of WBPU (Papaj, Mills, & Jamali, 2014). On the other hand, one downside of utilizing the above approach is the appearance of the agglomerate particles on the coated surface owing to low compatibility additive nano-microspheres and PU.

The adjustment of hydrazine hydrate and 2-[(2-aminoethyl) amino]ethyl sulfonic acid sodium (A-95) in WBPU has demonstrated an innovative method for producing AG coating on glass substrates. Different characterization methods were employed to evaluate the morphology and structure of WBPU coating. 60° glossiness and transmittance measured the AG properties of coated glass slides. A first and subsequent decrease in 60° glossiness and transmittance was observed as the content of the A-95 was increased. Under optimum conditions, these values were satisfactory. Results confirmed in the presence of nano-spheres, the coated glasses had low-gloss and high light-transmission properties and suggested its high ability on various glass surfaces (Zhu, Huang, & Peng, 2016b).

4.3 Functionalized WBPU-Based Graphite-Reinforced Composites

As a chemical medium for polymeric materials, vegetable oils have attracted many researchers and become an interesting field of research due to their universal nature, cheap price and biodegradable nature. The attention has been paid particularly, to investigate the aptness of vegetable oil-based polymers (VOBPs) as possible biomaterials. Renewable resource polymers based on their chain fundamentals have a principal interest. Thus, the modification is required to optimize their use by adding different additives as fillers. Due to the extraordinary electrical, thermal and mechanical properties of composites (carbon-based) such as carbon nanotubes (Sattar, Kausar, & Siddiq, 2014; Yang et al., 2017), carbon black (Király & Ronkay, 2015), graphene (Wu et al., 2018; Xia et al., 2016) and activated carbon (Olad & Gharekhani, 2015; Xia et al., 2016) have gained interest, making them potential candidates for multidisciplinary usage. The mechanical properties of the polymer matrix are the most prominent characteristics which are affected by reinforcement (Chakraborty et al., 2017; Pielichowska, Bieda, & Szatkowski, 2016).

Reinforced composites of WBPU-based oxidized graphite (WPUG) were synthesized successfully via the casting method. These composites were prepared by reacting synthesized vegetable oil followed by methylene diphenyl diisocyanate (MDI) crosslinking and finally particle reinforcement of graphite (Fig. 2).

Fourier transform infrared (FTIR) and field emission electron microscopy (FESEM) were employed to evaluate WPUG composites morphology-structure relationship. The result showed random homogeneous graphite particle distribution within the matrix of WBPG and resulted in the development of interconnected interface in the composites which finally encouraged approximately 100% increase in TS and 400% increase in modulus. Such findings were strongly associated with viscoelastic properties where the composites displayed a constructive approach to graphite adding. It was also confirmed that the rise in graphite content, impacting the presence of soft and hard segments in the

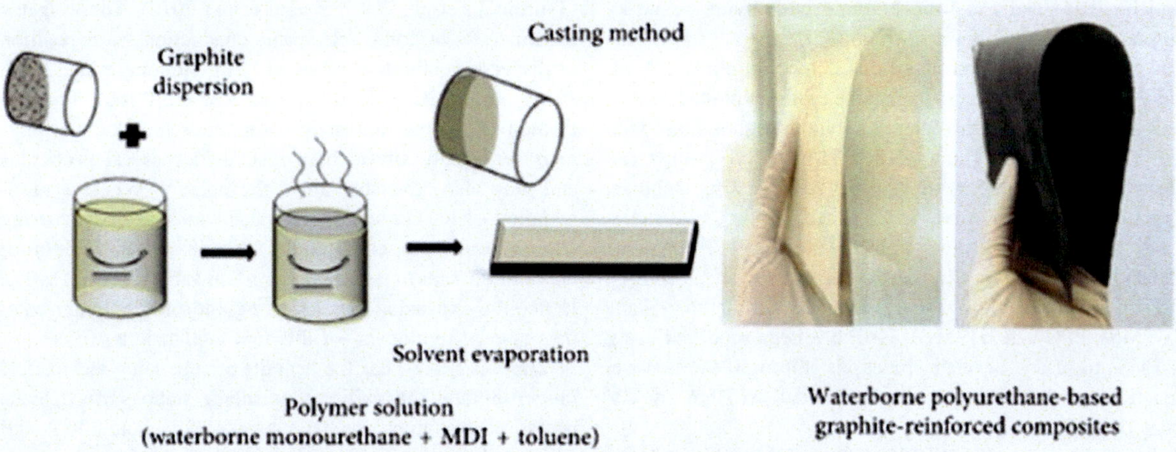

Fig. 2 Preparation diagram of WBPU-based graphene reinforced composites (Abdullah, Mohd Rus, & Abdullah, 2019)

system and therefore offering clearer explanations for mechanical, optical and electrical properties. Therefore, further studies of optical properties feature declining values in the optical energy band gap which subsequently led to electrical conductivity. The composites were found to have positive electrical conductivity by means of the two-point probing method. Basically, in the high content of graphite particles, the internal insulating WBPU gaps appeared to get smaller, leading to lower resistivity values with increasing filler content, supporting electrical conductivity occurrence. This shown the development of potential electric properties as alternative composites based on petroleum for the generation of energy from renewable resources, greener ways to consume energy (Abdullah et al., 2019).

4.4 Polydimethylsiloxane (PDMS)-Modified WBPUs

PDMS has many applications because of its special properties, primarily due to its usual structure consisting of Si–O bond (inorganic) and graft methyl (CH_3) group (organic). The fascinating properties include biocompatibility, higher thermal stability and molecular flexibility, Tg and surface energy, excellent resistance to water and oxidation, superior electrical insulation, chemical inactivity and many more (Chen et al., 2016; Dundua, Franzka, & Ulbricht, 2016; Li et al., 2016; Planes et al., 2016; Seethapathy & Gorecki, 2012). For two decades, the preparation of PDMS-modified WBPU has drawn the attention of numerous researchers and the modified approach has proven to be an efficient and feasible means of improving some of the main WBPU properties. PDMS-modified WBPU have the advantages of both PDMS and WBPU, but it should solve the incompatibility issue. Chemical modification receives much more research attention compared to physical modification, as it can improve compatibility from the molecular level. Diverse methods of polymerisation result in PDMS-modified WBPU having varied properties and structures. PDMS-grafted WBPU have improved surface properties than PDMS-blocked WBPU.

While WBPU's chemistry and technology continue to advance and the demand for a greener planet grows greater than ever before, WBPU production has gradually increased and its applications have grown significantly. Currently, although the PDMS-modified WBPU has made substantial progress and has been marketed in certain fields, still there are several aspects to the challenges. Firstly, there is a need for more kinds of pure, cheap, reactive PDMS having diverse molecular weights, which requires greater efforts from both engineers and researchers in the silicone field. Secondly, the solitary study of PDMS-modified WPU is not sufficient to satisfy the highly flexible and particular requirements for a variety of end uses; instead, an additional study must be conducted to compound with other suitable additives in order to acquire the end products. Thirdly, the challenge among various researchers is to combine the PDMS modification technique creatively and efficiently with other modification technologies to create new WBPUs which have unique properties and can find wider applications (Ji et al., 2017).

4.5 WBPU/hydroxyapatite Chemical Hybrids

Hydroxyapatite (HAP) has a composition and structure similar to that of the human bone mineral phase (Lee et al., 2007; Zhao & Zhang, 2008). This is therefore biocompatible and bioactive and can form bonding with living tissue by osteoconducting (Corcione, Striani, & Frigione, 2013). This has also been used for filling bone defects in clinics. A variety of HAP hydroxy groups can be easily exploited via the sol–gel reaction to form interchain siloxane bondings with vinyl alkoxysilane. The modified HAP vinyl groups can form covalent bonding to organic polymers to stabilize, functionalize and improve performance.

A variety of chemical hybrids of WBPU/HAP were developed utilizing 2-hydroxyethyl acrylate (HEA), dimethylbutanicacid (DMBA) and vinyltrimethoxysilane (VTMS), PPG and H12MDI. HAP has been modified using VTMS to carry acrylate functions through the sol–gel reaction and has been chemically bonded to HEA by UV cure (Fig. 3).

HAP thus provided WBPU with biodegradability, internal and interchain crosslinks in addition to filler reinforcement to increase mechanical, surface and thermal properties. Contact angle (CA), water swell resistance, hardness, Tg, tensile modulus and strength improved with a maximum HAP inclusion of 2%. In addition, beyond this value, considerable HAP particles were aggregated in the water phase and do not participate in the cross-linking process. With a physical blend, the results were less pronounced than the chemical hybrid (Kim & Kim, 2019).

4.6 WBPU-Acrylate modified Epoxy Resin

A cationic WBPU-acrylate modified epoxy acrylic resin was synthesized using acrylic acid, bisphenol A resin, 2,4-toluene diisocyanate (TDI), PEG and HEA. *N,N*-dimethylbenzylamine and hydroquinone were utilized as catalysts and inhibitors, respectively. The chemical titration method was employed to evaluate the change in acid value as well as the NCO group. The process was considered the best process for the synthesis of monoesters of epoxy acrylate with optimum reaction conditions for WBPU-acrylate modified epoxy acrylic resin (Liu et al., 2020).

4.7 Antistatic WBPU Coating

WBPUs are commonly used in the production of industrial coatings and adhesives for a variety of substrates (Harjunalanen & Lahtinen, 2003; Saeed & Shabir, 2013). The resins are insulating and may cause electrostatic discharge (ESD), injurious to electronics contacted. Thus, ESD must be avoided in PUs (He et al., 2017; Shen et al., 2016; Wang et al., 2016; Zia et al., 2015) in electrical or electronic devices. In that regard, the application of antistatic agents, conductive polymers, or conductive fillers will minimize their surface resistance (Rs) to values below 1011 sq^{-1} (Tsurumaki et al., 2015; Zheng et al., 2012). These agents minimize Rs by ion or electronic conduction via deposition or dispersal methods (Chiu et al., 2011; Luong et al., 2013; Weng et al., 2013; Zvonkina & Soucek, 2016). The utilization of internal antistatic compounds (conductive polymers) adversely affects base resin's mechanical properties and may also adversely affect the colours. Whereas minimization of Rs via adsorbing water was observed by using external antistatic compounds (phosphates, sulfonium or ammonium salts) (Haas, Amberg-Schwab, & Rose, 1999). Hence the expansion of everlasting antistatic polymer coatings is very significant for different applications.

Zirconia (ZrO_2) has the benefits of enhancing mechanical properties and the ability to disperse static surface loads across the different antistatic agents (Spanhel, 2006) and found to be enough potential in harsh atmospheres.

A new robust antistatic WBPU-ZrO_2 nanoparticles (NPs) coating was specially developed via sol–gel method. WBPU resin bonding to the substrate including uniform distribution of ZrO_2 NPs successfully improved the dissipation of ESD and adhesion of the coating. The Rs declined with the increase of ZrO_2 material. To achieve antistatic application, WBPUs with over 1.3 wt% of ZrO_2 NPs were utilized. In the meantime, adhesion increased approximately by 90% to 3 wt% ZrO_2 content compared to smooth resin, followed by a decrease. Ultimately, WBPU film with 6 wt% ZrO_2 NPs demonstrated $9.1 \times 109\ \Omega/\ Rs^2$ with an increased adhesive resistance of 3.4 ± 0.1 MPa. In addition, the homogeneous supply of customized/modified ZrO_2 NPs offer 27 nm surface roughnesses which can avoid dust from accumulation on the soil. The simple means of preparing green WBPU-ZrO_2 coating having superior antistatic, as well as mechanical properties provided a beneficial method to develop industrial antistatic surfaces to prevent accumulation of dust or electrical insulation (Yousefi, Dolati, & Najafkhani, 2020).

4.8 Fluorinated WBPU

Different methods have been adopted for modifying WBPU's ST or hydrophobic properties by blending and copolymerization using fluorinated components. Initially research (Guo, Wan, & Fu, 2010) was focused chiefly on researching the fluorine-based materials (additive type) having hydrophobic nature to improve the spreading and wetting properties of WBPU, whereas weak compatibility heterogeneous dispersion cannot be overlooked. Fluorine based monomers after copolymerization provide better resistivity towards water and solvent, lower ST and also preserve original mechanical as well as physical characteristics of WBPU and has simulated significant research interest. Past experiments reported fairly lesser surface-free

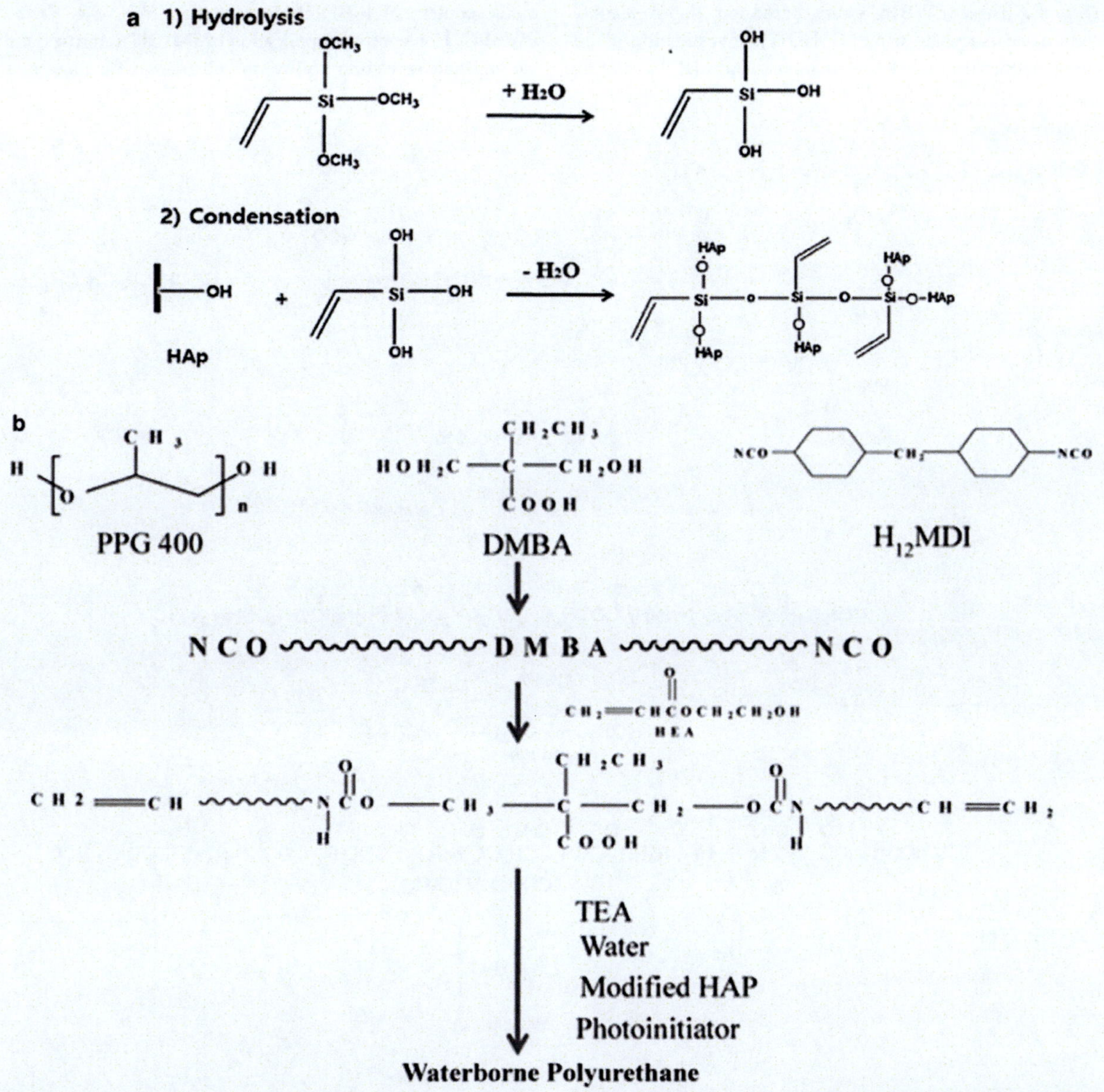

Fig. 3 **a** Scheme of chemical modification of HAP with VTMS, **b** synthesis scheme of WBPU/HAP hybrid. Reprinted from Progress in Organic Coatings, 128, H. A. Kim, B. K. Kim, Synthesis and properties of waterborne polyurethane/hydroxyapatite chemical hybrids, 69–74, Copyright (2018), with permission from Elsevier

energy and can be integrated into PU with fluorinated PEG (Ge et al., 2009) and chain extensors (Tan et al., 2004; Wang & Wei, 2005).

During the process, the segments of fluorine are generally linked to the major macromolecular chain, resulting in low enrichment of the surface and free fluorine migration. For this reason, in the latter case, fluorine-based WBPU (FWBPU) as a short side-chain was synthesized as a chain extender with 2,2,2-trifluoroethanol, but the CA of the water solid film was found to be around 90° and the changes cannot be attained further. The reason of the cause is the very short fluoroalkyl chains that are unable to orient on the surface well (Jiang et al., 2006).

Since fluorinated diols and isocynates are deficient, having longer fluoroalkyl side chains, research-based on fluorinated WBPU with convenient ST and high WR and SR properties have grown gradually. A series of FWBPU having longer fluoroalkyl chains, were prepared successfully

(Fig. 4). These FWBPU were based on fluoro alcohol-terminated isocynate trimer (F-HDIT). Investigation of different properties of FWBPU demonstrated that the increase in the content of F-HDIT also enhances WR, SR, EB of FWBPU. However, the application of F-HDIT improved the wetting and spreading ability on substrates with non-polar

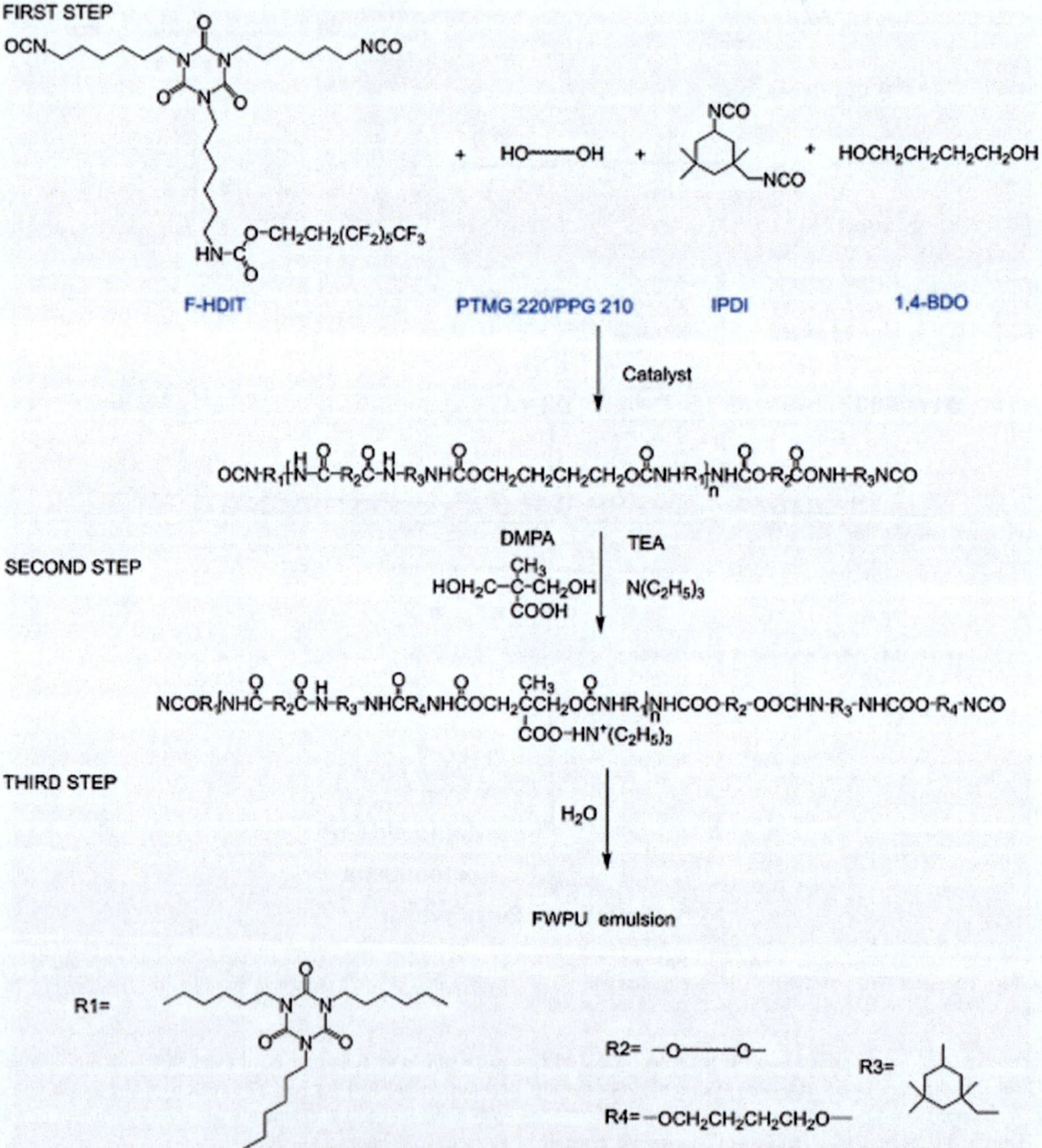

Fig. 4 Synthesis scheme of WBPU/HAP. Reprinted from Progress in Organic Coatings, 131, J. Wena, Z. Suna, H. Fana, Y. Chena, J. Yana, Synthesis and characterization of a novel fluorinated waterborne polyurethane, 291–300, Copyright (2019), with permission from Elsevier

nature and the ST of FWPU film was reduced. In the meantime, the CAs for water as well as methylene iodide were increased from 54.0°/34.1° to 121.8°/90.7° respectively, showing good hydrophobic and solvent-repellent properties. Ultimately, by controlling the fluorine content, a new approach was obtained for preparing FWDPU with lower surface-free energy and higher wetting capacity (Wena et al., 2019).

4.9 Hyperbranched WBPU

A hyperbranched polymeric substance is an OH functional polyester (aliphatic) consisting polyol core that results in core–shell structure having a large amount of OH groups at the peripheral surface. In combination with the dense structure, these reactive groups make them less susceptible to entanglement and thus less viscous and easily reachable for a reaction (Asif, Hu, & Shi, 2009; Han et al., 2012; Yates & Hayes, 2004). Hyperbranched polyester offers extremely lesser viscosity and remarkably fast drying for alkyds, drastically improves toughening for amine-curing epoxies and quick hexakis(methoxymethyl)melamine healing in WBPU (Pettersson, 1996). The effects of CEs (Asif et al., 2005a), structure composition (Asif & Shi, 2004; Florian et al., 2010), kinetics of polymerization (Asif, Huang, & Shi, 2005b) and final groups have been reported for the hyperbranched WBPUs. A hyperbranched 2,2-bis(hydroxymethyl) propionic acid polyester-16-hydroxyl (HBP), was incorporated (chemically) into WBPU as a chain extender for NCO-terminated PU prepolymers. After incorporation of HBP, the particle size decreased initially during the dispersion stage due to the effect of viscosity and increased by HBP extension due to HBP particles swelling. The intensity to break, decomposition temperature, initial modulus, and CA of the dispersion cast increased. The addition of HBP increases WBPU's thermal stability at lower temperatures but worsens it at higher temperatures. Particularly, when the NCO index was unity, the largest increases in those properties were obtained (Park, Bae, & Kim, 2013).

4.9.1 Polyurethane Acrylate (PUA) Modified WBPU

Acrylic polymers have high resistance to wear, flexible mechanical properties and low cost, but poor resistance to solvent and abrasion. A combination of PU with acrylate (AC) is expected to be successful in improving the efficiency of the resulting materials and thus, studies on PU modification by AC monomers have seen considerable progress in recent years (Chen & Chen, 2003; Hirose, Zhou, & Naga, 2000; Jeevananda & Siddaramaiah, 2003). The synthesis of PU/polyacrylate composite latex particles is the best way to achieve both systems. PU–AC interpenetration networks were developed and further improved material performance by using epoxy resin to form ternary latex IPNs (Lee et al., 2001).

IDPI, hydroxyethyl methyl acrylate (HEMA), dimethylolpropionic acid and N220 polyester polyols were employed for the preparation of WBPU and a sequence/series of PUA composites. The final modified WBPU have been obtained after the addition of initiating agent and mixed acrylates (methyl acrylate and butyl acrylate) in diverse mass ratios. HEMA was added at the end of PU chains to develop materials having outstanding characteristics. This was achieved via the reaction of the OH group (in HEMA) and NCO group (at the end of PU chain). Thermal stability, optical transparency and structures of PU as well as PUA were demonstrated using Ultra-violet, FTIR and DSC techniques. After modification, all of the WBPUDs were more stable. The findings showed that mechanical strength and SR of PUA film were enhanced in comparison to the PU film and was due to the intermolecular H-bonding and crosslinking in PUA film. In addition, PUA hybrids were found to have high decomposition temperature and low Tg in comparison to the PU. PUA hybrids obtained have outstanding integrated properties and immense prospective for meeting the extremely diverse demands of modern technologies like adhesives, rubbers, coatings (plastic and wood), wood and leather finishing (Qiu et al., 2010).

5 Conclusion

A broad variety of manufacturing uses such as textiles, skin care, houses, automobiles and bio-pharmaceuticals make use of PUs and are also of essence multipurpose. WBPU 's property of being safe and non-flammable has contributed to a new development in the creation of environmentally friendly coatings as well as adhesives. Many of the characteristics of WBPU depend on the chemical structure of the polymer since the dispersion of WBPU requires dispersion of the polymeric chain of PUs in water. The structure and characteristics of WBPUs are mainly determined by the key components used in their manufacture (diols, polyols, DIs, additives, neutralizers and CEs). Nevertheless, as a result of the deep work done to determine the role of these building components, very different and contradictory findings are reported in the literature. Such variations also result in the inculcation of different synthesis procedures and functional groups, rendering the comparison analysis difficult. This chapter was intended to include a description of the WBPU additive technologies and the related materials.

Acknowledgements Dr. Pallavi Jain is grateful to SRM Institute of Science & Technology, Modinagar for providing the facilities to research.

References

Abdullah, N. M., Mohd Rus, A. Z., & Abdullah, M. F. L. (2019). Functionalized waterborne polyurethane-based graphite-reinforced composites. *Advances in Materials Science and Engineering, 2019*, 8710370.

Akindoyo, J. O., Beg, M. D. H., Ghazali, S., Islam, M. R., Jeyaratnam, N., & Yuvaraj, A. R. (2016). Polyurethane types, synthesis and applications-a review. *RSC Advances, 6*, 114453–114482.

Akesson, B., Bengtsson, M., & Floren, I. (1986). Visual disturbances after industrial triethylamine exposure. *International Archives of Occupational and Environmental Health, 57*, 297–309.

Akram, N., Zia, K. M., Saeed, M., Mansha, A., & Khan, W. G. (2018). Morphological studies of polyurethane based pressure sensitive adhesives by tapping mode atomic force microscopy. *Journal of Polymer Research, 25*, 1591–1596.

Akram, N., Zia, K. M., Saeed, M., Usman, M., Khan, W. G., & Bashir, M. A. (2019). Investigation of non-adhesive behaviour of waterborne polyurethane dispersions. *Journal of Polymer Research, 26*, 45.

Al-Deyab, S. S., Al-Hazmi, A. M., & El-Newehy, M. H. (2010). Synthesis and characterization of organotin containing copolymers: Reactivity ratio studies. *Molecules, 15*, 1784–1797.

Alvarez, G. A., Fuensant, M., Orozco, V. H., Giraldo, L. F., & Martín-Martínez, J. M. (2018). Hybrid waterborne polyurethane/acrylate dispersion synthesized with bisphenol A-glicidylmethacrylate (Bis-GMA) grafting agent. *Progress in Organic Coatings, 118*, 30–39.

Arnoldus, R. (1990). Waterborne Coating, Surface Coating. In A. D. Wilson, J. W. Nicholson, & H. J. Prosser (Eds.), *Surface coating* (Vol. 3, pp. 93–127). Elsevier Applied Science.

Asif, A., Hu, L., & Shi, W. (2009). Synthesis, rheological and thermal properties of waterborne hyperbranched polyurethane acrylate dispersions for UV curable coatings. *Colloid and Polymer Science, 287*, 1041–1049.

Asif, A., Huang, C. Y., & Shi, W. F. (2005b). Photopolymerization of waterborne polyurethane acrylate dispersions based on hyperbranched aliphatic polyester and properties of the cured films. *Colloid and Polymer Science, 283*, 721–730.

Asif, A., Shi, W., Shen, X., & Nie, K. (2005a). Physical and thermal properties of UV curable waterborne polyurethane dispersions incorporating hyperbranched aliphatic polyester of varying generation number. *Polymer, 46*, 11066–11078.

Asif, A., & Shi, W. (2004). UV curable waterborne polyurethane acrylate dispersions based on hyperbranched aliphatic polyester: Effect of molecular structure on physical and thermal properties. *Polymers for Advanced Technologies, 15*, 669–675.

Bai, C. Y., Zhang, X. Y., Dai, J. B., & Zhang, C. Y. (2007). Water resistance of the membranes for UV curable waterborne polyurethane dispersions. *Progress in Organic Coatings, 59*, 331–336.

Barikani, M., Ebrahimi, M. V., & Mohaghegh, S. M. S. (2007). Influence of diisocyanate structure on the synthesis and properties of ionic polyurethane dispersions. *Polymer-Plastics Technology and Engineering, 46*, 1087–1092.

Behl, M., Razzaq, M. Y., & Lendlein, A. (2010). Multifunctional shape-memory polymers. *Advanced Materials, 22*, 3388–3410.

Bhavsar, R., Raj, R., & Parmar, R. (2013). Studies of sedimentation behaviour of high pigmented alkyd primer: A rheological approach. *Progress in Organic Coatings, 76*, 852–857.

Chakraborty, A., Agresti, A., Pizzoferrato, R., Matteis, F. D., Orsini, A., & Medaglia, P. G. (2017). Study of structural and optical properties of low temperature photo-activated ZnO-rGO composite thin film. *Materials Research Bulletin, 91*, 227–231.

Chattopadhyay, D. K., & Raju, K. V. S. N. (2007). Structural engineering of polyurethane coatings for high performance applications. *Progress in Polymer Science, 32*, 352–418.

Chen, S., & Chen, L. (2003). Structure and properties of polyurethane/polyacrylate latex interpenetrating networks hybrid emulsions. *Colloid and Polymer Science, 282*, 14–20.

Chen, J. K., Guo, H., Ding, P., Pan, R. Z., Wang, W. B., Xuan, W. P., Wang, X. Z., Jin, H., Dong, S. R., & Luo, J. K. (2016). Transparent triboelectric generators based on glass and polydimethylsiloxane. *Nano Energy, 30*, 235–241.

Chen, H., Jiang, X., He, L., Zhang, T., Xu, M., & Yu, X. (2002). Novel biocompatible waterborne polyurethane using l-lysine as an extender. *Journal of Applied Polymer Science, 84*, 2474–2480.

Chiu, H. T., Chang, C. Y., Chen, C. L., Chiang, T. Y., & Guo, M. T. (2011). Preparation and characterization of UV-curable organic/inorganic hybrid composites for NIR cutoff and antistatic coatings. *Journal of Applied Polymer Science, 120*, 202–211.

Chiu, S. C., Yu, H. C., & Li, Y. Y. (2010). High electromagnetic wave absorption performance of silicon carbide nanowires in the gigahertz range. *Journal of Physical Chemistry C, 114*, 1947–1952.

Corcione, C. E., Striani, R., & Frigione, M. (2013). UV-cured siloxane-modified methacrylic system containing hydroxyapatite as potential protective coating for carbonate stones. *Progress in Organic Coatings, 76*, 1236–1242.

Cunha, E., & Paiva, M. C. (2019). Composite films of waterborne polyurethane and few-layer graphene-enhancing barrier, mechanical and electrical properties. *Journal of Composites Science, 3*, 35.

Deka, A., & Dey, N. (2019). Rheological studies of two component high build epoxy and polyurethane based high performance coatings. *Journal of Coatings Technology and Research, 10*, 305–315.

Delebecq, E., Pascault, J. P., Boutevin, B., & Ganachaud, F. (2013). On the versatility of urethane/urea bonds: Reversibility, blocked isocyanate and non-isocyanate polyurethane. *Chemical Reviews, 113*, 80–118.

Dundua, A., Franzka, S., & Ulbricht, M. (2016). Improved antifouling properties of polydimethylsiloxane films via formation of polysiloxane/polyzwitterion interpenetrating networks. *Macromolecular Rapid Communications, 37*, 2030–2036.

Dieterich, D., Keberle, W., & Witth, H. (1970). Polyurethane ionomers, a new class of block polymers. *Angewandte Chemie International Edition Engineering, 9*, 40–50.

Feng, S. X., Lunney, P., & Wargo, R. (1999). Effects of additives on the performance of two-component waterborne polyurethane coatings. *Journal of Coatings Technology, 71*, 143–149.

Florian, P., Jena, K. K., Allauddin, S., Narayan, R., & Raju, K. V. S. N. (2010). Preparation and characterization of waterborne hyperbranched polyurethane-urea and their hybrid coatings. *Industrial and Engineering Chemistry Research, 49*, 4517–4527.

Gao, N., Zhang, Z., & Dong, Q. (2013). Preparation and properties of two-component and double-crosslinking waterborne polyurethane-acrylic dispersions. *Open Journal of Organic Polymer Materials, 3*, 27–33.

García-Pacios, V., Colera, M., Iwata, Y., & Martín-Martínez, J. M. (2013). Incidence of the polyol nature in waterborne polyurethane dispersions on their performance as coatings on stainless steel. *Progress in Organic Coatings, 76*, 1726–1729.

Garcia-Pacios, V., Jofre-Reche, J. A., Costa, V., Colera, M., & Martin-Martinez, J. M. (2013). Coatings prepared from waterborne polyurethane dispersions obtained with polycarbonates of 1,6-hexanediol of different molecular weights. *Progress in Organic Coatings, 76*, 1484–1493.

Ge, Z., Zhang, X., Dai, J., Li, W., & Luo, Y. (2009). Synthesis, characterization and properties of a novel fluorinated polyurethane. *European Polymer Journal, 45*, 530–536.

Guo, S., Ding, Z., Pang, X., Wang, W., Yin, Y., & You, G. (2018b). Study on the functionalized graphene modified waterborne polyurethane materials. In Albu, L., & Deselnicu, V. (Eds.),

Proceedings of International Conference on Advanced Materials and Systems (pp. 99–104). Bucharest, Romania.

Guo, L., Huang, S., & Qu, J. (2018a). Synthesis and properties of high functionality hydroxyl-terminated polyurethane dispersions. *Progress in Organic Coatings, 119*, 214–220.

Guo, Z., Lee, S. E., Kim, H., Park, S., Hahn, H. T., Karki, A. B., & Young, D. P. (2009). Fabrication, characterization and microwave properties of polyurethane nanocomposites reinforced with iron oxide and barium titanate nanoparticles. *Acta Materialia, 57*, 267–277.

Guo, Y. H., Li, S. C., Wang, G. S., Ma, W., & Huang, Z. (2012). Waterborne polyurethane/poly(*n*-butyl acrylate-styrene) hybrid emulsions: Particle formation, film properties and application. *Progress in Organic Coatings, 74*, 248–256.

Guo, W. J., Wan, J. Q., & Fu, H. Q. (2010). Adhesion mechanism of multi-modified waterborne polyurethane adhesive for polyolefin films. *China Adhesive, 22*, 178–189.

Haas, K. H., Amberg-Schwab, S., & Rose, K. (1999). Functionalized coating materials based on inorganic-organic polymers. *Thin Solid Films, 351*, 198–203.

Han, W., Lin, B., Yang, H., & Zhang, X. (2012). Synthesis and properties of UV-curable hyperbranched polyurethane acrylate oligomers containing carboxyl groups. *Polymer Bulletin, 68*, 1009–1022.

Harjunalanen, T., & Lahtinen, M. (2003). The effects of altered reaction conditions on the properties of anionic poly(urethane-urea) dispersions and films cast from the dispersions. *European Polymer Journal, 39*, 817–824.

He, H., Yan, Y., Qiu, Z., & Tan, X. (2017). A novel antistatic polyurethane hybrid based on nanoscale ionic material. *Progress in Organic Coatings, 113*, 110–116.

Hepburn, C. (1982). *Polyurethane elastomers* (pp. 355–406). Applied Science Publisher.

Hirose, M., Zhou, J. H., & Nagai, K. (2000). The structure and properties of acrylic- polyurethane hybrid emulsions. *Progress in Organic Coating, 38*, 27–34.

Honarkar, H. (2017). Waterborne polyurethanes: A review. *Journal of Dispersion Science and Technology, 39*, 507–516.

Honarkar, H., Barmar, M., & Barikani, M. (2016). New sulfonated waterborne polyurethane dispersions: Preparation and characterization. *Journal of Dispersion Science and Technology, 37*, 1219–1225.

Hourston, D. J., Williams, G., Satguru, R., Padget, J. D., & Pears, D. (1999). The influence of the degree of neutralization, the ionic moiety and the counterion on water-dispersible polyurethanes. *Journal of Applied Polymer and Science, 74*, 556–566.

Jana, D., Samanta, S., Maikap, S., & Cheng, H. M. (2016). Evolution of complementary resistive switching characteristics using $IrO_x/GdO_x/Al_2O_3/TiN$ structure. *Applied Physics Letters, 108*, 011605.

Jeevananda, T., & Siddaramaiah, H. (2003). Synthesis and characterization of polyaniline filled PU/PMMA interpenetrating polymer networks. *European Polymer Journal, 39*, 569–578.

Ji, X., Wang, H., Ma, X., Hou, C., & Ma, G. (2017). Progress in polydimethylsiloxane-modified waterborne polyurethanes. *RSC Advances, 7*, 34086–34086.

Jin, J., Huang, S. Z., Li, Y., Tian, H., Wang, H. E., Yu, Y., Chen, L. H., Hasan, T., & Su, B. L. (2015). Hierarchical nanosheet-constructed yolk–shell TiO_2 porous microspheres for lithium batteries with high capacity, superior rate and long cycle capability. *Nanoscale, 7*, 12979–12989.

Jiang, W. C., Huang, Y., Gu, G. T., Meng, W. D., & Qing, F. L. (2006). A novel waterborne polyurethane containing short fluoroalkyl chains: Synthesis, characterization and its application on cotton fabrics surface. *Applied Surface Science, 253*, 2304–2309.

Kang, S. Y., Ji, Z., Tseng, L. F., Turner, S. A., Villanueva, D. A., Johnson, R., Albano, A., & Langer, R. (2018). Design and synthesis of waterborne polyurethanes. *Advanced Materials, 30*, 1706237.

Kästner, U. (2001). The impact of rheological modifiers on water-borne coatings. *Colloids and Surfaces a: Physicochemical and Engineering Aspects, 183*, 805–821.

Kim, H. A., & Kim, B. K. (2019). Synthesis and properties of waterborne polyurethane/hydroxyapatite chemical hybrids. *Progress in Organic Coatings, 128*, 69–74.

Király, A., & Ronkay, F. (2015). Temperature dependence of electrical properties in conductive polymer composites. *Polymer Testing, 43*, 154–162.

Kuan, H. C., Ma, C. C. M., Chang, W. P., Yuen, S. M., Wu, H. H., & Lee, T. M. (2005). Synthesis, thermal, mechanical and rheological properties of multiwall carbon nanotube/waterborne polyurethane nanocomposite. *Composites Science and Technology, 65*, 1703–1710.

Kwak, Y. S., Kim, H. D., & Lee, J. B. (2005). Comparison of the properties of waterborne polyurethane-ureas containing different tri block glycols for water vapor permeable coatings. *Colloid and Polymer Science, 283*, 880–886.

Lee, S. K., & Kim, B. K. (2009). High solid and high stability waterborne polyurethanes via ionic groups in soft segments and chain termini. *Journal of Colloid and Interface Science, 336*, 208–214.

Lee, H. J., Kim, S. E., Choi, H. W., Kim, C. W., Kim, K. J., & Lee, S. C. (2007). The effect of surface-modified nano-hydroxyapatite on biocompatibility of poly(ε-caprolactone)/hydroxyapatite nanocomposites. *European Polymer Journal, 43*, 1602–1608.

Lee, J. S., Shin, J. H., Kim, B. K., & Kang, Y. S. (2001). Modification of aqueous polyurethanes by forming latex interpenetrating polymer networks with polystyrene. *Colloid Polymer Science, 79*, 959–965.

Lewandowski, K., Krepski, L. R., Mickus, D. E., Roberts, R. R., Heilmann, S. M., LarsonWK, P. M. D., Koecher, S. D., Johnson, S. A., Mcgurran, J. R., Chris, S. V., Pather, K. A., & Thakur, M. (2002). Synthesis and properties of waterborne self-crosslinkable sulfo-urethane silanol dispersions. *Journal of Polymer Science: Part a: Polymer Chemistry, 40*, 3037–3045.

Lei Wa, F. C., Zhou, X., Li, J., Yang, R., Zhang, Z., & Liuc, D. (2017). Morphology and thermal properties of polyurethane elastomer based on representative structural chain extenders. *Journal of Materials Science and Technology, 33*, 1424–1432.

Li, Q. A., & Sun, D. C. (2007). Synthesis and characterization of high solid content aqueous polyurethane dispersion. *Journal of Applied Polymer Science, 105*, 2516–2524.

Li, J., Zheng, W., Zeng, W., Zhang, D., & Peng, X. (2014). Structure, properties and application of a novel low-glossed waterborne polyurethane. *Applied Surface Science, 307*, 255–262.

Li, Y. Q., Zhu, W. B., Yu, X. G., Huang, P., Fu, S. Y., Hu, N., & Liao, K. (2016). Multifunctional wearable device based on flexible and conductive carbon sponge/polydimethylsiloxane composite. *ACS Applied Materials & Interfaces, 8*, 33189–33196.

Liu, H., Li, C., & Sun, X. S. (2017). Soy-oil-based waterborne polyurethane improved wet strength of soy protein adhesives on wood. *International Journal of Adhesion and Adhesives, 73*, 66–74.

Liu, Y., Liang, H. Q., Sihan Li, S., Liu, D., Long, Y. J., Liang, G. D., & Zhu, F. M. (2019). Preparation of waterborne polyurethane with high solid content and elasticity. *Journal of Polymer Research, 26*, 146.

Liu, H., Xue, X., Zhang, W., & Lin, F. (2020). Synthesis of waterborne polyurethane acrylate–modified epoxy resin. *IOP Conference Series: Material Science and Engineering, 770*, 012062.

Liu, W. K., Zhao, Y., Wang, R., Luo, F., Li, J. S., Li, H. J., & Tan, H. (2018). Effect of chain extender on hydrogen bond and microphase

structure of biodegradable thermoplastic polyurethanes. *Chinese Journal of Polymer Science, 36*, 520–524.
Luong, N. D., Korhonen, J. T., Soininen, A. J., Ruokolainen, J., Johansson, L. S., & Seppala, J. (2013). Processable polyaniline suspensions through in situ polymerization onto nanocellulose. *European Polymer Journal, 49*, 335–344.
Madbouly, S. A., & Otaigbe, J. U. (2009). Recent advances in synthesis, characterization and rheological properties of polyurethanes and POSS/polyurethane nanocomposites dispersions and films. *Progress in Polymer Science, 34*, 1283–1332.
Mahdi, E. M., & Tan, J. C. (2016). Dynamic molecular interactions between polyurethane and ZIF-8 in a polymer-MOF nanocomposite: Microstructural, thermo-mechanical and viscoelastic effects. *Polymer, 97*, 31–43.
Mathew, A., Kurmvanshi, S., Mohanty, S., & Nayak, S. K. (2018). Sustainable production of polyurethane from castor oil, functionalized with epoxy- and hydroxyl-terminated poly (dimethyl siloxane) for biomedical applications. *Journal of Materials Science, 53* (5), 3119–3130.
Meng, Q. B., Lee, S. I., Nah, C., & Lee, Y. S. (2009). Preparation of waterborne polyurethanes using an amphiphilic diol for breathable waterproof textile coatings. *Progress in Organic Coatings, 66*, 382–386.
Mishra, V., Desai, J., & Patel, I. K. (2018). (UV/Oxidative) dual curing polyurethane dispersion from cardanol basedpolyol: Synthesis and characterization. *Industrial Crops and Products, 11*, 165–178.
Noreen, A., Zia, K. M., Zuber, M., Tabasum, S., & Saif, M. J. (2016). Recent trends in environmentally friendly water-borne polyurethane coatings: A review. *Korean Journal of Chemical Engineering, 33*, 388–400.
Olad, A., & Gharekhani, H. (2015). Preparation and electrochemical investigation of the polyaniline/activated carbon nanocomposite for supercapacitor applications. *Progress in Organic Coatings, 81*, 19–26.
Osterhold, M., & Wagner, G. (2002). Methods for characterizing the mar resistance. *Progress in Organic Coatings, 45*, 365–371.
Papaj, E. A., Mills, D. J., & Jamali, S. (2014). Effect of hardener variation on protective properties of polyurethane coating. *Progress in Organic Coatings, 77*, 2086–2090.
Pardoa, J. I., Suna, P., Benthem, R. A. T. M., & Estevesa, A. C. C. (2018). Design of self-dispersible charged-polymer building blocks for waterborne polyurethane dispersions. *European Polymer Journal, 101*, 324–331.
Park, J. H., Bae, S. Y., & Kim, B. K. (2013). Hyperbranched waterborne polyurethanes. *Polymer Bulletin, 70*, 859–869.
Peng, C., & Joy, A. (2015). Self-emulsion polymerization of baylis-Hillman-derived α-hydroxymethyl-substituted acrylates. *Journal of Polymer Science and Polymer Chemistry, 53*, 1743–1747.
Pettersson, B. (1996). Hyperbranched polymers: Unique design tools for multi-property control in resins and coatings. *Pigment & Resin Technology, 25*, 4–14.
Pielichowska, K., Bieda, J., & Szatkowski, P. (2016). Polyurethane/graphite nano-platelet composites for thermal energy storage. *Renewable Energy, 91*, 456–465.
Planes, M., Brand, J., Lewandowski, S., Remaury, S., Sole, S., Le Coz, C., Carlotti, S., & Sebe, G. (2016). Improvement of the thermal and optical performances of protective polydimethylsiloxane space coatings with cellulose nanocrystal additives. *ACS Applied Materials & Interfaces, 8*, 28030–28039.
Poussard, L., Lazko, J., Mariage, J., Raquez, J. M., & Dubois, P. (2016). Biobased waterborne polyurethanes for coating applications: How fully biobased polyols may improve the coating properties. *Progress in Organic Coatings, 97*, 175–183.
Qiu, F. X., Zhang, J. L., Wu, D. M., & Yang, D. Y. (2010). Waterborne polyurethane and modified polyurethane acrylate composites. *Plastics, Rubber and Composites, 39*, 454–459.
Rahman, M. M., & Lee, W. K. (2010). Properties of isocyanate-reactive waterborne polyurethane adhesives: Effect of cure reaction with various polyol and chain extender content. *Journal of Applied Polymer Science, 114*, 3767–3773.
Rahman, M. M., Kim, H. D., & Lee, W. K. (2009). Properties of waterborne polyurethane adhesives: Effect of chain extender and polyol content. *Journal of Adhesion Science and Technology, 23*, 177–193.
Rosthauser, J. W., & Nachtkamp, K. (1987). Waterborne polyurathenes. In Klempner D., & Frisch, K. C. (Eds.), *Advances in urethane science and technology* (pp. 121–162). Lancaster, Technomic.
Saeed, A., & Shabir, G. (2013). Synthesis of thermally stable high gloss water dispersible polyurethane/polyacrylate resins. *Progress in Organic Coatings, 76*, 1135–1143.
Santamaria-Echart, A., Fernandes, I., Saralegi, A., Costa, M. R. P. F. N., Barreiro, F., Corcuera, M. A., & Eceiza, A. (2016). Synthesis of waterborne polyurethane-urea dispersions with chain extension step in homogeneous and heterogeneous media. *Journal of Colloid and Interface Science, 476*, 184–192.
Saralegi, A., Etxeberria, A., Fernández-d'Arlas, B., Mondragon, I., Eceiza, A., & Corcuera, M. A. (2013). Effect of H_{12}MDI isomer composition on mechanical and physico-chemical properties of polyurethanes based on amorphous and semicrystalline soft segments. *Polymer Bulletin, 70*, 2193–2210.
Sattar, R., Kausar, A., & Siddiq, M. (2014). Advances in thermoplastic polyurethane composites reinforced with carbon nanotubes and carbon nanofibers: A review. *Journal of Plast Film & Sheeting, 31*, 186–224.
Seethapathy, S., & Gorecki, T. (2012). Applications of polydimethylsiloxane in analytical chemistry: A review. *Analytica Chimica Acta, 750*, 48–62.
Shah, S. A. A., Imran, M., Lian, Q., Shehzad, F. Q., Athir, N., Zhang, J., & Cheng, J. (2018). Curcumin incorporated polyurethane urea elastomers with tunable thermomechanical properties. *Reactive and Functional Polymers, 128*, 97–103.
Shen, W., Feng, L., Liu, X., Luo, H., Liu, Z., Tong, P., & Zhang, W. (2016). Multiwall carbon nanotubes-reinforced epoxy hybrid coatings with high electrical conductivity and corrosion resistance prepared via electrostatic spraying. *Progress in Organic Coatings, 90*, 139–146.
Si, H. Y., Liu, H., Shang, S. B., Song, J., Liao, S. L., Wang, D., & Song, Z. Q. (2016). Preparation and properties of maleopimaric acid-based polyester polyol dispersion for two-component waterborne polyurethane coating. *Progress in Organic Coatings, 90*, 309–316.
Spanhel, L. (2006). Colloidal ZnO nanostructures and functional coatings: A survey. *Journal of Solgel Science and Technology, 39*, 7–24.
Szycher, M. (2013). *Szycher's handbook of polyurethane* (2nd ed., pp. 417–447). CRC Press.
Tan, H., Xie, X., Li, J., Zhong, Y., & Fu, Q. (2004). Synthesis and surface mobility of segmented polyurethanes with fluorinated side chains attached to hard blocks. *Polymer, 45*, 1495–1502.
Tanaka, H., & Kunimura, M. (2002). Mechanical properties of thermoplastic polyurethanes containing aliphatic polycarbonate soft segments with different chemical structures. *Polymer Engineering and Science, 42*, 1333–1349.
Tao, Q. Y., Xing, H. L., Guo, W. M., & Zhou, G. E. (2013). Study on synthesis of waterborne polyurethane for thermal infrared stealth coating. *Paint and Coatings Industry, 43*, 31–38.

Tsurumaki, A., Tajima, S., Iwata, T., Scrosati, B., & Ohno, H. (2015). Antistatic effects of ionic liquids for polyether-based polyurethanes. *Electrochimica Acta, 175*, 13–17.

Wang, L. F., & Wei, Y. H. (2005). Effect of soft segment length on properties of fluorinated polyurethanes. *Colloids and Surfaces. b, Biointerfaces, 41*, 249–255.

Wang, J., Zhang, C., Du, Z., Li, H., & Zou, W. (2016). Functionalization of MWCNTs with silver nanoparticles decorated polypyrrole and their application in antistatic and thermal conductive epoxy matrix nanocomposite. *RSC Advances, 6*, 31782–31789.

Wei, H., Ding, D., Wei, S., & Guo, Z. (2013). Anticorrosive conductive polyuratene multiwalled carbon nanotube nanocomposites. *Journal of Materials Chemistry A, 1*, 10805–10813.

Wena, J., Suna, Z., Fana, H., Chena, Y., & Yana, J. (2019). Synthesis and characterization of a novel fluorinated waterborne polyurethane. *Progress in Organic Coatings, 131*, 291–300.

Weng, C. J., Chen, Y. L., Jhuo, Y. S., Yi-Li, L., & Yeh, J. M. (2013). Advanced antistatic/anticorrosion coatings prepared from polystyrene composites incorporating dodecylbenzenesulfonic acid-doped SiO_2@polyaniline core-shell microspheres. *Polymer International, 62*, 774–782.

Wu, G., Xu, X., He, X., & Yan, Y. (2018). Preparation and characterization of graphene oxide-modified sapium sebiferum oil-based polyurethane composites with improved thermal and mechanical properties. *Polymers, 10*, 13.

Xia, C., Zhang, S., Ren, H., Shi, S. Q., Zhang, H., Cai, L., & Li, J. (2016). Scalable fabrication of natural-fiber reinforced composites with electromagnetic interference shielding properties by incorporating powdered activated carbon. *Materials, 9*, 1–9.

Xu, F., Qian, B. R., Hu, Z., Chen, W. D., Zhuang, Z. Y., Zhu, B. Y., Zhang, H. Q., & Zhu, K. (2013). A novel route to emulsifier-free, waterborne hydroxyl functional polyacrylate with low VOC level and its application in 2K-WPU coatings. *Journal of Macromolecular Science Part A, 50*, 555–561.

Yang, L. W., Zhang, X. S., Liu, H. T., & Zu, M. (2017). Thermal resistant, mechanical and electrical properties of a novel ultrahigh-content randomly-oriented CNTs reinforced SiC matrix composite-sheet. *Composites. Part b, Engineering, 119*, 10–17.

Yates, C. R., & Hayes, W. (2004). Synthesis and applications of hyperbranched polymers. *European Polymer Journal, 40*, 1257–1281.

Yen, M. S., & Kuo, S. C. (1997). PCL-PEG-PCL triblock copolydiol-based waterborne polyurethane. I. Effects of the soft-segment composition on the structure and physical properties. *Journal of Applied Polymer Science, 65*, 883–892.

Yen, M. S., & Kuo, S. C. (2015). PCL-PEG-PCL triblock ester-ether copolydiol-based waterborne polyurethane. II. Effect of NCO/OH mole ratio and DMPA content on the physical properties. *Journal of Applied Polymer Science, 67*, 1301–1311.

Yousefi, E., Dolati, A., & Najafkhani, H. (2020). Preparation of robust antistatic waterborne polyurethane coating. *Progress in Organic Coating, 139*, 105450.

Zhanga, H. H., Niua, R., Guan, X. B., Xua, D. H., & Shia, T. F. (2015). Rheological properties of waterborne polyurethane paints. *Chinese Journal of Polymer Science, 33*, 1750–1756.

Zhao, C. X., & Zhang, W. D. (2008). Preparation of waterborne polyurethane nanocomposites: Polymerization from functionalized hydroxyapatite. *European Polymer Journal, 44*, 1988–1995.

Zheng, F., Jiang, P., Hu, L., Bao, Y., & Xia, J. (2019). Functionalization of graphene oxide with different diisocyanates and their use as a reinforcement in waterborne polyurethane composites. *Journal of Macromolecular Science Part A, 56*, 1071–1081.

Zheng, A., Xu, X., Xiao, H., Li, N., Guan, Y., & Li, S. (2012). Antistatic modification of polypropylene by incorporating Tween/modified Tween. *Applied Surface Science, 258*, 8861–8866.

Zhou, X., Li, Y., Fang, C., Li, S., Cheng, Y., Lei, W., & Meng, X. (2015). Recent advances in synthesis of waterborne polyurethane and their application in water-based ink: A review. *Journal of Materials Science and Technology, 31*, 708–722.

Zhou, X., Fang, C., Lei, W., Du, J., Huang, T., Li, Y., & Cheng, Y. (2016). Various nanoparticle morphologies and surface properties of waterborne polyurethane controlled by water. *Science and Reports, 6*, 34574.

Zhu, J., Huang, H., & Peng, X. (2016b). Preparation and characterization of antiglare waterborne polyurethane. *RSC Advances, 6*, 102368–102372.

Zhu, Z., Li, R., Zhang, C., & Gong, S. (2018). Preparation and properties of high solid content and low viscosity waterborne polyurethane–acrylate emulsion with a reactive emulsifier. *Polymers, 10*, 154.

Zhu, R., Wang, Y., Zhang, Z., Ma, D., & Wang, X. (2016a). Synthesis of polycarbonate urethane elastomers and effects of the chemical structures on their thermal, mechanical and biocompatibility properties. *Heliyon, 2*, e00125.

Zia, F., Zia, K. M., Zuber, M., Kamal, S., & Aslam, N. (2015). Starch based polyurethanes: A critical review updating recent literature. *Carbohydrate Polymers, 134*, 784–798.

Zvonkina, I. J., & Soucek, M. D. (2016). Inorganic-organic hybrid coatings: Common and new approaches. *Current Opinion in Chemical Engineering, 11*, 123–127.

Waterborne Polyurethanes in Sustainability Development

Felipe M. de Souza and Ram K. Gupta

Abstract

The strive for more efficient materials, larger production, and diminish of cost along with the increasing environmental and health concerns led the scientific community and industry to shift gears to craft a more sustainable approach for the future. This chapter discusses the importance of waterborne polyurethanes as sustainable polymers that present high mechanical and thermal properties as well as resistance against external agents making them very suitable for applications in biomedical, corrosion and weather protective coatings, adhesives, sensors, electrical, electronics, etc. The use of bio-renewable sources has proven to be a smart way to address the main challenges of the status quo since most of these biomaterials replenish in a short time and can fit in most synthetic routes to acquire the targeted property. Materials such as soybean, corn, canola, castor, jatropha, and linseed oil can be easily converted into polyols to react with diisocyanates using green approaches that use water as the main solvent. The final products are a large number of bio-renewable waterborne polyurethanes with competitive properties and low cost that represent the future of the market.

Keywords

Waterborne polyurethanes • Polyols • Isocyanate • Green polymer • Green chemistry

1 Introduction

Around 1920, there was a perception among scientists that a high molecular weight compound (macromolecule) could not be synthesized in a laboratory. This partial perspective led the chemist Herman Staudinger to propose the "*Macromolecule hypothesis*", which stated that one single compound could be composed of 10,000 atoms or even more. This idea was heavily criticized during that time, but Staudinger was able to prove it by simply performing hydrogenation of isoprene. According to the belief of the colloid concept, natural rubber was an aggregate of isoprene units due to the interactions of double bonds, and therefore, after hydrogenation, the interactions should be disrupted. However, Staudinger obtained a rubber similar to the natural rubber showing that the double bonds were disrupted to form covalent bonds between each molecule. As a visual explanation, "polymer" could be described as a long chain of paper clips linked together, each paper clip representing what is called "monomer". Derived from the Greek words: "*mono*"—one, "*poly*"—many, and "*meros*"—part. Despite the strong scientific evidence, Staudinger's work started to get recognizing only after 1953 when he received the Nobel Prize in Chemistry. His theory and achievements officially gave birth to polymer science, which within 70 years of research has become one of the most influential areas of study that shape humankind to what we know. It is hard to find a sector where polymers are not used since they can be found in packing, buildings, electrical, electronics, automotive, medical, agriculture, furniture, bedding, consumer items, utilities, and more. The major applications of polymers in different sectors are shown in Fig. 1 (Grand View Research, 2020). With a wide range of applications, polymers are one of the fastest-growing industries with over 568 billion USD market in 2019 and are predicted to grow at 3.5% per year.

Based on the applications, various types of polymers are used. Polypropylene, low and high-density polyethylene,

F. M. de Souza
Kansas Polymer Research Center, Pittsburg State University, Pittsburg, KS 66762, USA

R. K. Gupta (✉)
Department of Chemistry, Pittsburg State University, Pittsburg, KS 66762, USA
e-mail: rgupta@pittstate.edu

Inamuddin et al. (eds.), *Sustainable Production and Applications of Waterborne Polyurethanes*, Advances in Science, Technology & Innovation, https://doi.org/10.1007/978-3-030-72869-4_5

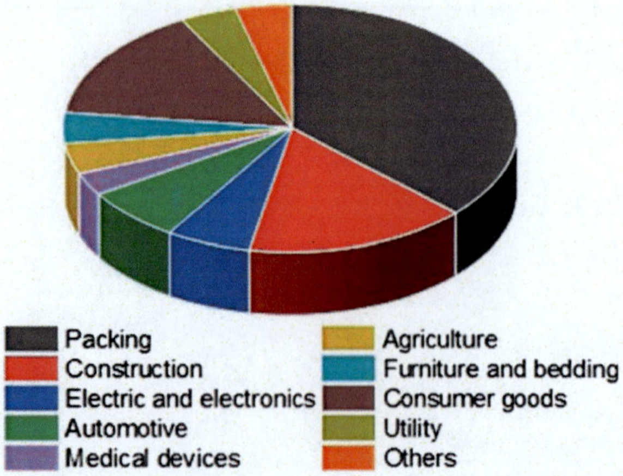

Fig. 1 Worldwide distribution of plastics production in 2019 (Grand View Research, 2020)

and polyvinyl chloride are some of the largest groups of polymers used in these applications (Fig. 2). Recently, polyurethane-based polymers are attracting considerable interest in both consumer and industrial applications making them one of the fastest-growing polymeric materials for industrial applications. The polyurethane market was over 65 billion USD in 2018 and is predicted to grow at 7.0% per year. They are being used in different areas such as furniture, construction, electronics, automotive, footwear, packing, and many others.

Polyurethanes are extremely versatile materials as noted by their broad range of applications. In general, polyurethanes are synthesized by an addition reaction where the hydroxyl groups (–OH) of a polyol react with isocyanates (–N=C=O) to form the urethane linkage (–NH–C(O)–O–) as a repeating unit. The wide variety of applications of polyurethanes is due to the ability to tune their properties by chemical modifications and/or formulation. By modifying the chemical structure of both polyol and diisocyanate, polyurethanes with a range of properties such as rigid, flexible, elastomers, thermoplastics, thermosetting, coatings, adhesives, binders, and waterborne can be realized (Akindoyo et al., 2016). Rigid and mechanically strong polyurethanes can be prepared by implementing low-mobility groups by using a small main chain, aromatic rings, and cross-linking sites (Charlon et al., 2014). Whereas, flexible polyurethanes are obtained by using polyols with a long, aliphatic, flexible chains with some side groups which add a plasticizing effect (Ionescu, 2006). Elastomers present structures with long chains that are lightly cross-linked show high stretchability (Ionescu, 2006). Thermoplastics are another important class of polymers used in footwear, automotive, and construction industries due to their stability against weather, impact, and abrasion (Palaskar et al., 2010). The thermosetting is highly cross-linked materials that grant them enhanced thermal and mechanical properties. They decompose before melting, unlike thermoplastics, hence making them very suitable for high-temperature applications (Dodiuk & Goodman, 2014). Coatings are other important applications of polymers as they increase the lifespan of materials by increasing resistance to corrosion and scratch (Szycher & Szycher, 2012). Adhesives are heavily used in automotive interiors, assembling of footwear, and many other products.

The variety of applications of polyurethanes is due to the versatility of materials that can be used in their synthesis along with a wide range of available processing techniques. Most of the starting materials used for the preparation of polymers for industrial applications are derived from petrochemicals. It is important to reinforce that the polymers derived from the petrochemicals show good properties and these industries have created many jobs and new technologies that made human life easier. However, due to the price instability, concerns regarding depletion of petrochemicals, and growing concerns about environmental protection, polymer industries are looking for sustainable and cost-effective ways to prepare polymers (Zhu et al., 2016).

Researchers are developing many green approaches to prepare polymers for a sustainable future which could be adopted by industries. One of the approaches is to use

Fig. 2 Largest polymeric materials produced in 2017 (Geyer et al., 2017)

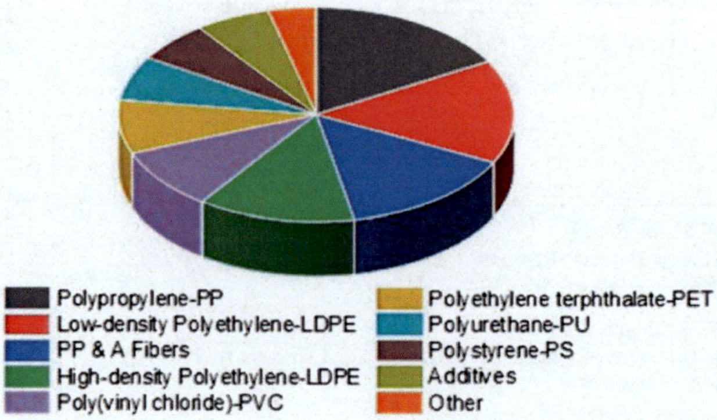

chemicals from renewable resources such as crops which can be grown every year. Some examples of such achievements are the conversion of soybean oil into a bio-polyol through epoxidation followed by ring-opening making it suitable for polyurethane application in industrial scale (Mu et al., 2012). Other successful procedures include blended flame-retardant polyurethanes foams based on polyol from corn oil, synthesized through a facile thiol-ene reaction (Ramanujam et al., 2019). Castor oil-based polyol shows to be a suitable low-cost and effective material for polyurethane as well (Petrović et al., 2008). Following the green and sustainable synthetic approaches, several bio-derived materials can be used for the preparation of polyurethanes such as cardanol, rapeseed oil, palm oil, sunflower oil, lignin, limonene, phellandrene, carvone, and many others (Arniza et al., 2015; Biermann et al., 2010; Dworakowska et al., 2012; Elbers et al., 2017; Lee & Deng, 2015; Souza et al., 2020; Wu et al., 2019). Apart from the polyols from sustainable resources, there has been some development for making polyurethanes without diisocyanate, a component to make polyurethanes along with polyols. Isocyanate-free routes allow us to develop polyurethanes from totally renewable resources (Lee & Deng, 2015).

The synthesis of polymeric materials requires several organic compounds and solvents. The presence of volatile organic components (VOC) in polymeric materials is one of the main concerns as it induces respiratory problems as well as environmental hazards (Saalah et al., 2015). To work around this situation, a new area in polyurethanes emerged named waterborne polyurethanes. They consist of organic reactions that can be performed in water dispersions, obtained by attaching polar groups such as carboxylic acids or tertiary amine groups into the main chain of the polyurethane. These pendant polar groups can be neutralized by salts creating enough polarity to form a colloidal system with water that acts mostly as an emulsifier and chain extender in some cases (Nanda et al., 2005). They are usually synthesized by procedures that involve mild temperature and relatively low synthesis time showing great potential for industrial applications (Jin et al., 2005). Several different properties can be achieved by varying the quantities of polyols, isocyanates, charged species (ionomers), and chain extenders.

One of the key components of polyurethanes is polyols. They normally receive more attention because they can define the properties of the final polyurethane and are readily available than isocyanates (Engels et al., 2013; Thomson, 2000). The size of the polyols chain determines the final properties of polyurethanes; for example, a long flexible chain provides flexible polyurethanes, while short chain creates rigid polyurethanes. For waterborne polyurethanes, polyols with long chains are commonly used so that polyurethanes can be dispersed in water easily. The presence of hydrophilic groups is one of the most crucial factors for high-quality waterborne polyurethanes. The hydrophilic groups could be in the main chain such as in polyether and polyester-based polyols or as pendant groups such as carboxylic acids and amines (Thomson, 2000). Both of these structures increase the dispersibility of polyurethanes in aqueous media, which grants a more stable colloidal system and reducing the VOC in the polymers (Saetung et al., 2012). The properties of waterborne polyurethanes can be further modified by the introduction of chain extenders (Chan-Chan et al., 2010). Chain extenders are an option to increase the molecular weight of the chain and influence the crystallinity of the polyurethane (Blackwell et al., 1982). They present a general structure of aliphatic or cyclic diols or diamines such as ethylene glycol, 1,4-butanediol, cyclohexane dimethanol, 1,6-butanediol, 1,4 butanediamine, and even water (Domanska & Boczkowska, 2014). Generally, chain extenders are added after the synthesis of the "prepolymer" to cap the isocyanate end groups with hydroxyl groups to improve dispersion in water and their environmental aspect (Chan-Chan et al., 2010). The diisocyanates represent the hard segment of the polyurethanes, and there is a demand to prepare isocyanate-free, water-dispersible polyurethanes. An isocyanate-free route was reported by performing a reaction with a cyclic carbonate derived from bio-precursor followed by a reaction with a diamine that yields a hydroxy polyurethane as described in Fig. 3. This type of alternative route provided polyurethane with hydrophilic behavior, hence making it suitable for water-based coatings (Kathalewar et al., 2014).

Apart from the environmental concerns of using petrochemical-based compounds, bio-based materials offer reduced cost, enabling sustainable production of commercial polyurethanes foam, elastomers, coatings, and adhesives (Desroches et al., 2012; Petrović, 2008; Pfister et al., 2011). Ford motors is using soybean oil-derived polyurethanes in almost all of their cars. Other important characteristics of these compounds are their tendency to be biodegradable and biocompatible, increasing their field of applications such as medical tissue, drug delivery, wound healer, and sealants (Lligadas et al., 2007, 2013). The presence of unsaturation (double bonds) is the main reactive site in vegetable, fruit, and flower oil that allows the conversion of biomaterials into polyols by chemically introducing hydroxyl groups. Many natural oils attend this requirement by presenting a high unsaturation level and ester groups due to their triglyceride form. The list of compounds that can be used for waterborne polyurethanes are composed of cardanol, corn, castor, cottonseed, tung, soybean, sunflower, peanut, safflower, olive, rapeseed, linseed, jatropha oil, and more (Lligadas et al., 2013; Zhou et al., 2015).

As previously mentioned, one of the important environmental aspects brought with waterborne polyurethanes is the

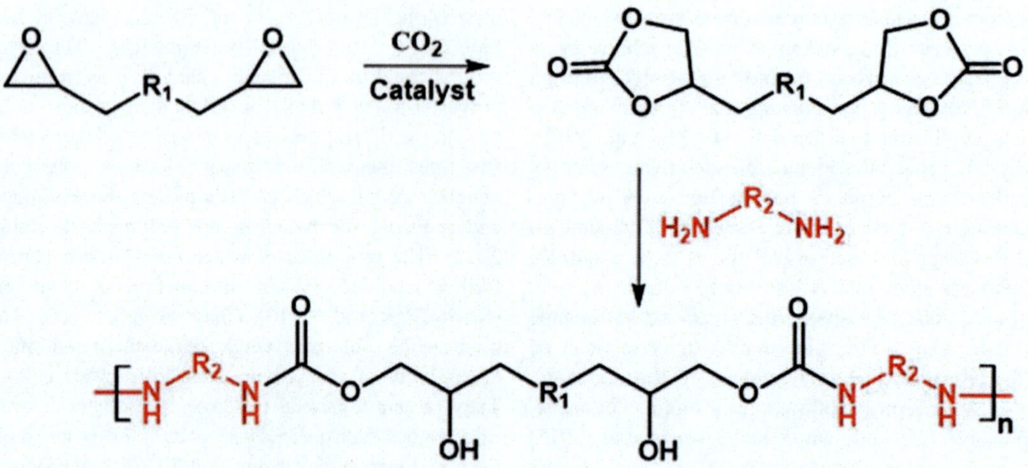

Fig. 3 Isocyanate-free route to obtain hydroxypolyurethanes

decrease in the use of volatile organic compounds. This process can be achieved by introducing pendant charged groups into the main chain to improve the water dispersibility, hence diminishing the use of toxic organic solvents. The charge of these groups defines the type of polyurethanes. It is called anionic when the pendant group has a negative charge or cationic if it has a positive charge (Coutinho et al., 2001; Lu & Larock, 2010a). The latest reports on the synthesis of anionic waterborne polyurethanes showed procedures that can be divided into three steps to facilitate understanding. First, a block copolymer, usually mentioned as "prepolymer" that can be synthesized using (a) an oligo polyol, (b) building block compound containing a carboxylic acid group as a pendant group, and (c) the diisocyanate that forms the urethane linkage between the polyols and represents the hard segment of the polyurethane. The second step is end-capping the isocyanates groups by reaction with a chain extender. The third step is adding a tertiary amine-based compound to form organic salts with the carboxylate originated from the carboxylic acids (Chen et al., 2014; Coutinho et al., 2001; Gaddam & Palanisamy, 2016). Figure 4 describes a general approach for the synthesis of an anionic waterborne polyurethane.

Polyurethane synthesis is known to be versatile and can be performed in many innovative ways. For example, a green approach was used to synthesize cottonseed oil-based anionic waterborne polyurethane (Gaddam & Palanisamy, 2016). In a sustainable process, the cottonseed oil was first epoxidized followed by a ring-opening reaction to synthesize hydroxylated cottonseed oil (HCSO). The carboxylic acid group was inserted using maleic anhydride through an esterification reaction on the hydroxyl groups of the polyol. The obtained prepolymer was then reacted with the diisocyanate. The isocyanate terminating groups were then capped with the chain extender hexane-1,6-diol. To improve the dispersity in water, tri ethyl amine was used. The synthetic routes are shown in Fig. 5. This synthetic route allows many biomaterials to be used in similar ways to obtain anionic waterborne polyurethane dispersions including many oils that are produced in large scales such as soybean, castor, and rapeseed oil (Lu & Larock, 2011; Lu et al., 2005a, 2005b). The synthesis of polyurethanes using these approaches is not only sustainable (through the use of renewable biomaterials) but also provides polyurethanes with thermophysical and mechanical properties similar to polyurethanes prepared using petrochemical-based counterparts (Lu & Larock, 2010a).

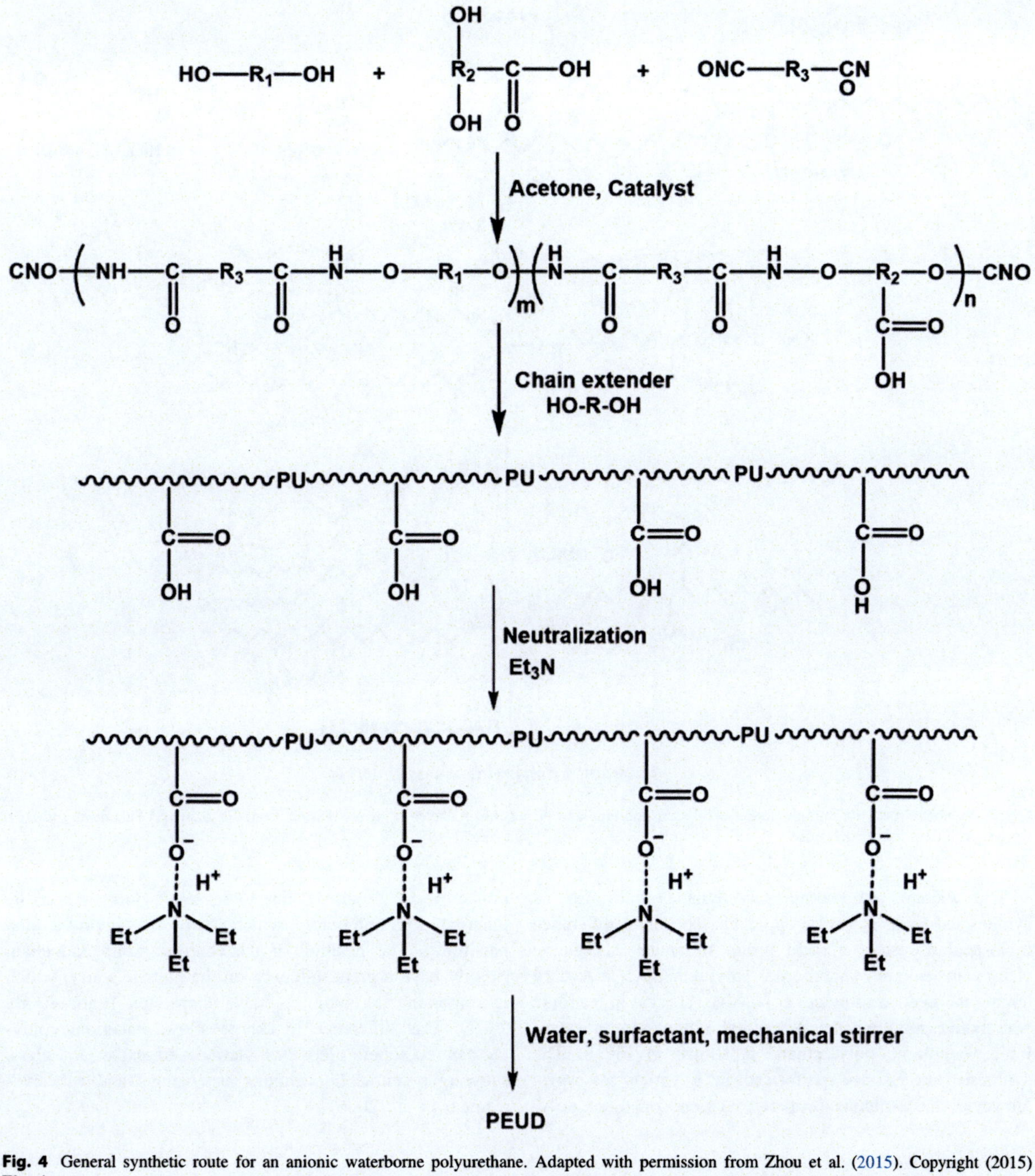

Fig. 4 General synthetic route for an anionic waterborne polyurethane. Adapted with permission from Zhou et al. (2015). Copyright (2015) Elsevier

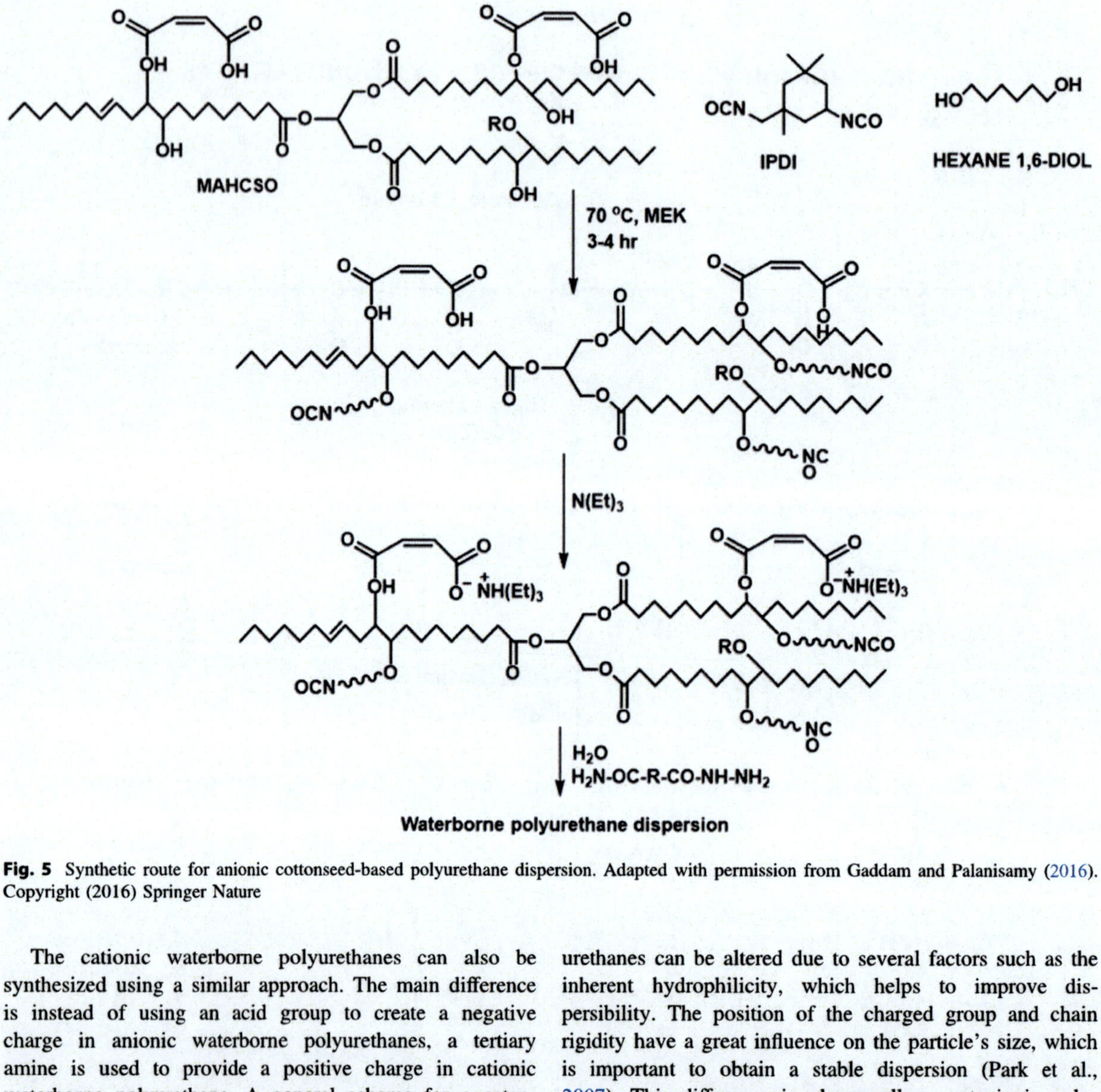

Fig. 5 Synthetic route for anionic cottonseed-based polyurethane dispersion. Adapted with permission from Gaddam and Palanisamy (2016). Copyright (2016) Springer Nature

The cationic waterborne polyurethanes can also be synthesized using a similar approach. The main difference is instead of using an acid group to create a negative charge in anionic waterborne polyurethanes, a tertiary amine is used to provide a positive charge in cationic waterborne polyurethane. A general scheme for a catanionic waterborne polyurethane is shown in Fig. 6. The synthesis can also be performed in a variety of ways. However, the structure–property relation for these polyurethanes can be altered due to several factors such as the inherent hydrophilicity, which helps to improve dispersibility. The position of the charged group and chain rigidity have a great influence on the particle's size, which is important to obtain a stable dispersion (Park et al., 2007). This difference in charge allows catanionic polyurethane to adhere properly to surfaces of leather and glass, showing potential applications as coatings and adhesives (Sundar et al., 2006).

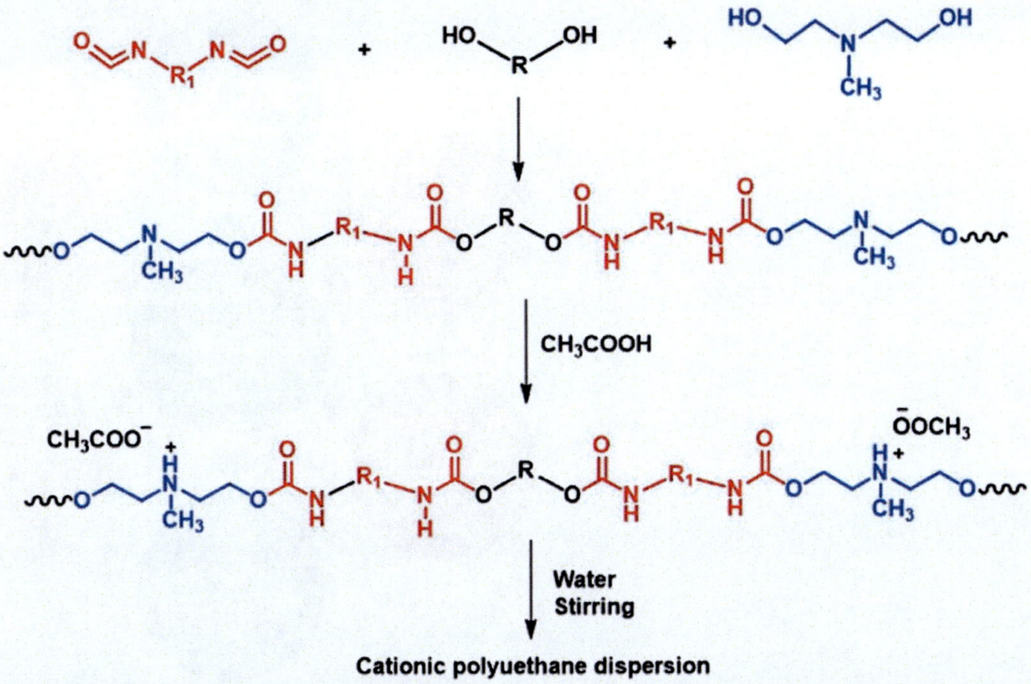

Fig. 6 General reactions for catanionic waterborne polyurethane

2 Sustainable Routes for Polyurethane Dispersion

Waterborne polyurethanes became an important area in both research and industry because of their green synthetic approach and a broad range of properties such as quick drying time, brightness, transparency, low toxicity and resistance to abrasion, temperature, and impact (Mumtaz et al., 2013). Due to these unique features, water-dispersible polyurethanes established themselves as sustainable and efficient materials especially for the adhesives and coatings sector covering from textile, which requires flexible coatings, to wood and metal surfaces (Choi et al., 2010). In the following section, some of the synthetic routes for waterborne polyurethanes are discussed and shown in Fig. 7.

2.1 Prepolymer Emulsification Process

The prepolymer emulsification process is one of the most important methods as it uses water instead of organic solvents, bringing a sustainable and non-toxic route for the synthesis of waterborne polyurethane dispersions and their green applications. The process starts with the preparation of a prepolymer, which is usually a copolymer that contains a polyol, an ionic building block monomer, and an isocyanate as an end group. The properties of waterborne polyurethanes largely depend on the nature of polyol (petrochemical, bio-based) and their characteristics (chain length, functionality, flexibility, hydrophilicity, and others) (Remya et al., 2016). The building block, a monomer with the ionic group, is responsible for improving the dispersion of the polyurethane in water. These internal emulsifiers improve the interactions with water, causing a decrease in particle size that leads to more stable dispersions (Szycher & Szycher, 2012). Some examples of the compounds commonly used are dimethylolpropionic acid and N-methyldiethanolamine to synthesize anionic and cationic polyurethanes, respectively. The ionic group in the backbone must then be neutralized. For the case of a carboxylic acid in the backbone, an amine-based compound can be used for neutralization and vice-versa. The neutralized prepolymer can be dispersed in water, usually under mild temperatures (40–70 °C) and intense stirring (600–750 rpm) (Madbouly & Otaigbe, 2009). To further improve the molecular weight as well as the dispersibility of the polyurethane, the isocyanate end groups can be reacted with diamine or dihydroxyl-based compounds that make the final polyurethane dispersion environmentally friendly and even more dispersible in aqueous media (Madbouly & Otaigbe, 2009). Green waterborne polyurethane dispersion using soybean oil as a starting material was prepared through the prepolymer emulsification process. The use of this bio-renewable source improved the

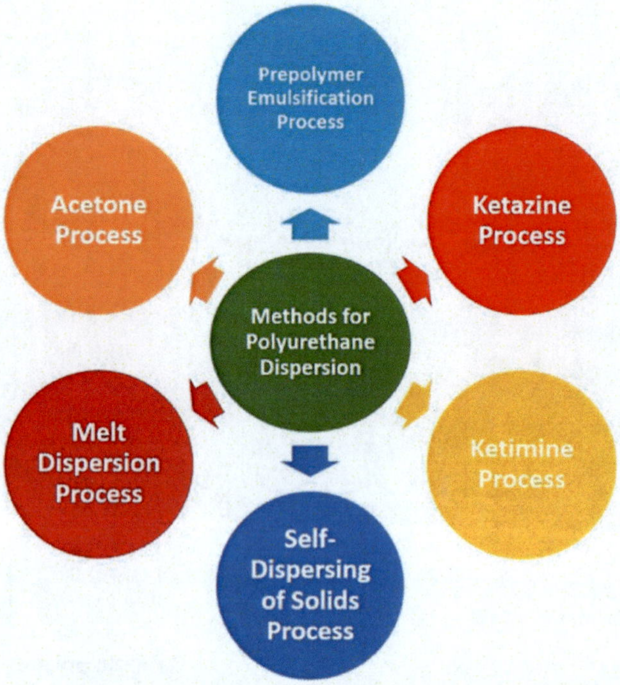

Fig. 7 Techniques for preparation of polyurethane dispersion

overall properties of the polyurethane, making it comparable to a polypropylene glycol-based polyurethane (Lu & Larock, 2007). It encourages the use of bio-renewable sources as they are abundant, low cost, present similar synthesis and properties, and can usually be replenished in about six months, which allows a stable schedule for large-scale productions (Lu & Larock, 2007).

2.2 Acetone Process

The acetone process consists of the preparation of the prepolymer in acetone, which is inert, water-soluble, and can be easily removed by distillation or rotary evaporation. Other organic hydrophilic solvents such as tetrahydrofuran, methyl, and ethyl ketone can also be used. Since prepolymer has isocyanate end groups, the chain extension can be performed using a sulfonated diamine compound that provides polyurethane and polyurea segments (Szycher & Szycher, 2012). A solvent exchange procedure is then performed in which water is added into the mixture and acetone is later removed. This process allows proper control of the viscosity and molecular weight of starting materials that do not disperse well in water initially (Kim, 1996). It yields products with exceptional properties that are reproducible and can be applied in conditions where acetone solubility of the final coating is not an issue (coating is not resistant to acetone) (Remya et al., 2016). Nevertheless, the properties of the polyurethane dispersions can be modified in many ways since they can be influenced by polyurethane/solvent ratio, the temperature of solvent exchange, rate of water addition, use of different catalysts, agitation rate, and others (Yang et al., 1995). The employment of bio-based materials into the synthesis has shown to be an effective way to improve many properties of polyurethane dispersion. However, it is also important to mention that the large amounts of acetone in dispersion require distillation which may increase the cost of the process. Despite the requirement of distillation, the acetone process is largely used in industry for coatings and adhesives (Yang et al., 2002). A reported example was the synthesis of a polyurethane dispersion that used poly(tetramethylene ether) and isophorone diisocyanate. Cardanol oil was used as a chain extender, and it decreased the particle size providing better film formation with soft and flexible properties (Suresh & Harikrishnan, 2014). The synthesis schematic is given in Fig. 8. Meanwhile, butanediol used as a chain extender for the same polyurethane yielded a larger size of particles that decrease the quality of the film and became brittle. It shows that the research in applying bio-derived materials can deliver promising and sustainable routes to obtain materials with even better properties if compared with petrochemical-based (Suresh & Harikrishnan, 2014).

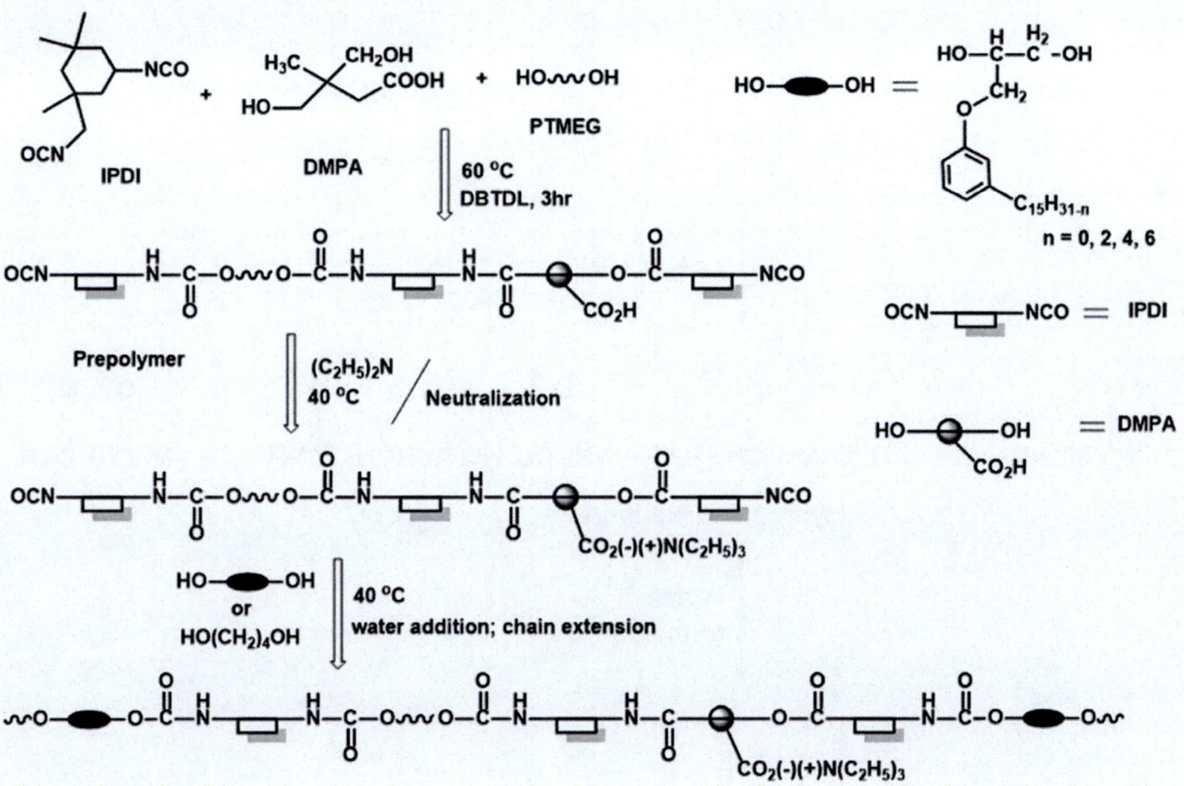

Fig. 8 Schematic of synthetic procedure for polyurethane dispersion using cardanol oil as a chain extender. Adapted by permission from Springer Nature: (Suresh & Harikrishnan, 2014). Copyright (2014) Springer Nature

2.3 Melt Dispersion Process

The melt dispersion process is applied for the prepolymer with terminal isocyanate which reacts with urea or ammonia leading to biuret groups. It can be applied to both ionic or non-ionic prepolymers, and the process is performed at higher temperatures than other methods (around 130 °C) (Szycher & Szycher, 2012). Water is used as a solvent to form the dispersion around 90 °C and to decrease the viscosity (Kim, 1996). Formaldehyde is also added as a reagent along with water for a methylation reaction, which enables control of molecular weight by introducing hydroxyl to the end of the chain (Remya et al., 2016). Some researchers have used an organic solvent and catalyst-free route for the synthesis of waterborne polyurethanes by using 2,4-diaminobenzenesulfonate and water as a solvent to decrease the viscosity (Xiao et al., 2016). The waterborne polyurethane obtained from this method presented similar mechanical properties as well as water resistance when compared with petrochemical-based. Figure 9 represents a melt dispersion process as an example using urea, sulfonate sodium, and methylation reaction as chain extending procedures (Szycher & Szycher, 2012).

2.4 Self-Dispersing of Solids Process

Self-dispersing of solids process is a procedure that adopts a mix of two polymers, generally the same, with the difference that one contains ionic groups, usually carboxylic acids, and the other does not. This type of system allows both polymers to disperse in each other, while water can be used as a solvent which helps to decrease the use of volatile organic compounds and improve solvent resistance. It is a procedure that applied in paints industries to make paints less toxic and odorless. Some approaches functionalize the isocyanate with hydrophilic groups to improve its dispersibility in aqueous media. The general procedure for a self-dispersing of solids is described in Fig. 10 (Kong et al., 2010).

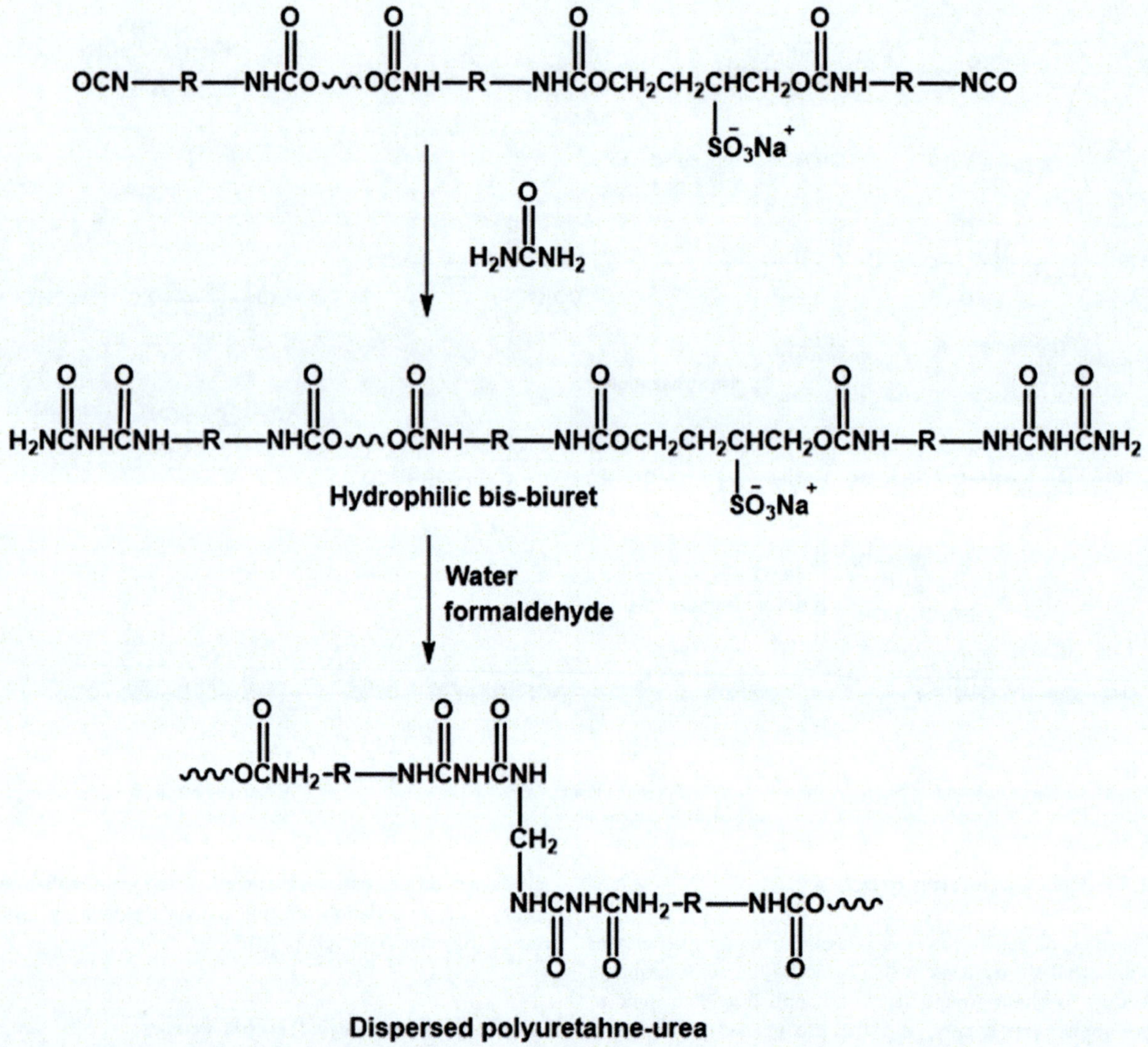

Fig. 9 General hot-melt process to obtain a polyurethane-urea dispersion. Copyright (2012) from Szycher's Handbook of Polyurethanes by Szycher and Szycher (2012). Reproduced by permission of Tylor and Francis Group, LLC, a division of Informa plc

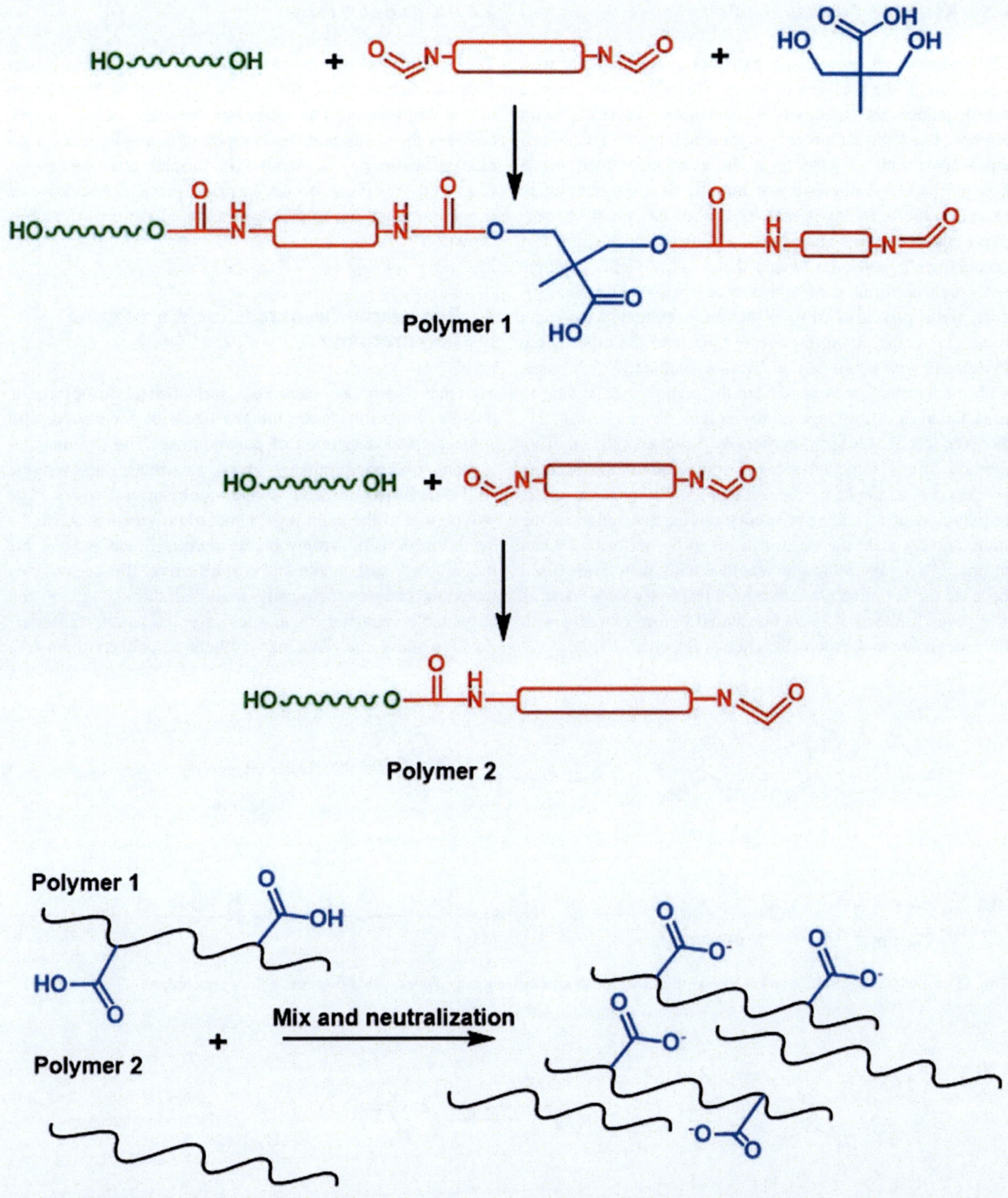

Fig. 10 General synthetic procedure for self-dispersing waterborne polyurethane

2.5 Ketimine Process

The amine used as a chain extender reacts rapidly with isocyanate in the presence of water. The addition of ketones could reduce the reactivity by reacting reversibly with amines. The ketimine process is generated by the reaction of an amine with a ketone; it is also mentioned as a ketone-blocked diamine. It acts initially as a dormant chain extender due to its low reactivity toward isocyanate in regular conditions. This characteristic allows proper control of prepolymer's molecular weight that usually yields coatings with high performance (Remya et al., 2016). For this process, water presents two roles that are executed at the same time. One is dispersing the prepolymer, and the other is the hydrolysis with ketimine that forms a diamine, which reacts with the terminal isocyanate from the prepolymer leading to urea linkages at the edge of the polymeric chain (Fig. 11) (Reeves et al., 2015; Szycher & Szycher, 2012). This method brings some advantages compared to the acetone process. For example, by the fact that less or no co-solvent is required, which yields polyurethane-urea that tends to the more solvent resistant. Also, it tends to be performed when aromatic isocyanates (more reactive with water) are used because the ketimine presents a high reactivity with water as well, which allows it to be consumed before reacting with the isocyanate end group (Szycher & Szycher, 2012).

2.6 Ketazine Process

The ketazine process is very similar to the ketimine procedure and also works on the logic of masking the NCO group of the prepolymer. The difference between these two processes is the chain extender compound that is formed. In the case of ketimine, a diamine is formed after hydrolysis (Fig. 11); however, for the ketazine process, hydrazine is formed to mask the isocyanate groups of the prepolymer as shown in Fig. 12.

3 Renewable Resources for Waterborne Polyurethane

Materials such as catalysts, surfactants, diisocyanates, polyols, and solvents are the key elements for a successful synthesis and dispersion of polyurethane. The commercialization of polyurethanes was possible due to the well-established research which took several years. The polyol, one of the main ingredients of polyurethanes, can be synthesized using renewable resources which makes this process very sustainable and cost-effective. Bio-derived raw materials can be chemically modified through green and sustainable synthetic routes such as epoxidation, transesterification, thiol-ene, and many others to convert them into

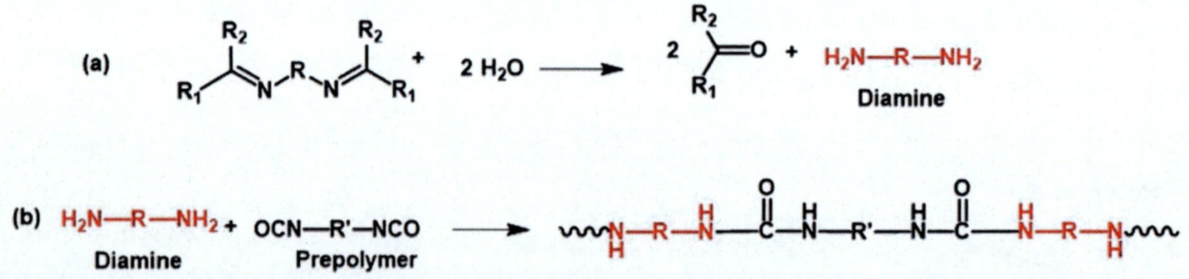

Fig. 11 a Ketimine hydrolysis and b follow-up reaction with diisocyanate prepolymer with a diamine to form urea linkage

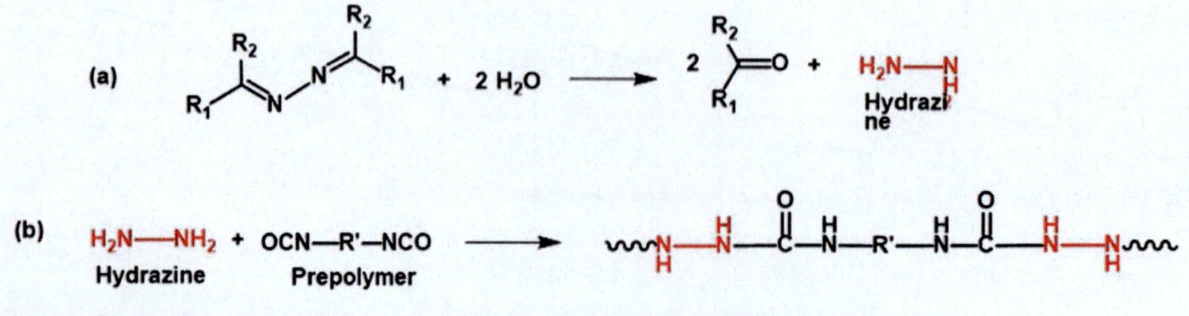

Fig. 12 a Ketazine hydrolysis and b reaction with diisocyanate prepolymer with hydrazine yielding urea linkages

polyols for polyurethanes (Arniza et al., 2015; Dai et al., 2009; Liu & Chung, 2017; Mosiewicki et al., 2009; Petrović et al., 2005). The properties of polyols such as functionality, chain length, presence of pendant groups, and flexibility are responsible for defining most of the properties of polyurethanes (Ionescu, 2006). Some examples of currently used bio-derived raw materials for waterborne polyurethanes are corn, soybean, castor, cardanol oils, etc. (Garrison et al., 2014a; Kosheeladevi et al., 2016; Liang et al., 2018; Lu & Larock, 2008; Saalah et al., 2015; Suresh & Harikrishnan, 2014).

Castor oil is one of the most widely used raw materials for polyurethanes due to the presence of both hydroxyl groups and unsaturation in their chemical structure that makes them very suitable for functionalization using several green methods to synthesize a variety of polyols (Babb, 2012). Also, the castor plant easily grows in most types of terrain, even those with a low concentration of nutrients making them cost-effective crops for commercial applications as castor oil is non-edible oil in general (Haynes et al., 2001; Milliano et al., 2010). Castor oil is mostly harvested in India for polyurethane industries as a starting material for polyol. The polyurethanes prepared using castor oil yield coatings and elastomers having water and weather-resistance characteristics, which extend the lifetime of materials or components (Hablot et al., 2008; Mittal et al., 1991; Valero et al., 2009). Cationic waterborne polyurethanes based on castor oil also show higher shelf-life as they presented the same properties after one year of storage. Some properties such as glass transition temperature (T_g) and tensile strength can be tuned by varying the percentage of cross-link. The use of a cationic chain extender such as N-methyl diethanolamine (MDEA) improved both water and solvent resistance of the coating and showed anti-bacterial properties. These properties make them suitable for applications in medical equipment and food processing (Liang et al., 2018). Epoxidation followed by a ring-opening reaction was performed to introduce hydroxyl groups at double bonds in castor oil to prepare a waterborne polyurethane (Sardari et al., 2019). Sodium 2,4-diamino benzene sulfonate (SDBS) was used as a chain extender to add ionic centers to the polymeric chain, which decrease particle size and viscosity. The increase in the quantity of castor oil in the waterborne polyurethane also increased the Tg, glossiness, hardness, and adhesion of the coatings due to the controlled functionality and regular distribution of particles size, showing promising applications for the inclusion of castor oil into large-scale production (Sardari et al., 2019).

As previously mentioned, many bio-derived compounds can be used as vital components for waterborne polyurethane to endorse sustainable production and applications. Most of the bio-renewable materials contain double bonds in their structures that allow chemical functionalization through many routes. The general factors that make bio-oils attractive are the environmentally friendly routes, biodegradability, health safety, availability, low or zero use of volatile organic compounds, and low cost (Lu & Larock, 2009). Soybeans are one of the most important bio-renewable materials produced worldwide reaching around 348 million tonnes in 2018 with 35% of that amount produced in the United States alone, followed by Brazil (33%), Argentina (10.69%), China (4.07%), and India (3.95%) (Agricultural Marketing Resource Center, 2020). Hence, soybean oil fits as a very attractive source for bio-renewable materials and has already shown great potential and versatility to obtain waterborne polyurethanes with a wide range of properties for many applications.

Figure 13 shows a general route for the synthesis of soybean oil-based cationic polyurethane dispersions (Lu & Larock, 2010b). The obtained polyurethane showed high mechanical properties with Young's moduli in range of 30–550 MPa, and elongation at break percentages exceeds 200%, and tensile strength between 5 and 23 MPa. It was observed that the properties were highly influenced by the functionality of the polyol to improve the coating and stability of the dispersion, decrease the viscosity to ease processability, and improve the mechanical and thermal properties. As the functionality increases, the polyurethane changes from an elastomeric to a flexible plastic, which can define different fields of applications for the polyurethane (Lu & Larock, 2010b). The result obtained for this soybean-based cationic polyurethane dispersion matches and in some cases even surpasses the properties of other petrochemical-based polyurethanes, showing potential applications for coating on glass or lead due to its positively charged polymeric chain (Lu & Larock, 2010b). Other studies have also demonstrated that soybean oil-based anionic polyurethane can sweep a variety of properties in a similar fashion. Other studies have also demonstrated that soybean oil-based anionic polyurethane can sweep a variety of properties due to awide range of OH functionality (2.4 to 4). Higher the functionality higher the tendency to cross-link, which leads to rigid coating. But an interesting feature provided by most of the bio-renewable materials is being able to control the functionality by using a proper ratio of reagents to obtain the targeted property, which is one of the reasons these materials are so attractive along with procedures that are facile and well-established (Lu & Larock, 2008).

Canada contributes around 70% of the worldwide production of rapeseed oil (Shahbandeh, 2020). Rapeseed oil contains erucic acid, and more than 2% of erucic acid in rapeseed oil is not suitable for human consumption as it may cause heart diseases (Agricultural Marketing Resource Center, 2020). Therefore, it is mostly used as a starting material for various applications including the lubricant

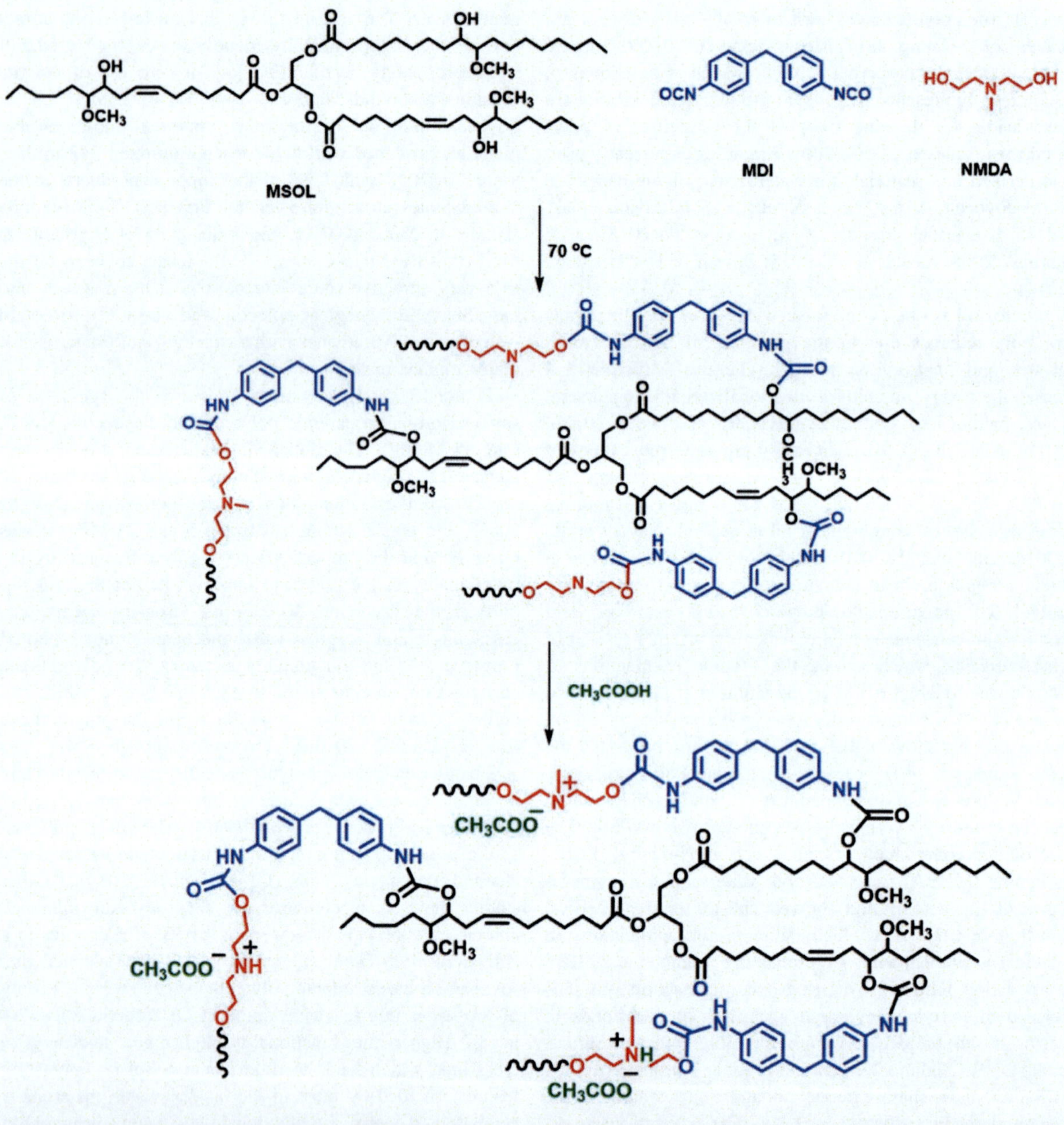

Fig. 13 Synthetic route for a soybean-based cationic polyurethane. Adapted with permission from Lu and Larock (2010b). Copyright (2010) Elsevier

market, biodiesel, and polyurethanes due to the presence of unsaturation in its structure which allows functionalizing the double bond using several green synthetic routes (Nieschlag & Wolff, 1971; Philipp & Eschig, 2012; Vicente et al., 2005). Sustainability has one of its solids foundations on finding applications for such materials that were initially considered as not suitable food and convert them into value-added products. With this keystone in perspective, a previous report obtained a rapeseed-based polyurethane that can be used as a biomedical scaffold for bone tissue to improve regeneration in damaged areas of the human body. A block prepolymer obtained from rapeseed oil-based polyol, synthesized through epoxidation followed by ring-opening with diethylene glycol (Fig. 14), was mixed in different ratios with poly(ε-caprolactone) diol to react with hexamethylene diisocyanate. Water was used as a chain extender to

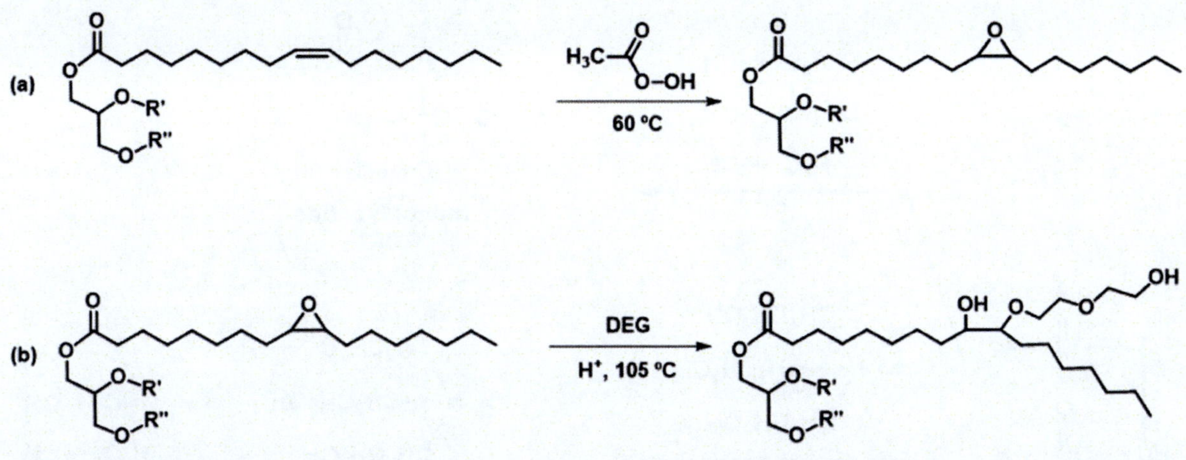

Fig. 14 Epoxidation followed by ring-opening of rapeseed oil to obtain a polyol. Adapted with permission from Zieleniewska et al. (2014). Copyright (2014) Elsevier

implement urea linkage into the structure to increase the biocompatibility, and sodium chloride was added to increase the porosity of the microstructure, which aids cells to grow within the polymer to prompt tissue regeneration (Zieleniewska et al., 2014). For biomedical applications, it is required that materials must present a safe way to leave the body to avoid allergic reactions or rejection by the body. Polyurethane-urea-based polymers are known to undergo hydrolysis when submitted to water or body fluids and go through the degradation that allows a gradual and safe release of the bio-based polymers (Santerre et al., 1994, 2005).

However, the use of waterborne polyurethane as a VOC-free coating has different requirements. For coating applications, polyurethane coating should be able to withstand aggressive environment conditions such as radiation, corrosion, chemical attack, and temperature amplitude to effectively protect the material that is underneath the coating. To meet these requirements, a rigid polyurethane coating was prepared using rapeseed oil-based polyols (Stirna et al., 2014). In this work, rapeseed oil-based polyol (RO) was synthesized by amidization with diethanolamine (DEA) and esterification with triethanolamine and glycerin (GL) as shown in Fig. 15 (Stirna et al., 2013). The produced polyurethane could be coated on materials as VOC-free coating using the spray technique. It showed remarkable mechanical properties as well as resistance against external agents even after exposure for a long period (Stirna et al., 2014). Rapeseed oil is one of the many cases where raw materials that were considered waste were found to be important components for many products with a broad range of applications as discussed. It shows that renewable materials present a great potential for sustainable production turning waste into value-added products through green methodologies and profitable investments.

Linseed oil extracted from the flax plant is another renewable resource for many applications. It can be used for wood finish since it can be soaked into the pores of the wood improving its aspect and durability (Tripathi et al., 2013). Linseed oil has provided significant improvement to paint industries as it is environmentally friendly, enhance the colors of pigments, and facilitate the mixing of components in the paint. It is also used as a supplement diet which helps against high blood pressure, heart disease, arthritis, control levels of cholesterol, and many others (Tripathi et al., 2013). Linseed oil contains linolenic acid (48%), linoleic acid (15%), and oleic acid (23%) which makes it very suitable for functionalization using sustainable routes and thus can be used for a wide-range application (Vereshchagin & Novitskaya, 1965). A representation of these fatty acids into a triglyceride is shown in Fig. 16 (Chen et al., 2002).

Linseed oil is used as an ionic center for waterborne polyurethane. For the synthesis of waterborne polyurethane, it was converted into a polyol through epoxidation and ring-opening with diethylene glycol followed by saponification with KOH to obtain free fatty acids chains (Chen et al., 2014). Since the obtained material presented both carboxylic acid and hydroxyl groups (functionality between 2 and 4), it was suitable as an ionic center to improve dispersibility in aqueous media. In general, 2,2-bis(hydroxymethyl)-propionic acid (DMPA) is used for this purpose; however, it is not a bio-derived material, hence by using a fatty acid chain originated from linseed oil for the same purpose adds a bio-renewable aspect to the polyurethane dispersion. Due to the presence of a higher concentration of both linoleic and linolenic fatty acids in linseed oil, it is very suitable for coating applications as it easily undergoes oxidation, self-polymerization, and cross-linking which makes coating to dry rapidly (Wicks & Wicks, 2005).

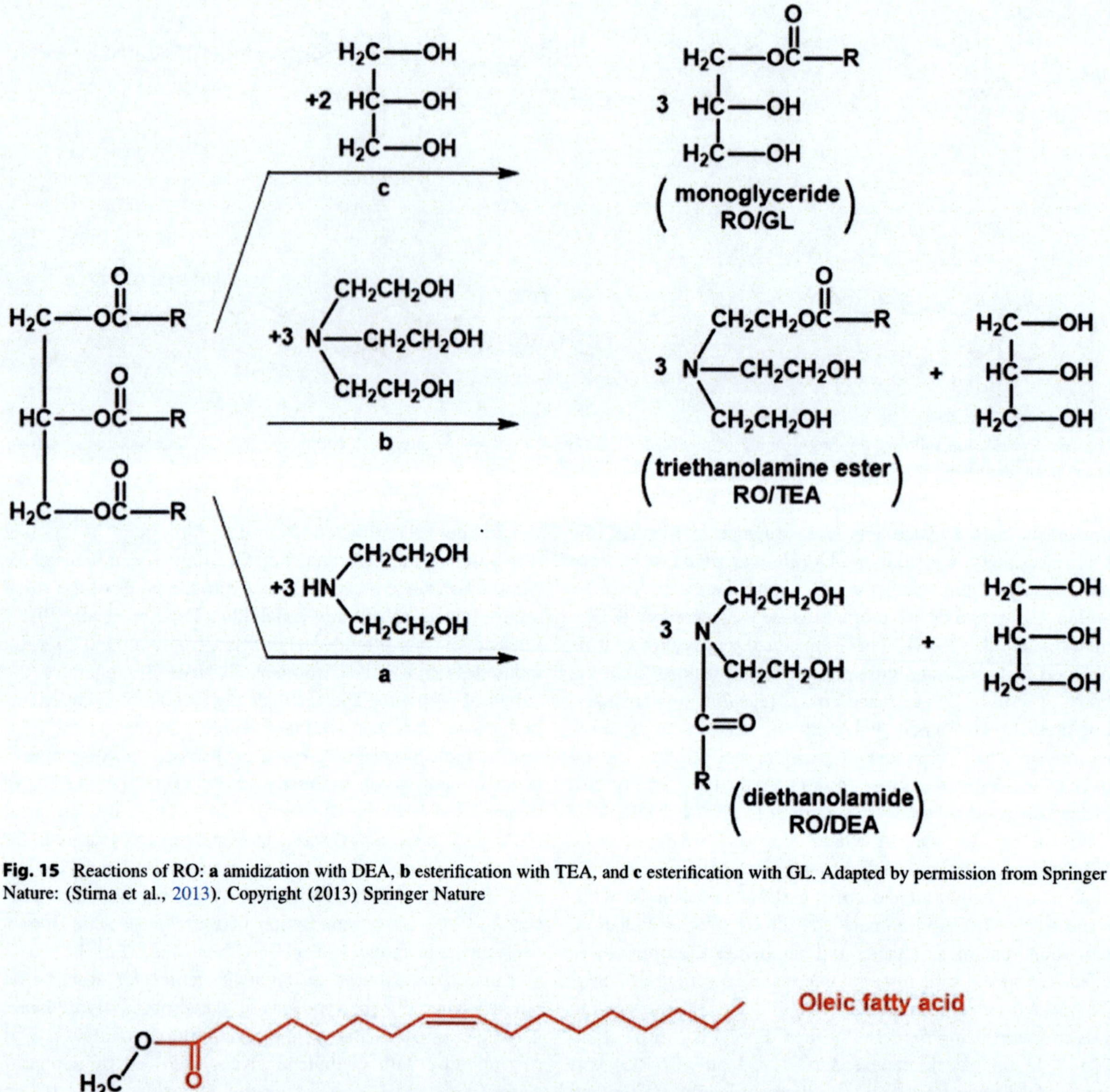

Fig. 15 Reactions of RO: **a** amidization with DEA, **b** esterification with TEA, and **c** esterification with GL. Adapted by permission from Springer Nature: (Stirna et al., 2013). Copyright (2013) Springer Nature

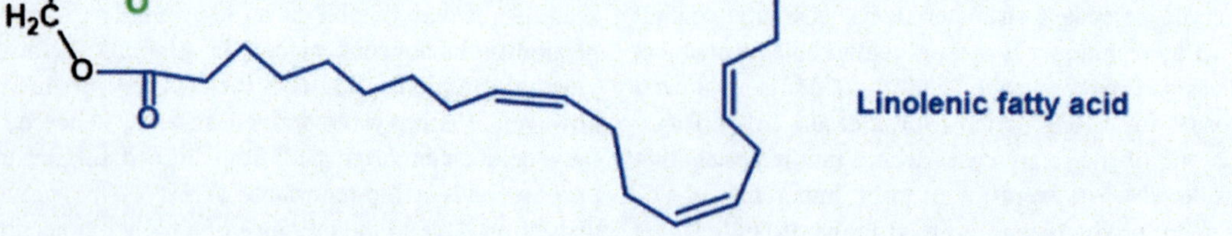

Fig. 16 Representation of linseed oil. Adapted with permission from Chen et al. (2002). Copyright (2002) Elsevier

Ammonolysis reaction was used to functionalize the linseed oil for waterborne polyurethanes. The final product displayed high thermal stability (decomposition started ~ 294 °C) along with hydrolysis resistance and fast drying of the coating (Chen et al., 2011).

Corn is the most produced grain in the USA accounting for around 96% of feed grains that are being cultivated. The production and price of corn have been increasing since 1926 which indicates that it is a stable and profitable market that has been growing constantly despite any economic recession. The high demand for corn is due to its versatile applications in energy, food, and industry. The majority of produced corn is primarily used as a source for ethanol production (40%), animal feed (36%), and food and industrial applications (24%). The use of corn for energy production through the extraction of ethanol is already a sustainable route with high demands. However, the acres that have been used to plant the crop are mostly constant (100 million) throughout the 20s and up to recent date, while the yield of production has been constantly increasing. The advance in agricultural technology played an important role in that improvement, for example, more effective machinery, fertilizers, pesticides along with a larger variety of seeds, and genetic modification (Castleberry et al., 1984). These facts encourage a deeper implementation of corn into research to obtain value-added products such as corn oil-based rigid and flexible polyurethanes (Ramanujam et al., 2019; Souza et al., 2012). Corn oil present very similar features if compared with soybean oil such as functionality, the concentration of fatty acids, the yield of extraction, and many others (Meier et al., 2007). These characteristics make it suitable for use in the synthesis of waterborne polyurethanes. Corn oil-based waterborne polyurethanes were synthesized and compared with soybean and several other oils. The results showed that corn oil-based waterborne polyurethanes provide similar or better properties in regards to mechanical properties and thermal stability making corn oil a promising starting renewable material for industrial applications (Garrison et al., 2014a).

Another biomaterial that is constantly growing in both research and industrial applications is jatropha oil which is mostly grown in Asia (Kumar & Sharma, 2008). Jatropha oil presents a high level of unsaturation (~ 80%) because of the higher content of oleic (~ 43%) and linoleic (~ 37%) fatty acids which provide several opportunities to functionalize for applications in polyurethanes (Sarin et al., 2007). Additionally, jatropha oil is non-edible oil making it viable for value-added industrial applications in various areas such as in polymers, green fuel, etc. Jatropha oil has been used for the production of biodiesel, elastomers, adhesives, coatings, and resins (Aung et al., 2014; Gogoi et al., 2014; Hazmi et al., 2013; Sugita & Mas'ud ZA, 2013). Waterborne polyurethanes synthesized using jatropha oil show high thermal stability around 286 °C for the first decomposition stage, an elastomeric behavior with elongation at break in a range of 85–325% with a potential application in wood coating (Saalah et al., 2015). Peanut oil is a commodity that remained stagnated from 1994 to 2013 finding most use as a frying oil due to its flavor and healthy properties (List, 2016). Because of its relatively higher price among other bio-renewable oils and superb properties for deep fat frying, peanut oil may not be as competitive for waterborne polyurethane applications (Warner et al., 1994). Despite that, previous reports that have converted it into a bio-polyol to synthesize polyurethanes showed that polyurethanes presented higher thermal stability if compared with polypropylene ones (Javni et al., 2000).

Tung oil has the capability to self-polymerize when exposed to air which makes it very suitable for coating. The coating on wood usually provides a golden color and improves resistance to weather and aesthetics aspects (Poth, 2001). Since it is non-edible and already present applications as wood coating, tung oil is a promising candidate for waterborne polyurethane (Wang et al., 2014; Yang, 2015; Zhengxiang et al., 2012). Polyurethanes based on tung oil provide better properties such as elongation, tensile strength, water repellency, and thermal stability compared to polyurethanes synthesized using other bio-oils and petrochemical (Man et al., 2019). Tung oil was also used in cationic dispersed polyurethanes for anti-bacterial properties (Man et al., 2019). Tung oil contains a high level of conjugation which allows them self-cross-linking and rapid drying on coating which enhances mechanical properties, thermal stability, and overall resistance to external agents (Yoo & Youngblood, 2017). Due to these factors, tung oil is a proper example of sustainability; it can be easily extracted and used to synthesize waterborne polyurethanes through facile and effective methods yielding superior properties if compared with most materials in the field. Even though it is already used in industrial applications, its potential has not been fully explored since it still has a relatively lower number of reports if compared with other biomaterials (Huang et al., 2014; Liang et al., 2019). Cationic dispersed polyurethanes based on tung oil provide superior anti-bacterial properties (Man et al., 2019).

Cottonseed oil is mostly used as frying oil due to its property of enhancing flavor and relatively low cost but also being used for various applications (Daniel et al., 2005). It is also implemented in cosmetics due to its both anti-inflammatory and anti-aging properties (Oskoueian et al., 2011). For this, food-grade cottonseed oil must be refined to remove gossypol, which is a pigment naturally produced by the plant with the role to protect it against insects; however, these days it is studied as a potential antifertility agent for the man (Dodou, 2005; Stipanovic et al., 1986). Reports suggest that many versatile routes can

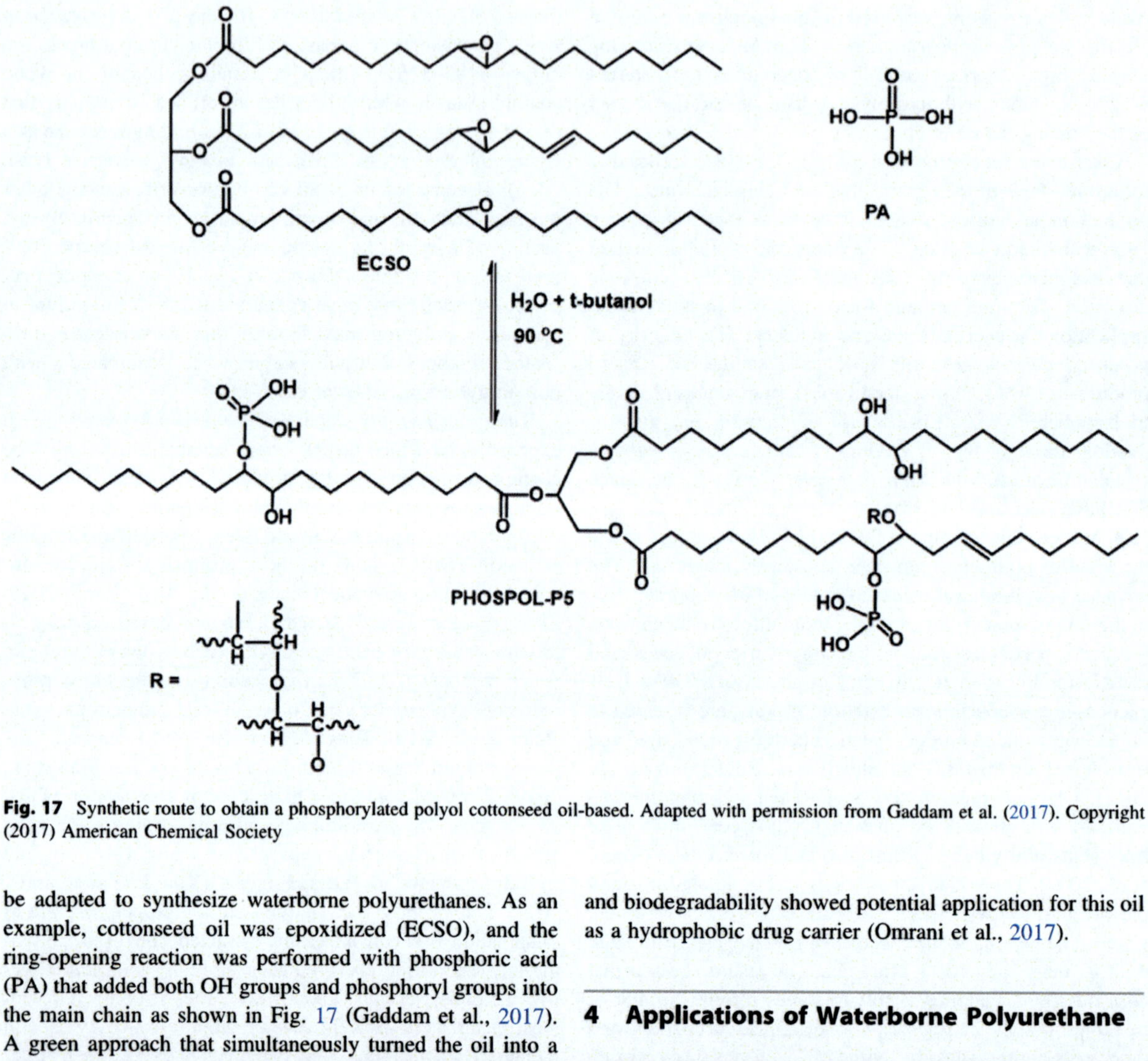

Fig. 17 Synthetic route to obtain a phosphorylated polyol cottonseed oil-based. Adapted with permission from Gaddam et al. (2017). Copyright (2017) American Chemical Society

be adapted to synthesize waterborne polyurethanes. As an example, cottonseed oil was epoxidized (ECSO), and the ring-opening reaction was performed with phosphoric acid (PA) that added both OH groups and phosphoryl groups into the main chain as shown in Fig. 17 (Gaddam et al., 2017). A green approach that simultaneously turned the oil into a polyol and in an ionic center, since the phosphorylated oil was able to replace the commonly used dimethylol propanoic acid. A siloxane-based compound, 3-aminopropyltriethanoxysiloxane, was used as a cross-linker, responsible for improving some of the properties such as thermal stability and an increase in mechanical properties.

Sunflower oil is mostly used for frying and as an emollient in the cosmetics industry. The demand for sunflower oil is slowly growing due to its healthy and flavor-enhancing properties. To attend those requirements, selective breeding and genetic manipulation are currently being performed to increase the amount of sunflower oil that can be extracted from the plant (Biresaw et al., 2014). By presenting a similar chemical structure with other polyunsaturated oils, sunflower oil can be naturally used for the synthesis of waterborne polyurethane in a variety of procedures. The biocompatibility and biodegradability showed potential application for this oil as a hydrophobic drug carrier (Omrani et al., 2017).

4 Applications of Waterborne Polyurethane

Due to the availability of versatile synthetic routes, tunable and unique properties, waterborne polyurethanes find their applications in several areas such as adhesives, coatings, biomedical, electrical, electronic, and many others. The process and the products are very sustainable as virtually any bio-polyols can be used for the synthesis of waterborne polyurethanes. Waterborne polyurethanes for adhesive applications must meet some requirements such as adhesion strength, durability, weather, and solvent stability. Coatings and adhesives play an important role in improving the durability and life span of the materials that they are applied to protect against corrosion, water uptake, temperature oscillation, mechanical shock, etc. Cationic waterborne polyurethanes are also suitable for coating on glass and leather (Kim, 1996; Lu & Larock, 2010a, 2010b). The adhesion takes place due to electrostatic attraction between the substrate and cationic waterborne polyurethanes. In general, glass and leather have a negatively charged surface which combines with the positively charged centers at the backbone of cationic waterborne polyurethanes, providing a strong bonding. This strong bonding allows the waterborne polyurethane to be used as coatings and adhesives to increase the longevity of these materials by protecting them against external agents such as radiation, thermal amplitude, chemical attack, and others.

Several bio-based materials such as soybean, linseed, castor, jatropha, and many more are being used to make waterborne polyurethanes for coating and adhesive applications (Liang et al., 2018; Saalah et al., 2015; Vereshchagin & Novitskaya, 1965; Xia et al., 2012). Recent studies have demonstrated remarkable anticorrosion properties and low levels of water absorption for waterborne polyurethane coatings that used grafted nanosheets of graphene oxide as a source of hydroxyl groups (Wen et al., 2019). The enhanced corrosion resistance and lower water uptake of the composite were directly related to the stability of the graphene oxide dispersion that guaranteed the formation of a continuous coating, the inherent impedance of the composite itself, and the partial hydrophobic chain. Other work implemented polyaniline blended with a polyurethane water dispersion matrix to obtain corrosion-resistant coatings (Gurunathan et al., 2013). The mechanical and thermal behavior are other important properties that define the field of applications for any materials. Waterborne polyurethanes also present satisfactory properties on these terms. The authors reported that linseed oil-based polyurethanes presented high thermal stability along with versatile synthesis procedures (Chen et al., 2011).

Cationic waterborne polyurethanes also extant anti-microorganism property. This, in particular, enlarges the field of applications for these polymeric materials to be used in bathtubs, boats, submarines, medical devices, and other related areas to prevent the growth of fungus, mold, bacteria, etc. (Liang et al., 2018; Lu & Larock, 2010b; Zhu et al., 2016). The high anti-microorganism property of cationic waterborne polyurethanes is due to the presence of positive charges in the backbone which effectively neutralizes the negative charges in the cell wall of microorganisms (Lu & Larock, 2010b; Qi et al., 2015). Also, soybean-based waterborne polyurethane presented simultaneous anti-bacterial and tunable thermal properties (Garrison et al., 2014b). These features make them suitable for food packing, thermal protective coatings, and so on.

Waterborne polyurethanes also find their wide applications in the biomedical field. For example, bio-renewable and biodegradable sources such as rapeseed oil-based polyol, poly(ε-caprolactone) diol, and water as a chain extender were used to obtain a biocompatible and biodegradable scaffold for bone tissue regeneration (Zieleniewska et al., 2014). Other biomedical applications of waterborne polyurethanes include heart valves, drug delivery, anticoagulant for body fluids, anti-bacterial dispersions, etc. (Jiang et al., 2007; Song et al., 2016; Unsworth et al., 1997). The use of bio-renewable sources for such applications yields value-added products with a lower cost and facile manufacture providing more accessibility for people that require related medical treatment. Waterborne polyurethanes also find their applications in specific areas such as shape memory, self-healing, sensors, and flame-retardant coatings (Chen et al., 2004; Díez-García et al., 2019; Gu & Luo, 2015; Wang et al., 2018; Zhang et al., 2019). The authors reported a biodegradable and biocompatible shape memory polyurethane that was functionalized with iron oxide nanoparticles (Eberhart et al., 2003). The polymers were stable in body fluids and could be seeded with specific cells to prompt cell growth which avoids more invasive surgeries and other medical procedures. The biggest advantage of waterborne polyurethanes synthesized using renewable resources is being biodegradable. Such waterborne polyurethanes could be consumed in the body without causing side effects or allergies.

Fractural defects and microcracks appear in coatings due to external provocations which hinder their wide applicability. Such issues can be overcome by introducing functional groups that can self-heal such defects in waterborne polyurethanes. The functional groups which are capable of self-healing could be easily introduced in waterborne polyurethanes due to the presence of unsaturation in biomaterials used to synthesize polyols. Zhang et al. introduced self-healing properties in waterborne polyurethanes by implementing disulfide bonds (Zhang et al., 2019). The fracture in coating healed completely in 15 min above room temperature. The effect of the addition of TiO_2 on the self-healing capability of waterborne poly(urethane-urea) was also investigated (Díez-García et al., 2019). It was observed that the addition of TiO_2 improved the thermal stability of the waterborne poly(urethane-urea); however, it decreased the self-healing capacity of the polymer which was correlated to the change in the morphology and interactions among various groups of the coating.

Waterborne polyurethanes are also used for some specific applications such as sensors and flame-retardant coatings. In general, waterborne polyurethanes are functionalized such as with conductive segments, which can turn them into effective sensors for gases. This process consists of an appreciable change in resistivity of the composite when in contact with a certain type of gases. This interaction leads to swelling of the polymeric structure causing it to rearrange the conductive segments allowing a proper measurement of the analyte gas (Chen et al., 2004). Flame-retardant materials are a huge ground of research that quickly evolved and brought components that release non-toxic or low toxic smoke as well as act as an efficient flame retardant to suppress the flame. Some known flame-retardant materials that are commonly explored are phosphorus and nitrogen-based compounds (Chen et al., 2012; Gu et al., 2015). The implementation of these materials into waterborne polyurethanes to obtain flame-retardant coatings showed remarkable results as explored by previous reports (Feng et al., 2016; Gu & Luo, 2015; Wang et al., 2019; Yin et al., 2017). The development of flame-retardant waterborne polyurethane coatings is very significant for society as it saves lives, eases the work of firemen, and helps the environment by decreasing the amount of smoke.

Sustainability is also efficiently obtaining energy to establish a balance between minimal and optimal usage to guarantee resources for future generations. The scientific community has provided a tremendous amount of effort into finding new materials to improve capacitance, energy efficiency, potency, and so on (Malini et al., 2009; Xie et al., 2016). Lithium ions and lithium-based inorganic compounds are largely used in electronic devices due to their high ionic conductivity and capacitance (Ritchie, 2001). However, they present some concerns such as the risk of fire hazards, instability at high temperatures, and toxicity (Cho et al., 2017). The solid polymer electrolytes (SPEs) were developed as an answer to these issues by providing the ability to fabricate flexible and light devices, use of eco-friendly solvents, and improved thermal stability (Long et al., 2016; Xue et al., 2015). Waterborne polyurethanes are applicable as solid polymer electrolytes due to the use of a non-toxic solvent (water), high mechanical durability (through the rigid domain), enough flexibility (due to the presence of chain extenders) for flexible devices and high ionic conduction (Wang & Min, 2010; Wang et al., 2010). The blending of waterborne polyurethane with other polymers used for SPE prompted an improvement in mechanical, thermal, and ionic conduction properties (Bao et al., 2018). The intermolecular interaction between the components is the key factor that allows the application of waterborne polyurethanes as solid polymer electrolytes. A general scheme of their mechanism is given in Fig. 18.

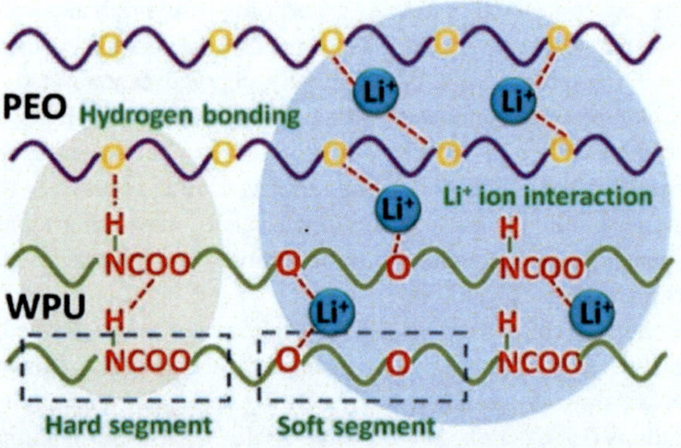

Fig. 18 Interactions between polymers and lithium ions for a solid polymer electrolyte (SPE). Adapted by permission from Bao et al. (2018). Copyright (2018) Elsevier

The increasing use of electric and electronic equipment for fast information trade and telecommunications has brought some issues such as electromagnetic interferences. Carbon nanotubes (CNT) were found to be useful for shielding electromagnetic waves, because of their high conductivity and flexibility that enable a good amount of wave absorption (Kim et al., 2004). However, one of the inherent issues of working with CNT is their tendency to form clusters, which often leads to a decay of the desired properties. To counter this, waterborne polyurethanes are being used to allow proper levels of percolation between the chains to turn it into stable dispersions. The attractive intermolecular interaction with polyurethane allows a high content of CNT to be used in the dispersion which enhances the shielding effectiveness (Hsiao et al., 2013; Zhang et al., 2010). Another allotrope of carbon such as graphene nanosheets dispersed into a polyurethane that contained a sulfonate pendant group allowing electrostatic interaction between the cationic and anionic segments found to be very effective for electromagnetic shielding applications (Hsiao et al., 2013).

Waterborne polyurethanes are also used for light-emitting electrochemical cells (Wang et al., 2004). Electrical conduction in waterborne polyurethanes happens due to the intercalation of conjugated and ionic patterns. The properties of waterborne polyurethanes can be tuned by changing the ratio of ionic centers on the polyurethane. Also, both thermal and mechanical properties can be optimized by varying the soft segment from the polyol, hard segment from the isocyanate, and by the introduction of different metal ions to neutralize the ionic center on the polymeric chain (Wang et al., 2003; Yin et al., 2000). Waterborne polyurethanes also used in thermoelectric, a device that converts heat into electricity (Tritt et al., 2008). Besides the use in solar cells, recent research is finding ways to implement waterborne polyurethanes with thermoelectric composites into cotton to use the body heat to generate electrical current (Wu & Hu, 2016). Related to the optical and light-emitting properties, waterborne polyurethane can be used as polymeric matrixes to fabricate photocatalyst hybrid films via functionalization with TiO_2 nanoparticles. The proper particle size distribution and stable dispersion are required to synthesize effective waterborne polyurethane coatings for photovoltaic cells, photocatalysis, UV stability, and water treatment (Li et al., 2015; Qiu et al., 2015). Polyurethane functionalized with expandable graphite and calcium carbonate showed its enhanced efficiency for water purification (Calcagnile et al., 2012; Vásquez et al., 2019).

5 Concluding Remarks and Future Perspective

Throughout this chapter, it was notable that many bio-renewable materials can be used to design waterborne polyurethanes with suitable properties for large-scale applications. The versatility and ease of synthetic routes strongly encourage the scientific community to find new biomaterials that can be used for this purpose. The main assets regarding the use of bio-renewable sources are the inherent availability since these resources can be replenished around six months to a year. Relative low-cost, which not only prompts the economy of the industry but ultimately leads to a virtuous economic chain or low harm to the environment, which is important to preserve natural areas for the sake of biodiversity also pushing the boundaries of technology to improve the yield of production with the lands that are already being used for growing renewable resources. The implementation of synthetic routes that avoid the use of the volatile organic compound is the proper way to both prevent environmental issues and to guarantee a safer place for research and industry. The biomaterials and procedures are interchangeable, which means that any materials can be suitable for any method applied making it a simple yet versatile process. Most importantly, the biomaterials currently reported showed properties that most of the time either match or surpass if compared with petrochemical materials, which is an important factor to establish an industrial segment that can rely on a stable provider.

References

Agricultural Marketing Resource Center. (2020). No Title. In: Rapeseed. https://www.agmrc.org/commodities-products/grains-oilseeds/rapeseed. Accessed 6 Nov 2020

Akindoyo, J. O., Beg, M. D. H., Ghazali, S., Islam, M. R., Jeyaratnam, N., & Yuvaraj, A. R. (2016). Polyurethane types, synthesis and applications—A review. *RSC Advances, 6*, 114453–114482. https://doi.org/10.1039/c6ra14525f

Arniza, M. Z., Hoong, S. S., Idris, Z., Yeong, S. K., Hassan, H. A., Din, A. K., & Choo, Y. M. (2015). Synthesis of transesterified palm olein-based polyol and rigid polyurethanes from this polyol. *Journal of the American Oil Chemists Society, 92*, 243–255. https://doi.org/10.1007/s11746-015-2592-9

Aung, M. M., Yaakob, Z., Kamarudin, S., & Abdullah, L. C. (2014). Synthesis and characterization of Jatropha (*Jatropha curcas* L.) oil-based polyurethane wood adhesive. *Industrial Crops and Products, 60*, 177–185.

Babb, D. A. (2012). Polyurethanes from renewable resources. *Advances in Polymer Science, 245*, 315–360. https://doi.org/10.1007/12-2011-130

Bao, J., Qu, X., Qi, G., Huang, Q., Wu, S., Tao, C., Gao, M., & Chen, C. (2018). Solid electrolyte based on waterborne polyurethane and poly(ethylene oxide) blend polymer for all-solid-state lithium ion batteries. *Solid State Ionics, 320*, 55–63. https://doi.org/10.1016/j.ssi.2018.02.030

Biermann, U., Metzger, J. O., & Meier, M. A. R. (2010). Acyclic triene metathesis oligo-and polymerization of high oleic sun flower oil. *Macromolecular Chemistry and Physics, 211*, 854–862.

Biresaw, G., Laszlo, J. A., Evans, K. O., Compton, D. L., & Bantchev, G. B. (2014). Synthesis and tribological investigation of lipoyl glycerides. *Journal of Agriculture and Food Chemistry, 62*, 2233–2243. https://doi.org/10.1021/jf404289r

Blackwell, J., Nagarajan, M. R., & Hoitink, T. B. (1982). Structure of polyurethane elastomers: Effect of chain extender length on the structure of MDI/diol hard segments. *Polymer (Guildf), 23*, 950–956. https://doi.org/10.1016/0032-3861(82)90392-5

Calcagnile, P., Fragouli, D., Bayer, I. S., Anyfantis, G. C., Martiradonna, L., Cozzoli, P. D., Cingolani, R., & Athanassiou, A. (2012). Magnetically driven floating foams for the removal of oil contaminants from water. *ACS Nano, 6*, 5413–5419. https://doi.org/10.1021/nn3012948

Castleberry, R. M., Crum, C. W., & Krull CF (1984) Genetic yield improvement of U.S. maize cultivars under varying fertility and climatic environments. *Crop Science, 24*. https://doi.org/10.2135/cropsci1984.0011183X002400010008x

Chan-Chan, L. H., Solis-Correa, R., Vargas-Coronado, R. F., Cervantes-Uc, J. M., Cauich-Rodríguez, J. V., Quintana, P., & Bartolo-Pérez, P. (2010). Degradation studies on segmented polyurethanes prepared with HMDI, PCL and different chain extenders. *Acta Biomaterialia, 6*, 2035–2044. https://doi.org/10.1016/j.actbio.2009.12.010

Charlon, M., Heinrich, B., Matter, Y., Couzigné, E., Donnio, B., & Avérous, L. (2014). Synthesis, structure and properties of fully biobased thermoplastic polyurethanes, obtained from a diisocyanate based on modified dimer fatty acids, and different renewable diols. *European Polymer Journal, 61*, 197–205. https://doi.org/10.1016/j.eurpolymj.2014.10.012

Chen, J. B., Guo, Q., Sun, J. L., Shao, X. L., & Nie, Z. J. (2011). A study on linseed oil modified waterborne polyurethane coatings. *Materials Science Forum, 686*, 528–532. https://doi.org/10.4028/www.scientific.net/MSF.686.528

Chen, M.-J., Shao, Z.-B., Wang, X.-L., Chen, L., & Wang, Y.-Z. (2012). Halogen-free flame-retardant flexible polyurethane foam with a novel nitrogen-phosphorus flame retardant. *Industrial and Engineering Chemistry Research, 51*, 9769–9776. https://doi.org/10.1021/ie301004d

Chen, R., Zhang, C., & Kessler, M. R. (2014). Anionic waterborne polyurethane dispersion from a bio-based ionic segment. *RSC Advances, 4*, 35476–35483. https://doi.org/10.1039/c4ra07519f

Chen, J., Soucek, M. D., Simonsick, W. J., & Celikay, R. W. (2002). Synthesis and photopolymerization of norbornyl epoxidized linseed oil. *Polymer (Guildf), 43*, 5379–5389. https://doi.org/10.1016/S0032-3861(02)00404-4

Chen, S. G., Hu, J. W., Zhang, M. Q., Li, M. W., & Rong, M. Z. (2004) Gas sensitivity of carbon black/waterborne polyurethane composites. *Carbon N Y, 42*, 645–651. https://doi.org/10.1016/j.carbon.2004.01.002

Cho, J., Jung, Y.-C., Lee, Y. S., & Kim, D.-W. (2017). High performance separator coated with amino-functionalized SiO_2 particles for safety enhanced lithium-ion batteries. *Journal of Membrane Science, 535*, 151–157. https://doi.org/10.1016/j.memsci.2017.04.042

Choi, H. Y., Bae, C. Y., & Kim, B. K. (2010). Nanoclay reinforced UV curable waterborne polyurethane hybrids. *Progress in Organic Coatings, 68*, 356–362. https://doi.org/10.1016/j.porgcoat.2010.03.015

Coutinho, F. M. B., Delpech, M. C., & Alves, L. S. (2001). Anionic waterborne polyurethane dispersions based on hydroxyl-terminated polybutadiene and poly(propylene glycol): Synthesis and characterization. *Journal of Applied Polymer Science, 80*, 566–572.

https://doi.org/10.1002/1097-4628(20010425)80:4%3c566::AID-APP1131%3e3.0.CO;2-H

Dai, H., Yang, L., Lin, B., Wang, C., & Shi, G. (2009). Synthesis and characterization of the different soy-based polyols by ring opening of epoxidized soybean oil with methanol, 1,2-ethanediol and 1,2-propanediol. *Journal of the American Oil Chemists Society, 86*, 261–267. https://doi.org/10.1007/s11746-008-1342-7

Daniel, D. R., Thompson, L. D., Shriver, B. J., Wu, C.-K., & Hoover, L. C. (2005). Nonhydrogenated cottonseed oil can be used as a deep fat frying medium to reduce trans-fatty acid content in french fries. *Journal of the American Dietetic Association, 105*, 1927–1932. https://doi.org/10.1016/j.jada.2005.09.029

De Souza, V. H. R., Silva, S. A., Ramos, L. P., & Zawadzki, S. F. (2012). Synthesis and characterization of polyols derived from corn oil by epoxidation and ozonolysis. *Journal of the American Oil Chemists Society, 89*, 1723–1731. https://doi.org/10.1007/s11746-012-2063-5

Desroches, M., Escouvois, M., Auvergne, R., Caillol, S., & Boutevin, B. (2012). From vegetable oils to polyurethanes: synthetic routes to polyols and main industrial products. *Polymer Reviews, 52*, 38. https://doi.org/10.1080/15583724.2011.640443

Díez-García, I., Eceiza, A., & Tercjak, A. (2019). Self-healable nanocomposites with enhanced thermal stability by incorporation of TiO_2 nanoparticles to waterborne poly(urethane-urea) matrices based on amphiphilic triblock copolymers. *Journal of Physical Chemistry C, 123*, 21290–21298. https://doi.org/10.1021/acs.jpcc.9b06184

Dodiuk, H., & Goodman, S. H. (2014) *Handbook of thermoset plastics* (3rd edn.). William Andrew.

Dodou, K. (2005). Investigations on gossypol: Past and present developments. *Expert Opinion on Investigational Drugs, 14*, 1419–1434. https://doi.org/10.1517/13543784.14.11.1419

Domanska, A., & Boczkowska, A. (2014). Biodegradable polyurethanes from crystalline prepolymers. *Polymer Degradation and Stability, 108*, 175–181. https://doi.org/10.1016/j.polymdegradstab.2014.06.017

Dworakowska, S., Bogdal, D., & Prociak, A. (2012). Microwave-assisted synthesis of polyols from rapeseed oil and properties of flexible polyurethane foams. *Polymers (basel), 4*, 1462–1477. https://doi.org/10.3390/polym4031462

Eberhart, R. C., Su, S.-H., Nguyen, K. T., Zilberman, M., Tang, L., Nelson, K. D., & Frenkel, P. (2003). Review: Bioresorbable polymeric stents: Current status and future promise. *Journal of Biomaterials Science, Polymer Edition, 14*, 299–312. https://doi.org/10.1163/156856203321478838

Elbers, N., Ranaweera, C. K., Ionescu, M., Wan, X., Kahol, P. K., & Gupta, R. K. (2017). Synthesis of novel biobased polyol via thiol-ene chemistry for rigid polyurethane foams. *Journal of Renewable Materials, 5*, 74–83. https://doi.org/10.7569/jrm.2017.634137

Engels, H.-W., Pirkl, H.-G., Albers, R., Albach, R. W., Krause, J., Hoffmann, A., Casselmann, H., & Dormish, J. (2013). Polyurethanes: Versatile materials and sustainable problem solvers for today's challenges. *Angewandte Chemie International Edition, 52*, 9422–9441. https://doi.org/10.1002/anie.201302766

Feng, J., Ge, Z., Chai, C., Wang, S., Yu, D., Wu, G., & Luo, Y. (2016). Flame retardant modification of waterborne polyurethane fabric coating agent with high hydrostatic pressure resistance. *Progress in Organic Coatings, 97*, 91–98. https://doi.org/10.1016/j.porgcoat.2016.03.020

Gaddam, S. K., & Palanisamy, A. (2016). Anionic waterborne polyurethane dispersions from maleated cotton seed oil polyol carrying ionisable groups. *Colloid and Polymer Science, 294*, 347–355. https://doi.org/10.1007/s00396-015-3787-1

Gaddam, S. K., Kutcherlapati, S. N. R., & Palanisamy, A. (2017). Self-cross-linkable anionic waterborne polyurethane-silanol dispersions from cottonseed-oil-based phosphorylated polyol as ionic soft segment. *ACS Sustainable Chemical & Engineering, 5*, 6447–6455. https://doi.org/10.1021/acssuschemeng.7b00327

Garrison, T. F., Zhang, Z., Kim, H. J., Mitra, D., Xia, Y., Pfister, D. P., Brehm-Stecher, B. F., Larock, R. C., & Kessler, M. R. (2014b). Thermo-mechanical and antibacterial properties of soybean oil-based cationic polyurethane coatings: Effects of amine ratio and degree of crosslinking. *Macromolecular Materials and Engineering, 299*, 1042–1051. https://doi.org/10.1002/mame.201300423

Garrison, T. F., Kessler, M. R., & Larock, R. C. (2014). Effects of unsaturation and different ring-opening methods on the properties of vegetable oil-based polyurethane coatings. *Polymer (Guildf), 55*, 1004–1011. https://doi.org/10.1016/j.polymer.2014.01.014

Geyer, R., Jambeck, J. R., & Law, K. L. (2017). Production, use, and fate of all plastics ever made—Supplementary Information. *Science Advances, 3*, 19–24. https://doi.org/10.1126/sciadv.1700782

Gogoi, P., Boruah, M., Bora, C., & Dolui, S. K. (2014). Jatropha curcas oil based alkyd/epoxy resin/expanded graphite (EG) reinforced bio-composite: Evaluation of the thermal, mechanical and flame retardancy properties. *Progress in Organic Coatings, 77*, 87–93. https://doi.org/10.1016/j.porgcoat.2013.08.006

Grand View Research. (2020). Plastic market size, share & trends analysis report by product forecasts, 2020–2027. https://www.grandviewresearch.com/industry-analysis/global-plastics-market. Accessed 6 Jul 2020

Gu, L., Ge, Z., Huang, M., & Luo, Y. (2015). Halogen-free flame-retardant waterborne polyurethane with a novel cyclic structure of phosphorus−nitrogen synergistic flame retardant. *Journal of Applied Polymer Science*, 132.https://doi.org/10.1002/app.41288

Gu, L., & Luo, Y. (2015). Flame retardancy and thermal decomposition of phosphorus-containing waterborne polyurethanes modified by halogen-free flame retardants. *Industrial and Engineering Chemistry Research, 54*, 2431–2438. https://doi.org/10.1021/ie5045692

Gurunathan, T., Rao, C. R. K., Narayan, R., & Raju, K. V. S. N. (2013). Synthesis, characterization and corrosion evaluation on new cationomeric polyurethane water dispersions and their polyaniline composites. *Progress in Organic Coatings, 76*, 639–647. https://doi.org/10.1016/j.porgcoat.2012.12.009

Hablot, E., Zheng, D., Bouquey, M., & Avérous, L. (2008). Polyurethanes based on castor oil: Kinetics, chemical, mechanical and thermal properties. *Macromolecular Materials and Engineering, 293*, 922–929. https://doi.org/10.1002/mame.200800185

Haynes, J., McLaughlin, J., Vasquez, L., & Hunsberger, A. (2001). Low-maintenance landscape plants for South Florida. Environ Hortic Dept, Florida Coop Ext Serv Univ Florida-IFAS Publ ENH854, Florida, USA, pp. 1–49

Hazmi, A. S. A., Aung, M. M., Abdullah, L. C., Salleh, M. Z., & Mahmood, M. H. (2013). Producing Jatropha oil-based polyol via epoxidation and ring opening. *Industrial Crops Products, 50*, 563–567. https://doi.org/10.1016/j.indcrop.2013.08.003

Hsiao, S.-T., Ma, C.-C.M., Tien, H.-W., Liao, W.-H., Wang, Y.-S., Li, S.-M., & Huang, Y.-C. (2013). Using a non-covalent modification to prepare a high electromagnetic interference shielding performance graphene nanosheet/water-borne polyurethane composite. *Carbon N Y, 60*, 57–66. https://doi.org/10.1016/j.carbon.2013.03.056

Huang, K., Liu, Z., Zhang, J., Li, S., Li, M., Xia, J., & Zhou, Y. (2014). Epoxy monomers derived from Tung oil fatty acids and its regulable thermosets cured in two synergistic ways. *Biomacromolecules, 15*, 837–843. https://doi.org/10.1021/bm4018929

Ionescu, M. (2006). Mihail Ionescu: Chemistry and technology of polyols for polyurethanes. *Polimeri, 26*, 218–218.

Javni, I., Petrović, Z. S., Guo, A., & Fuller, R. (2000). Thermal stability of polyurethanes based on vegetable oils. *Journal of Applied Polymer Science, 77*, 1723.

Jiang, X., Li, J., Ding, M., Tan, H., Ling, Q., Zhong, Y., & Fu, Q. (2007). Synthesis and degradation of nontoxic biodegradable waterborne polyurethanes elastomer with poly(ε-caprolactone) and poly(ethylene glycol) as soft segment. *European Polymer Journal, 43*, 1838–1846. https://doi.org/10.1016/j.eurpolymj.2007.02.029

Jin, Y.-Z., Hahn, Y. B., Nahm, K. S., & Lee, Y.-S. (2005) Preparation of stable polyurethane–polystyrene copolymer emulsions via RAFT polymerization process. *Polymer (Guildf), 46*, 11294–11300 . https://doi.org/10.1016/j.polymer.2005.10.049

Kathalewar, M., Sabnis, A., & D'Mello, D. (2014). Isocyanate free polyurethanes from new CNSL based bis-cyclic carbonate and its application in coatings. *European Polymer Journal, 57*, 99–108. https://doi.org/10.1016/j.eurpolymj.2014.05.008

Kim, B. K. (1996). Aqueous polyurethana dispersions. *Colloid & Polymer Science, 274*, 599–611. https://doi.org/10.1007/BF00653056

Kim, H. M., Kim, K., Lee, C. Y., Joo, J., Cho, S. J., Yoon, H. S., Pejaković, D. A., Yoo, J.-W., & Epstein, A. J. (2004). Electrical conductivity and electromagnetic interference shielding of multi-walled carbon nanotube composites containing Fe catalyst. *Applied Physics Letters, 84*, 589–591. https://doi.org/10.1063/1.1641167

Kong, X., Li, S., Qu, J., & Chen, H. (2010). Self-emulsifying hydroxy acrylic polymer dispersions for two component waterborne polyurethane coatings. *Journal of Macromolecular Science, Part A Pure and Applied Chemistry, 47*, 368–374. https://doi.org/10.1080/10601320903539330

Kosheeladevi, P. P., Tuan Noor Maznee, T. I., Hoong, S. S., Nurul'Ain, H., Mohd Norhisham, S., Norhayati, M. N., Srihanum, A., Yeong, S. K., Hazimah, A. H., Sendijarevic, V., & Sendijarevic, A. (2016). Performance of palm oil-based dihydroxystearic acid as ionizable molecule in waterborne polyurethane dispersions. *Journal of Applied Polymer Science, 133*, 1–10.https://doi.org/10.1002/app.43614

Kumar, A., & Sharma, S. (2008). An evaluation of multipurpose oil seed crop for industrial uses (*Jatropha curcas* L.): A review. *Industrial Crops and Products, 28*, 1–10. https://doi.org/10.1016/j.indcrop.2008.01.001

Lee, A., & Deng, Y. (2015). Green polyurethane from lignin and soybean oil through non-isocyanate reactions. *European Polymer Journal, 63*, 67–73. https://doi.org/10.1016/j.eurpolymj.2014.11.023

Li, K., Peng, J., Zhang, M., Heng, J., Li, D., & Mu, C. (2015). Comparative study of the effects of anatase and rutile titanium dioxide nanoparticles on the structure and properties of waterborne polyurethane. *Colloids and Surfaces A: Physicochemical and Engineering Aspects, 470*, 92–99. https://doi.org/10.1016/j.colsurfa.2015.01.072

Liang, H., Liu, L., Lu, J., Chen, M., & Zhang, C. (2018). Castor oil-based cationic waterborne polyurethane dispersions: Storage stability, thermo-physical properties and antibacterial properties. *Industrial Crops and Products, 117*, 169–178. https://doi.org/10.1016/j.indcrop.2018.02.084

Liang, B., Zhao, J., Li, G., Huang, Y., Yang, Z., & Yuan, T. (2019). Facile synthesis and characterization of novel multi-functional bio-based acrylate prepolymers derived from tung oil and its application in UV-curable coatings. *Industrial Crops and Products, 138*, 111585. https://doi.org/10.1016/j.indcrop.2019.111585

List, G. R. (2016). Chapter 15—Processing and food uses of peanut oil and protein. In H. T. Stalker & F. Wilson (Eds.), *Peanuts* (pp. 405–428). AOCS Press.

Liu, H., & Chung, H. (2017). Visible-light induced thiol-ene reaction on natural lignin. *ACS Sustainable Chemistry & Engineering, 5*, 9160–9168. https://doi.org/10.1021/acssuschemeng.7b02065

Lligadas, G., Ronda, J. C., Galià, M., & Cádiz, V. (2007). Poly(ether urethane) networks from renewable resources as candidate biomaterials: Synthesis and characterization. *Biomacromolecules, 8*, 686–692. https://doi.org/10.1021/bm060977h

Lligadas, G., Ronda, J. C., Galià, M., & Cádiz, V. (2013) Renewable polymeric materials from vegetable oils: A perspective. *Materials Today, 16*, 337–343 . https://doi.org/10.1016/j.mattod.2013.08.016

Long, L., Wang, S., Xiao, M., & Meng, Y. (2016). Polymer electrolytes for lithium polymer batteries. *Journal Materials Chemistry A, 4*, 10038–10069. https://doi.org/10.1039/C6TA02621D

Lu, Y., Tighzert, L., Dole, P., & Erre, D. (2005). Preparation and properties of starch thermoplastics modified with waterborne polyurethane from renewable resources. *Polymer (Guildf), 46*, 9863–9870. https://doi.org/10.1016/j.polymer.2005.08.026

Lu, Y., Tighzert, L., Berzin, F., & Rondot, S. (2005) Innovative plasticized starch films modified with waterborne polyurethane from renewable resources. *Carbohydrate Polymers, 61*, 174–182. https://doi.org/10.1016/j.carbpol.2005.04.013

Lu, Y., & Larock, R. C. (2007). New hybrid latexes from a soybean oil-based waterborne polyurethane and acrylics via emulsion polymerization. *Biomacromolecules, 8*, 3108–3114. https://doi.org/10.1021/bm700522z

Lu, Y., & Larock, R. C. (2008). Soybean-oil-based waterborne polyurethane dispersions: Effects of polyol functionality and hard segment content on properties. *Biomacromolecules, 9*, 3332–3340. https://doi.org/10.1021/bm801030g

Lu, Y., & Larock, R. C. (2009). Novel polymeric materials from vegetable oils and vinyl monomers: Preparation, properties, and applications. *Chemsuschem, 2*, 136–147. https://doi.org/10.1002/cssc.200800241

Lu, Y., & Larock, R. C. (2010a). Aqueous cationic polyurethane dispersions from vegetable oils. *Chemsuschem, 3*, 329–333. https://doi.org/10.1002/cssc.200900251

Lu, Y., & Larock, R. C. (2010b). Soybean oil-based, aqueous cationic polyurethane dispersions: Synthesis and properties. *Progress in Organic Coatings, 69*, 31–37. https://doi.org/10.1016/j.porgcoat.2010.04.024

Lu, Y., & Larock, R. C. (2011). Synthesis and properties of grafted latices from a soybean oil-based waterborne polyurethane and acrylics. *Journal of Applied Polymer Science, 119*, 3305–3314. https://doi.org/10.1002/app.29029

Madbouly, S. A., & Otaigbe, J. U. (2009). Recent advances in synthesis, characterization and rheological properties of polyurethanes and POSS/polyurethane nanocomposites dispersions and films. *Progress in Polymer Science, 34*, 1283–1332. https://doi.org/10.1016/j.progpolymsci.2009.08.002

Malini, R., Uma, U., Sheela, T., Ganesan, M., & Renganathan, N. G. (2009). Conversion reactions: A new pathway to realise energy in lithium-ion battery—Review. *Ionics (kiel), 15*, 301–307. https://doi.org/10.1007/s11581-008-0236-x

Man, L., Feng, Y., Hu, Y., Yuan, T., & Yang, Z. (2019). A renewable and multifunctional eco-friendly coating from novel tung oil-based cationic waterborne polyurethane dispersions. *Journal of Cleaner Production, 241*, 118341. https://doi.org/10.1016/j.jclepro.2019.118341

Meier, M. A. R., Metzger, J. O., & Schubert, U. S. (2007). Plant oil renewable resources as green alternatives in polymer science. *Chemical Society Reviews, 36*, 1788–1802. https://doi.org/10.1039/B703294C

De Milliano, J. W., Woolnough, A., Reeves, A., & Shepherd, D. (2010). Ecologically significant invasive species: A monitoring framework for natural resource management groups in Western Australia. Department of Agriculture and Food, Western Australia, Perth

Mittal, J. P., Dhawan, K. C., & Thyagraj, C. R. (1991). Energy scenario of castor crop under dryland agriculture of Andhra Pradesh. *Energy*

Conversion and Management, 32, 425–430. https://doi.org/10.1016/0196-8904(91)90003-2

Mosiewicki, M. A., Dell'Arciprete, G. A., Aranguren, M. I., & Marcovich, N. E. (2009). Polyurethane foams obtained from castor oil-based polyol and filled with wood flour. *Journal of Composite Materials, 43*, 3057–3072. https://doi.org/10.1177/0021998309345342

Mu, Y., Wan, X., Han, Z., Peng, Y., & Zhong, S. (2012). Rigid polyurethane foams based on activated soybean meal. *Journal of Applied Polymer Science, 124*, 4331–4338. https://doi.org/10.1002/app.35612

Mumtaz, F., Zuber, M., Zia, K. M., Jamil, T., & Hussain, R. (2013). Synthesis and properties of aqueous polyurethane dispersions: Influence of molecular weight of polyethylene glycol. *Korean Journal of Chemical Engineering, 30*, 2259–2263. https://doi.org/10.1007/s11814-013-0166-9

Nanda, A. K., Wicks, D. A., Madbouly, S. A., & Otaigbe, J. U. (2005). Effect of ionic content, solid content, degree of neutralization, and chain extension on aqueous polyurethane dispersions prepared by prepolymer method. *Journal of Applied Polymer Science, 98*, 2514–2520.

Nieschlag, H. J., & Wolff, I. A. (1971). Industrial uses of high erucic oils. *Journal of the American Oil Chemists Society, 48*, 723–727. https://doi.org/10.1007/BF02638529

Omrani, I., Babanejad, N., Shendi, H. K., & Nabid, M. R. (2017). Preparation and evaluation of a novel sunflower oil-based waterborne polyurethane nanoparticles for sustained delivery of hydrophobic drug. *European Journal of Lipid Science and Technology, 119*, 1600283. https://doi.org/10.1002/ejlt.201600283

Oskoueian, E., Abdullah, N., Hendra, R., & Karimi, E. (2011). Bioactive compounds, antioxidant, xanthine oxidase inhibitory, tyrosinase inhibitory and anti-inflammatory activities of selected agro-industrial by-products. *International Journal of Molecular Sciences, 12*.

Palaskar, D. V., Boyer, A., Cloutet, E., Alfos, C., & Cramail, H. (2010). Synthesis of biobased polyurethane from oleic and ricinoleic acids as the renewable resources via the AB-type self-condensation approach. *Biomacromolecules, 11*, 1202–1211. https://doi.org/10.1021/bm100233v

Park, S. H., Chung, I. D., Hartwig, A., & Kim, B. K. (2007). Hydrolytic stability and physical properties of waterborne polyurethane based on hydrolytically stable polyol. *Colloids and Surfaces A: Physicochemical and Engineering Aspects, 305*, 126–131. https://doi.org/10.1016/j.colsurfa.2007.04.051

Petrović, Z. S. (2008). Polyurethanes from vegetable oils. *Polymer Reviews, 48*, 109. https://doi.org/10.1080/15583720701834224

Petrović, Z. S., Zhang, W., & Javni, I. (2005). Structure and properties of polyurethanes prepared from triglyceride polyols by ozonolysis. *Biomacromolecules, 6*, 713–719. https://doi.org/10.1021/bm049451s

Petrović, Z. S., Cvetković, I., Hong, D., Wan, X., & Zhang, W. (2008). Polyester polyols and polyurethanes from ricinoleic acid. *Journal of Applied Polymer Science, 108*, 1184–1190. https://doi.org/10.1002/app.27783

Pfister, D. P., Xia, Y., & Larock, R. C. (2011). Recent advances in vegetable oil-based polyurethanes. *ChemSusChem, 4*, 703. https://doi.org/10.1002/cssc.201000378

Philipp, C., & Eschig, S. (2012). Waterborne polyurethane wood coatings based on rapeseed fatty acid methyl esters. *Progress in Organic Coatings, 74*, 705–711. https://doi.org/10.1016/j.porgcoat.2011.09.028

Poth, U. (2001). Drying oils and related products. *Ullmann's Encyclopedia of Industrial Chemistry*.

Qi, J., Lai, X., Wang, J., Tang, H., Ren, H., Yang, Y., Jin, Q., Zhang, L., Yu, R., Ma, G., Su, Z., Zhao, H., & Wang, D. (2015). Multi-shelled hollow micro-/nanostructures. *Chemical Society Reviews, 44*, 6749–6773. https://doi.org/10.1039/C5CS00344J

Qiu, S., Deng, F., Xu, S., Liu, P., Min, X., & Ma, F. (2015). Degradation of pollutant and antibacterial activity of waterborne polyurethane/doped TiO_2 nanoparticle hybrid films. *J Wuhan Univ Technol Sci Ed, 30*, 447–451. https://doi.org/10.1007/s11595-015-1169-7

Ramanujam, S., Zequine, C., Bhoyate, S., Neria, B., Kahol, P., & Gupta, R. (2019). Novel biobased polyol using corn oil for highly flame-retardant polyurethane foams. *C, 5*, 13. https://doi.org/10.3390/c5010013

Reeves, J. T., Visco, M. D., Marsini, M. A., Grinberg, N., Busacca, C. A., Mattson, A. E., & Senanayake, C. H. (2015). A general method for imine formation using $B(OCH_2CF_3)_3$. *Organic Letters, 17*, 2442–2445. https://doi.org/10.1021/acs.orglett.5b00949

Remya, V., Patil, D., Abitha, V., & Ajay Rane, R. M. (2016). Biobased materials for polyurethane dispersions. *Chemistry International, 2*, 158–167.

Ritchie, A. G. (2001). Recent developments and future prospects for lithium rechargeable batteries. *Journal of Power Sources, 96*, 1–4. https://doi.org/10.1016/S0378-7753(00)00673-X

Saalah, S., Abdullah, L. C., Aung, M. M., Salleh, M. Z., Awang Biak, D. R., Basri, M., & Jusoh, E. R. (2015) Waterborne polyurethane dispersions synthesized from jatropha oil. *Industrial Crops and Products, 64*, 194–200. https://doi.org/10.1016/j.indcrop.2014.10.046

Saetung, A., Kaenhin, L., Klinpituksa, P., Rungvichaniwat, A., Tulyapitak, T., Munleh, S., Campistron, I., & Pilard, J.-F. (2012). Synthesis, characteristic, and properties of waterborne polyurethane based on natural rubber. *Journal of Applied Polymer Science, 124*, 2742–2752. https://doi.org/10.1002/app.35318

Santerre, J. P., Labow, R. S., Duguay, D. G., Erfle, D., & Adams, G. A. (1994). Biodegradation evaluation of polyether and polyester-urethanes with oxidative and hydrolytic enzymes. *Journal of Biomedical Materials Research, 28*, 1187–1199. https://doi.org/10.1002/jbm.820281009

Santerre, J. P., Woodhouse, K., Laroche, G., & Labow, R. S. (2005). Understanding the biodegradation of polyurethanes: From classical implants to tissue engineering materials. *Biomaterials 26*, 7457–7470. https://doi.org/10.1016/j.biomaterials.2005.05.079

Sardari, A., Sabbagh Alvani, A. A., & Ghaffarian, S. R. (2019). Castor oil-derived water-based polyurethane coatings: Structure manipulation for property enhancement. *Progress in Organics Coatings, 133*, 198–205. https://doi.org/10.1016/j.porgcoat.2019.04.030

Sarin, R., Sharma, M., Sinharay, S., Malhotra, R. K. (2007) Jatropha–Palm biodiesel blends: An optimum mix for Asia. Fuel 86:1365–1371. https://doi.org/10.1016/j.fuel.2006.11.040

Shahbandeh, M. (2020). No Title. In: Rapeseed Prod. Vol. Worldw. by Ctry. 2019/2020. https://www.statista.com/statistics/263930/worldwide-production-of-rapeseed-by-country/#:~:text=Rapeseed production volume worldwide by country 2019%2F2020&text=Canada was the leading producer,metric tons in 2019%2F2020. Accessed 6 Nov 2020

Song, Y., Gao, Y., Wan, X., Luo, F., Li, J., Tan, H., & Fu, Q. (2016). Dual-functional anticoagulant and antibacterial blend coatings based on gemini quaternary ammonium salt waterborne polyurethane and heparin. *RSC Advances, 6*, 17336–17344. https://doi.org/10.1039/c5ra27081b

de Souza, F., Choi, J., Bhoyate, S., Kahol, P. K., Gupta, R. K. (2020). Expendable graphite as an efficient flame-retardant for novel partial bio-based rigid polyurethane foams. *C*, 1–13. https://doi.org/10.3390/c6020027

Stipanovic, R. D., Stoessl, A., Stothers, J. B., Altman, D. W., Bell, A. A., & Heinstein, P. (1986). The stereochemistry of the biosynthetic precursor of gossypol. *Journal of the Chemical Society, Chemical Communications*, 100–102.https://doi.org/10.1039/C39860000100

Stirna, U., Fridrihsone, A., Lazdiņa, B., Misāne, M., & Vilsone, D. (2013). Biobased polyurethanes from rapeseed oil polyols:

Structure, mechanical and thermal properties. *Journal of Polymers and the Environment, 21*, 952–962. https://doi.org/10.1007/s10924-012-0560-0

Stirna, U., Fridrihsone-Girone, A., Yakushin, V., & Vilsone, D. (2014). Processing and properties of spray-applied, 100% solids polyurethane coatings from rapeseed oil polyols. *Journal of Coatings Technology and Research, 11*, 409–420. https://doi.org/10.1007/s11998-013-9545-8

Sugita, P., & Mas'ud, Z. A. (2013). Synthesis and application of Jatropha oil based polyurethane as paint coating material. *Makara Journal of Science*, 134–140. https://doi.org/10.7454/mss.v16i2.1409

Sundar, S., Vijayalakshmi, N., Gupta, S., Rajaram, R., & Radhakrishnan, G. (2006). Aqueous dispersions of polyurethane–polyvinyl pyridine cationomers and their application as binder in base coat for leather finishing. *Progress in Organic Coatings, 56*, 178–184. https://doi.org/10.1016/j.porgcoat.2006.04.001

Suresh, K. I., & Harikrishnan, M. G. (2014). Effect of cardanol diol on the synthesis, characterization, and film properties of aqueous polyurethane dispersions. *Journal of Coatings Technology and Research, 11*, 619–629. https://doi.org/10.1007/s11998-014-9571-1

Szycher, M. (2012). Waterborne polyurethanes. In M. Szycher (Ed.), *Szychers handbook of polyurethanes* (2nd ed., pp. 449–494). CRC Press.

Thomson, T. (2000). *Design and applications of hydrophilic polyurethanes*. CRC Press.

The Food and Agriculture Organization. (2020). No Title. In: Prod. Quant. Soybeans by Ctry. http://www.fao.org/faostat/en/#data/QC/visualize. Accessed 6 Nov 2020

Tripathi, V., Abidi, A. B., Markerb, S., & Bilal, S. (2013). Linseed and linseed oil: Health benefits—A review. *International Journal Pharmacy and Biological Sciences, 3*, 434–442.

Tritt, T. M., Böttner, H., & Chen, L. (2008). Thermoelectrics: Direct solar thermal energy conversion. *MRS Bulletin, 33*, 366–368. https://doi.org/10.1557/mrs2008.73

Unsworth, W. D., Angelini, G. D., & Pistolesi, C. (1997). *Heart valve prosthesis*.

Valero, M. F., Pulido, J. E., Hernández, J. C., Posada, J. A., Ramírez, A., & Cheng, Z. (2009). Preparation and properties of polyurethanes based on castor oil chemically modified with yucca starch glycoside. *Journal of Elastomers and Plastics, 41*, 223–244. https://doi.org/10.1177/0095244308091785

Vásquez, L., Campagnolo, L., Athanassiou, A., & Fragouli, D. (2019). Expanded graphite-polyurethane foams for water-oil filtration. *ACS Applied Materials & Interfaces, 11*, 30207–30217. https://doi.org/10.1021/acsami.9b07907

Vereshchagin, A. G., & Novitskaya, G. V. (1965). The triglyceride composition of linseed oil. *Journal of the American Oil Chemists Society, 42*, 970–974. https://doi.org/10.1007/BF02632457

Vicente, G., Martínez, M., & Aracil, J. (2005). Optimization of Brassica carinata oil methanolysis for biodiesel production. *Journal of the American Oil Chemists Society, 82*, 899–904. https://doi.org/10.1007/s11746-005-1162-6

Wang, Z. X., Yuan, J. T., Liu, Q. L., Xiao, X. M., Li, Y. H., & Chen, H. (2014). Synthesis and characterization of a novel waterborne polyurethane modified by Tung Oil. *Applied Mechanics and Materials, 513–517*, 251–254. https://doi.org/10.4028/www.scientific.net/AMM.513-517.251

Wang, Y. J., Jeng, U. S., & Hsu, S. H. (2018). Biodegradable water-based polyurethane shape memory elastomers for bone tissue engineering. *ACS Biomaterials Science & Engineering, 4*, 1397–1406. https://doi.org/10.1021/acsbiomaterials.8b00091

Wang, H., Wang, S., Du, X., Wang, H., Cheng, X., & Du, Z. (2019). Synthesis of a novel flame retardant based on DOPO derivatives and its application in waterborne polyurethane. *RSC Advances, 9*, 7411–7419. https://doi.org/10.1039/C8RA09838G

Wang, S., & Min, K. (2010). Solid polymer electrolytes of blends of polyurethane and polyether modified polysiloxane and their ionic conductivity. *Polymer (Guildf), 51*, 2621–2628. https://doi.org/10.1016/j.polymer.2010.04.038

Wang, H.-L., Gopalan, A., & Wen, T.-C. (2003). A novel lithium single ion based polyurethane electrolyte for light-emitting electrochemical cell. *Materials Chemistry and Physics, 82*, 793–800. https://doi.org/10.1016/S0254-0584(03)00363-8

Wang, H.-L., Fu, C.-M., Gopalan, A., & Wen, T.-C. (2004). Frequency dependent conductivity of the thin film blend of electroluminescent poly(p-phenylene vinylene) with waterborne polyurethane as ionomer. *Thin Solid Films, 466*, 197–203. https://doi.org/10.1016/j.tsf.2004.01.098

Wang, S., Jeung, S., & Min, K. (2010). The effects of anion structure of lithium salts on the properties of in-situ polymerized thermoplastic polyurethane electrolytes. *Polymer (Guildf), 51*, 2864–2871. https://doi.org/10.1016/j.polymer.2010.04.022

Warner, K., Orr, P., Parrott, L., & Glynn, M. (1994). Effects of frying oil composition on potato chip stability. *Journal of the American Oil Chemists Society, 71*, 1117–1121. https://doi.org/10.1007/BF02675905

Wen, J. G., Geng, W., Geng, H. Z., Zhao, H., Jing, L. C., Yuan, X. T., Tian, Y., Wang, T., Ning, Y. J., & Wu, L. (2019). Improvement of corrosion resistance of waterborne polyurethane coatings by covalent and noncovalent grafted graphene oxide nanosheets. *ACS Omega, 4*, 20265–20274. https://doi.org/10.1021/acsomega.9b02687

Wicks, D. A., & Wicks, Z. W. (2005). Autoxidizable urethane resins. *Progress in Organic Coatings, 54*, 141–149. https://doi.org/10.1016/j.porgcoat.2004.12.006

Wu, S., Deng, D., Zhou, L., Zhang, P., & Tang, G. (2019). Flame retardancy and thermal degradation of rigid polyurethane foams composites based on aluminum hypophosphite. *Materials Research Express, 6*.https://doi.org/10.1088/2053-1591/ab41b2

Wu, Q., & Hu, J. (2016). Waterborne polyurethane based thermoelectric composites and their application potential in wearable thermoelectric textiles. *Composites. Part B, Engineering, 107*, 59–66.

Xia, Y., Zhang, Z., Kessler, M. R., Brehm-Stecher, B., & Larock, R. C. (2012). Antibacterial soybean-oil-based cationic polyurethane coatings prepared from different amino polyols. *Chemsuschem, 5*, 2221–2227. https://doi.org/10.1002/cssc.201200352

Xiao, Y., Fu, X., Zhang, Y., Liu, Z., Jiang, L., & Lei, J. (2016). Preparation of waterborne polyurethanes based on the organic solvent-free process. *Green Chemistry, 18*, 412–416. https://doi.org/10.1039/C5GC01197C

Xie, Y., Zou, H., Xiang, H., Xia, R., Liang, D., Shi, P., Dai, S., & Wang, H. (2016). Enhancement on the wettability of lithium battery separator toward nonaqueous electrolytes. *Journal of Membrane Science, 503*, 25–30. https://doi.org/10.1016/j.memsci.2015.12.025

Xue, Z., He, D., & Xie, X. (2015). Poly(ethylene oxide)-based electrolytes for lithium-ion batteries. *Journal of Materials Chemistry A, 3*, 19218–19253. https://doi.org/10.1039/C5TA03471J

Yang, X. (2015). Synthesis and properties of novel non-ionic polyurethane dispersion based on hydroxylated tung oil and alicyclic isocyanates. *Journal of Materials Science and Chemical Engineering, 3*, 88. https://doi.org/10.4236/msce.2015.31013

Yang, C.-H., Lin, S.-M., & Wen, T.-C. (1995). Application of statistical experimental strategies to the process optimization of waterborne polyurethane. *Polymer Engineering & Science, 35*, 722–730. https://doi.org/10.1002/pen.760350812

Yang, J.-E., Kong, J.-S., Park, S.-W., Lee, D.-J., & Kim, H.-D. (2002). Preparation and properties of waterborne polyurethane–urea anionomers. I. The influence of the degree of neutralization and counterion. *Journal of Applied Polymer Science, 86*, 2375–2383. https://doi.org/10.1002/app.11249

Yin, C., Zhao, Y., Yang, C., & Zhang, S. (2000). Single-ion transport light-emitting electrochemical cells: Designation and analysis of the fast transient light-emitting responses. *Chemistry of Materials, 12*, 1853–1856. https://doi.org/10.1021/cm0001408

Yin, X., Luo, Y., & Zhang, J. (2017). Synthesis and characterization of halogen-free flame retardant two-component waterborne polyurethane by different modification. *Industrial and Engineering Chemistry Research, 56*, 1791–1802. https://doi.org/10.1021/acs.iecr.6b04452

Yoo, Y., & Youngblood, J. P. (2017). Tung oil wood finishes with improved weathering, durability, and scratch performance by addition of cellulose nanocrystals. *ACS Applied Materials & Interfaces, 9*, 24936–24946. https://doi.org/10.1021/acsami.7b04931

Zhang, J. L., Wu, D. M., Yang, D. Y., & Qiu, F. X. (2010). Environmentally friendly polyurethane composites: Preparation, characterization and mechanical properties. *Journal of Polymers and the Environment, 18*, 128–134. https://doi.org/10.1007/s10924-010-0178-z

Zhang, M., Zhao, F., & Luo, Y. (2019). Self-healing mechanism of microcracks on waterborne polyurethane with tunable disulfide bond contents. *ACS Omega, 4*, 1703–1714. https://doi.org/10.1021/acsomega.8b02923

Zhengxiang, W., Jianlong, W., & Lizheng, G. (2012). Synthesis and characterization of Tung oil-based waterborne polyurethane. *Paint & Coatings Industry, 7*.

Zhou, X., Li, Y., Fang, C., Li, S., Cheng, Y., Lei, W., & Meng, X. (2015). Recent Advances in synthesis of waterborne polyurethane and their application in water-based ink: A review. *Journal of Materials Science and Technology, 31*, 708–722. https://doi.org/10.1016/j.jmst.2015.03.002

Zhu, Y., Romain, C., & Williams, C. K. (2016). Sustainable polymers from renewable resources. *Nature, 540*, 354–362. https://doi.org/10.1038/nature21001

Zieleniewska, M., Auguścik, M., Prociak, A., Rojek, P., & Ryszkowska, J. (2014). Polyurethane-urea substrates from rapeseed oil-based polyol for bone tissue cultures intended for application in tissue engineering. *Polymer Degradation and Stability, 108*, 241–249. https://doi.org/10.1016/j.polymdegradstab.2014.03.010

Properties and Characterization Techniques for Waterborne Polyurethanes

M. Ramesh, L. Rajeshkumar, D. Balaji, and M. Priyadharshini

Abstract

Polyurethanes are a class of polymeric materials possessing a broad range of properties and characteristics both chemical and physical when compared with monolithic materials. Due to a wide availability of monolithic materials commercially and due to the fact that the properties of those materials could be altered suitably, combination of polyurethanes to obtain customizable properties cater to the growing needs of contemporary technologies including elastomers, foams, coatings, glass, paper, wood and adhesives. Waterborne Polyurethanes (WPUs) are currently in the research limelight among various researchers due to its novelty, unique properties and wide scope for applicability in fields like caulking materials, paint additives, various fibers, emulsion polymerization media, dyes and primers for metals, pigment pastes and defoamers. WPUs are considered to be green elements which are characterized by non-flammability, non-toxicity and less environment degradability and for this reason these materials are taken up for research during recent times. Surface oxygen groups of WPUs are amended during the recent researchers for converting them into polymer nano-composites. Due to the unique phase characteristics, nano-structured organic and inorganic hybrid composites possess better and enhanced functional characteristics. Such unique phase characteristics arise due to exfoliation and interlayer collation which increases the interfacial bond between organic and inorganic phased thus improving the composite properties. Manufacturing of such hybrid materials requires special attention toward their miscibility since agglomeration and accretion are considered to be major problems that results due to poor interaction between polymer and dispersoid along with least homogeneity.

Keywords

Polyurethanes • Waterborne • Properties • Characterization • Nano-fillers • Nano-composites

1 Introduction

Polyurethanes (PU) are fabricated by stacking the rigid and flexible polymer layers alternatively and are considered to be much better in terms of their performance. Fabrication of Pus comprises of two steps: Formation of a thick viscous liquid called as prepolymer is the first step, which has its molecular weight in medium range, as a result of reaction between polyol and isocyanate. Secondly, the molecular weight of the prepolymer is shifted to higher values by making it to react with diamine or diol chain extender. Various final polymeric configurations can be obtained by producing the PUs by changing the reactants nature to such as catalysts and additives, di-isocyanates and polyols. Usually PUs are characterized by few note-worthy properties such as toughness, flexibility, abrasion and scratch resistance and resistance toward chemicals. PUs mostly comprises of a notable quantity of free isocyanates and a major quantity of volatile organic compounds (VOC) in most of the cases (Hepburn, 1992). But the presence of toxic organic compounds may pollute the environment. Hence, during late 1960s, aqueous PU dispersions were discovered and the WPU contains the PU particulates suspended in an aqueous medium, thus forming a binary colloidal solution. But PUs are discordant toward water and they have to be modified to behave as

M. Ramesh (✉)
Department of Mechanical Engineering,
KIT-Kalaignarkarunanidhi Institute of Technology, Tamil Nadu, Coimbatore, 641402, India

L. Rajeshkumar · D. Balaji
Department of Mechanical Engineering, KPR Institute of Engineering and Technology, Tamil Nadu, Coimbatore, 641407, India

M. Priyadharshini
Department of Computer Science and Engineering, VIT-AP, Amaravathi, Andhra Pradesh, India

Inamuddin et al. (eds.), *Sustainable Production and Applications of Waterborne Polyurethanes*, Advances in Science, Technology & Innovation, https://doi.org/10.1007/978-3-030-72869-4_6

waterborne compounds. During further developments, WPUs were manufactured from polymers containing many hydrophilic groups so as to have better solubility in water. This could be done by addition of hydrophilic monomers that contain ionic functions into the polymers. Such materials include carboxylate or sulfonate groups and quaternary ammonium. Resulting materials are termed as PU ionomers and internal emulsifiers respectively. WPUs are most widely applied in many areas due to the following advantages: Minimizing the release of solvent into atmosphere due to the organization's administrative regulations, high cost of the currently used solvents and the WPUs quality is most suited for various applications. Generally, WPUs are ecofriendly, non-polluting of air and water and non-toxic and non-flammable. Only product from the reaction involving WPUs is the evaporation of water, which is no way hazardous to the environment. WPUs find their majority of application in the areas including waterproof textiles, coatings, films for packaging, adhesives, biomaterials and membranes, ink binder, synthetic leathers, paper sizing and glass fibers (Arnolds, 1990).

Composite materials find more applications in the field of engineering and technology due to its superior characteristics like low price, less in weight, ease of manufacturability and corrosion resistant nature. Specifically, organic polymers have very high applicability in many high-end products which reoriented the research on organic polymers to newer directions and heights. Nevertheless, these materials are poor conductors of heat and so their applicability in electronic products and other heat involved devices are relatively low. Thus, organic polymers with better heat transfer and heat dissipation characteristics has to be developed (Dieterich et al., 1970; Rosthauser & Nachtkamp, 1987). Few researches stated that the analysis of thermal conductivity of organic polymers contains two different aspects: First aspect is the synthesis of organic polymers with ordered structures to enhance the phonon mobility, materials with high degree of crystallinity, materials with higher order polarization structure, enhanced pathway strategy to improve heat transfer or materials with conjugated radicle. Few such materials are polypyrrole (PPY), polyacetylene (PA) and polyaniline (PANI). Second, incorporation of fillers into the matrix for enhancing the heat transfer by the way of composing a heat transfer path with conductive matrix and fillers (Garcia-Pacios et al., 2013). Results of the experiments indicate that the heat transfer network is affected by the usage of less amount of fillers incorporated in polymers. This renders lower connectivity between the particles within the path of heat transfer, hinders the phonon transport and increases the heat resistance on the interfacial surface of the two elements and reduces the conduction rate between those particles. Heat network is formed between two particles in the path of heat transfer when the number of particles present in the network goes beyond the threshold of percolation. One such organic polymer incorporated with filler is PU whose characteristics ranges between a rigid thermoplastic to an elastomer which possess far better tensile strength, solvent resistance, physical properties and abrasion and tear resistance (Poussard et al., 2016).

Utilization of PU in higher end applications can be easily carried out by customizing its properties since PUs have a property of existing with adaptable chemical structures unlike the non-tailorable monomers. By the way, development of ecofriendly WPUs is also increasing these days due to the current day demands of minimizing the emissions from VOCs. WPUs are most widely used in coating industries and technologies where VOCs evolution during drying process is of major concern when conventional materials are used. Unlike traditional solvent-based polymers, WPUs offer numerous advantages like high molecular weight, good adaptability and low viscosity which make them suitable for electromagnetic shielding and coating on electronics devices that needs enhanced thermal conductivity (Zhou et al., 2015a).

Thermal properties of WPUs can be easily tailored by incorporating conductive fillers of nano-size into them which percolate them even at minimum loads. Graphene, which is the single laminate of the graphite sheet can act as an important nano-filler with carbon as base for not only enhancing the thermal conductivity but also the strength of the nano-composite. Graphene-based nano-materials and nano-platelets can be developed by researchers, as these materials are easy to process and are easily soluble, for applications in polymer-based nano-composites that were enhanced mechanically and thermally. Such carbon-based materials like graphene, single and multiwalled carbon nanotubes (CNTs) and graphite can be hybridized with WPUs for the enhancement of its thermal conductivity. Current chapter focuses on WPU filled with hybrid additives for enhancing the thermal conductivity along with chemical and physical amendments (Bai et al., 2007).

2 Bio-Based Water Polyurethanes

Generally, bio-based WPUs were manufactured by addition of natural filler into the polymer matrix. Few researchers used castor oil (CO) and tartaric acid (TA) as additives and cellulose nano-crystals (CNCs) as fillers to form a bio-based WPUs. Since CNC is a bio-based reinforcement, these materials were coated over the metal substrate to obtain numerous tailored properties. CNCs were emulsified with WPUs in various concentrations and the mixed suspensions were obtained as aqueous suspensions in various concentrations of 5% and 10% by weight on dry state. These suspensions were adjusted for their viscosity so that they could be used for coating of aluminum substrate by dip-coating. Such coatings were subjected to contact angle tests and knife and tape tests. Additionally, nano-scratch, nano-indentation and wear tests were carried out on the coated aluminum and the tests revealed the feasibility of utilizing nano-size filled WPUs as coatings to metal substrates (Meng et al., 2009).

2.1 Synthesis of WPUs

WPUs were manufactured in a two-step process: First, CO is mixed with Isophorone diisocyanate (IPDI) in a 250 ml flask at a temperature of 78 °C for a period of two and half an hour. The flask is attached with a mechanical stirrer, a nitrogen gas inlet and a condenser. Second, TA is mixed with 30% by weight of dimethyl formamide (DMF) while 20 ml of acetone is added to the mixture for controlling the viscosity of the emulsion. After this tetra ethanolamine (TEA) was added in equimolar ratio to TA into the final mixture and the temperature was brought down to 60 °C. In order to obtain an aqueous mixture, the above constituents were stirred aggressively at about 800 rpm and distilled water is added to it. If any unsettled organic solvents were present in the stirred mixture, they were removed from the system with the help of a vacuum evaporator at a temperature of 30 °C. Figure 1 shows the final chemical structure of CO-based WPUs for various types of emulsions. These WPUs were nomenclated as WPU-TA. Another configuration of WPUs was prepared by using polycaprolactone (PCL) which partly replaces CO in the mixture and serves as an alternative for 16% of –OH groups in CO. During this synthesis PCL was added during the initial step in the process while the other steps remain the same (Liu et al., 2017; Ramesh & Rajesh Kumar, 2018). These WPUs were nomenclated as WPU-TA-PCL. Few cationic and anionic series of WPUs were prepared based on castor oil with feed ratios as listed in Table 1.

WPU-TA samples were subjected to macroscopic examinations and they revealed that these samples were harder film coatings. Nano-indentation tests also supported the above results rendering quantitative data. On the other hand, WPU-TA-PCL samples were found to be more elastic than former sample owing to the addition of PCL into the mixture. This enhanced the possibility of using them as self-standing films with better mechanical characteristics. Incorporation of CNC into WPUs rendered better properties such as lower coefficient of friction, elastic recovery, high hardness, wear resistance and lower plastic deformation. WPU-TA-PCL samples with 5% CNC has better properties comparatively with better wear resistance, hardness and adhesion (Zhang et al., 2020).

Analysis of the effects of moisture imbibition of the PU films and its mechanical characteristics, effects of CO functional groups, particle size of WPUs due to chain extension were carried out by various researchers. WPUs exhibited better stability during storage for about 24 months. Resultant PUs exhibited low hydrophilicity, low moisture intake, better mechanical properties and high transmittance. Test results on WPUs indicated a Young's modulus of about 50 MPa, moisture intake of about less than 4%, and an elongation ranging between 15 and 300%. These WPUs were also tested by applying on animal skins and they displayed better hair-styling performance with practically no skin irritation. Hence, these castor oil-based PU can be utilized as potential substitutes for hair-styling commodities (Bloor et al., 2006).

WPUs were also synthesized by using trimethyol ethane (TME) which is a bio-based epoxy resin along with some chemicals likeneopentyl glycol, diethanolamine and N-benzylethanolamine. When these WPU polyols were subjected to testing their glass transition temperature and storage modulus increased with the presence of aromatic ring and larger hydroxyl value in the crosslinked polymer (Sim et al., 2005). Presence of tertiary amines affected the thermal stability of the crosslinked polymers while the WPU amino polyols without tertiary amines executed better thermal stability due to the absence of C-N bonds. WPUs combined with amino polyols, on the other hand, can render a better efficiency on development of coating film which also minimizes the VOC toxic emissions (Riffat & Ma, 2003).

2.2 WPU Synthesis from Cellulose Nano-Fibers

Interests were shown by some researchers in processing of WPUs using cellulose nano-fibers (CNF) because of its environmental friendliness and easily tailorable properties. Since CNFs are hydrophilic in nature, aqueous-based reinforcements can also be used for synthesis. Usage of CNFs poses greater advantages like better specific properties, naturally available and less in cost. Though CNFs need few compatibility alterations, their advantages mask this fact. CNFs were generally obtained from the carboxylation and mechanical disintegration of hardwood kraft pulp that was bleached by standard procedures. When the degrees of carboxylation along with number of passes and processing time was varied, various configurations of WPU incorporated with CNFs could be obtained. Analysis was carried out to evaluate the influence of degree of carboxylation over the mechanical properties of the WPU-CNF nano-composites. Unreinforced WPUs exhibited poor thermal, mechanical and thermo-mechanical characteristics than the CNF reinforced WPUs in spite of damage in cellulose structure and its crystallinity index caused by the carboxylation process (Saqr et al., 2008). WPUs reinforced with carboxylated CNF fibers exhibited 86% of breaking stress, 167% of modulus and

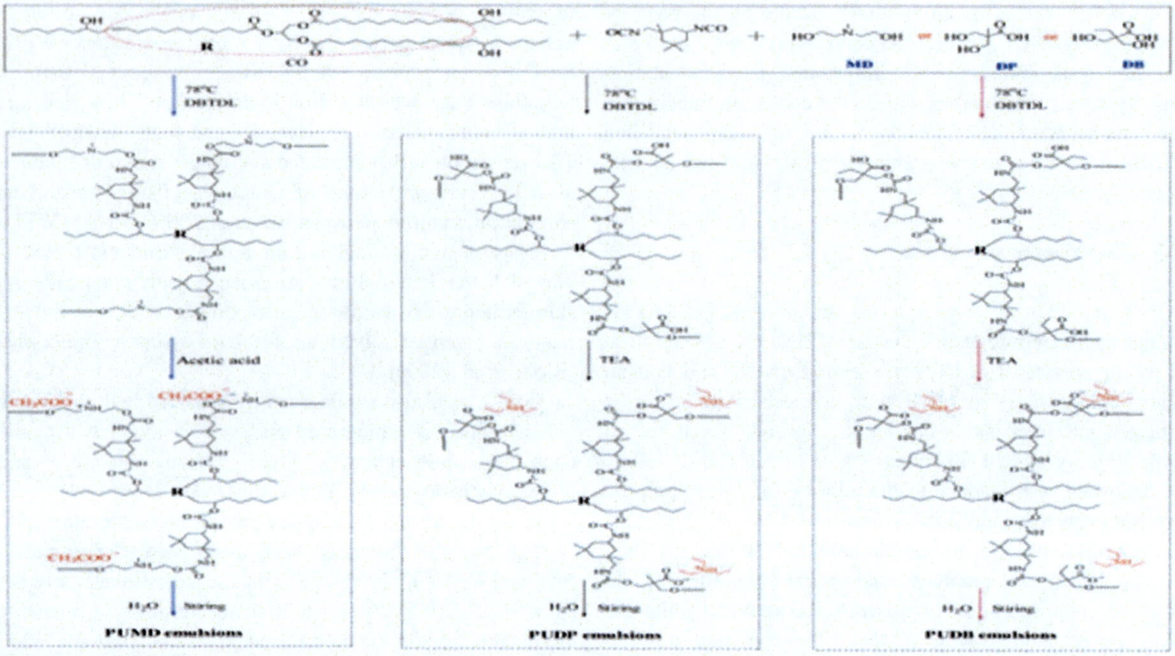

Fig. 1 Representation of castor oil-based WPUs (Zhang et al., 2020)

Table 1 Reaction conditions of WBPUs (Zhou et al., 2015b)

S. No.	Sample	Time and steps	Equivalent ratios	NCO/OH ratio	ACID/TEA ratio
1	WBPU-TA	2.5 h and 2 h 2 steps	CO/TA = 0.8	1.6	1.0
2	WBPU-TA-PCL	2.5 h and 2 h 2 steps	(CO + PCL)/TA = 0.8	1.6	1.0

377% of yield stress when compared with its counterparts. Interaction between the matrix and the reinforcement at the interface was also found to be improved as a result of presence of carboxyl groups on the CNF surfaces. It could be seen from the above test results that carboxylation process on CNF fibers have a positive effect on the WPU-CNF composites. Nevertheless, when the degree of carboxylation crosses certain limit, the crystal structure of CNF undergoes damage which in turn decreases the composite properties (Bozlar et al., 2010; Han & Fina, 2011).

3 Synthetic Water Polyurethanes

Many researchers prepared synthetic-based WPUs from various artificial sources such as silane-based inorganic compounds. Siloxane-functionalized polyether carbonate (Si-PEC) containing a maximum of 9.8% by weight of siloxane was synthesized along with WPUs and compounds of various molecular weights were prepared. Above prepared emulsion had a particle size ranging between 100 and 684 nm and this range is due to the variation of siloxane content. Carbonate unit of polyol enhanced the oxidation and mechanical resistance while the siloxane content rendered better thermal stability and cross-link density to WPUs. Si-PEC-WPU compounds displayed a good mechanical performance with an elongation of 684%, enhanced hydrophobic nature and a tensile strength of 17 MPa owing to the presence of siloxane and carbonate units. These compounds could be used as better substitutes for petroleum derived polyols as far as synthetic WPU application is concerned (Maldovan, 2013).

Experiments were conducted by many researchers in view of enhancing the hydrophobicity of cationic WPUs by manufacturing them with octadecyl side chains containing compounds such as 1-monostearoyl-rac-glycerol using cathodic electrode position technique (CED). Figure 2 shows the schematic of synthesis of CED coatings. Characterization of such synthesized compounds portrayed that there is an increase in the phase separation rate between hard and soft segments of WPUs due to the introduction of octadecyl chains into them. Water absorption and water contact angle analysis revealed that when the above emulsion was used as coatings then the water resistance of the substrates increase due to the formation of hydrogen bonds on the surface and low surface energy. CED coatings also possessed better flexibility, hardness, impact resistance and adhesion in the tests conducted by international standardization test of paints and varnishes. These coatings displayed an optical transmittance of 90% during optical transmittance tests and the thermal stability of the coatings was also improved (He et al., 2020).

4 Self-Healable Waterborne Polyurethanes

WPUs were generally prepared by a process called acetone process and in case of damage of the WPU coatings an inorganic resin or curing agent is infused which induces a crosslinking reaction resulting in curing process. Various characterization techniques could be used to analyze the incorporation of curing agent into the system quantitatively (Sanada et al., 2009). Few researchers tried to prepare a self-curing or self-healable WPUs from either organic or inorganic sources. WPUs were synthesized by developing a dynamic covalent bond within them through aromatic schiff base (ASB) under visible light. These ASB-WPUs have shown a better self-healing ability with an efficiency of 84% when exposed to LED table lamp for a period of 24 h at a temperature of around 25 ºC. These polymers, when maintained at room temperature, exhibited enhanced mechanical properties such as 14.35 MPs of tensile stress and 65 MJ/m^3 of toughness. These enhanced properties were possible primarily due to the imine metathesis present in the ASB bonds which were elicited by visible light (Xie et al., 2005). Meanwhile, interactions of H-bonds in the urethane were also responsible for such a better self-healing of the ASB-WPU composites. It was also a supporting fact that these ASB-WPU composites could be repaired and rehabilitated for obtaining better mechanical properties after damage under visible light. Next-generation smart materials could be easily prepared from the above-stated manufacturing process and it could possibly result in room temperature self-healing WPU composites which can be reprocessed under clean and safe visible light to restore their mechanical properties to the maximum possible extent (Im & Kim, 2012; Lee et al., 2009). Figure 3a shows the schematic of processing of polymers under visible light and the resulting ASB-WPU composites. Figure 3b shows the chemical structure of self-healing mechanisms happening under visible light.

Self-matting WPUs were prepared by some researchers by incorporating hydrophilic elements into hard and soft segments of WPUs. This could be easily done by addition of PCKL polyols that comprises of carboxylate groups into the WPUs for enhancement of their hydrophilicity and matting characteristics (Choi et al., 2011). Characterization techniques like FTIR and SEM were used for confirmation of chemical structure and WPU morphology respectively. Upon analysis of details pertaining to the manufacturing of WPUs, it was determined that hydrophilic units added to the WPUs influence the particle size as well as the surface gloss of WPUs. Surface gloss was found to be directly proportional to the increase in sulfonates or carboxylates contents while the particle size was inversely proportional to it. It was found from the experiments that when the dispersion particle size was about 3 microns, the size of coated surface was less than 1 mm. Thermo gravimetric analysis was carried out to

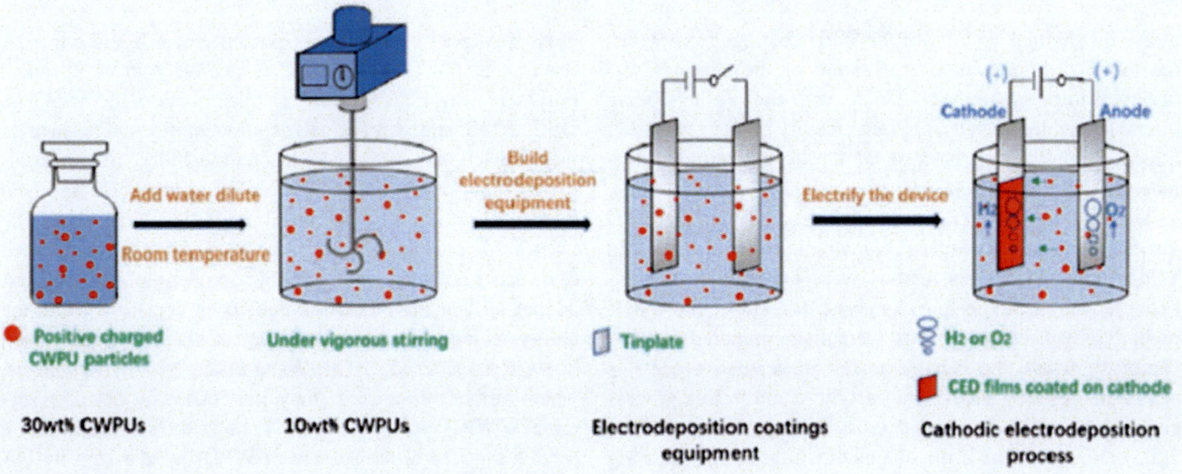

Fig. 2 Preparation of CED coatings and the cathodic electrode position process (He et al., 2020)

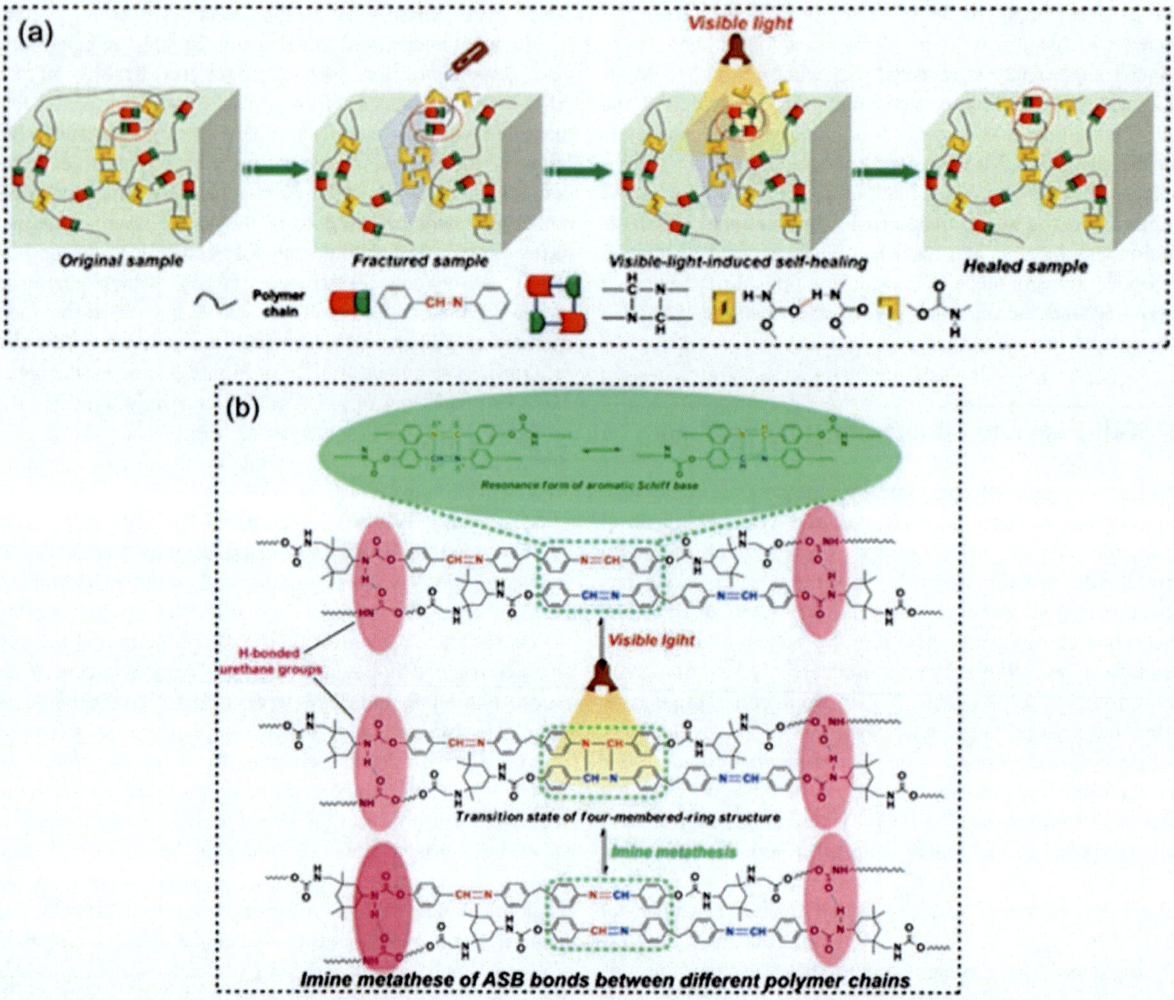

Fig. 3 **a** Schematic depicting the healing process of ASB-WPU polymers; **b** the proposed self-healing mechanisms (Fan et al., 2020)

assess the thermal stability of the WPU coatings from which the initial decomposition temperature of the coatings was observed to be greater than 285 °C with the coatings being thermally stable. Such coatings can readily be used in leather finishing applications because of its aforementioned attributes (Kim et al., 2003).

Diels–Alder (DA) chemistry was used to prepare WPU dispersions with ultraviolet (UV) curable resin. Crosslinked WPUs obtained by this method were expected to possess better recycling and healing characteristics. Hence the WPUs prepared by acetone process were then prepared by DA chemistry where the chain extender in acetone process is substituted by DA adduct with acrylic chain in this process enabling the UV curing and cross-linking (Jeong & Lee, 2003). It was found from the photo differential scanning calorimetry (PDSC) that DA adduct supports UV cross-linkages in the coating. Existence of DA and retro DA reaction at 60 °C and 120 °C respectively in cross-linked polymers was noticed from characterization methods like DSC, FTIR and rheological experimentations. Presence of these reactions enhanced the permeability of materials toward CO_2 and enhanced the ability to reprocess and heal by also sustaining its mechanical properties. Yet it should be taken into consideration that a balance has to be maintained at all times between acrylic double bond component and DA adduct so that the cross-link density is kept low while the ability to recycle enhances (Jeong et al., 2003; Lee et al., 2019; Ramesh et al., 2020; Wang et al., 2009). These facts, when further researched, may pave way to develop sustainable WPU coatings. Figure 4 shows the self-healing of a scratch induced at the surface of WPUs synthesized by DA chemistry.

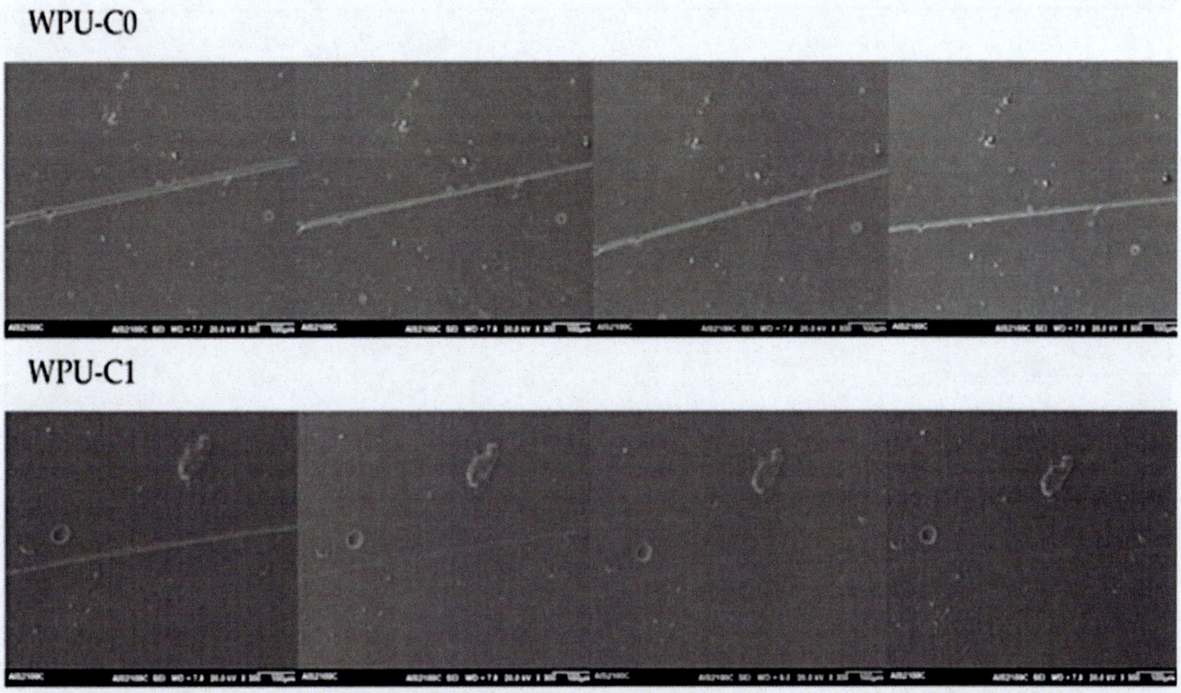

Fig. 4 SEM micrograph of self-healable WPUs with DA adduct (Lee et al., 2019)

5 Water Polyurethanes Adhesives

The challenge lies in developing the novel adhesives which are soft in nature is hard brought in the property of adhesion to the materials with least surface energy. By holding this as a key, WPU dispersions based on (PEO) poly (ethylene oxide) homopolymer and (PEO-b-PPO-b-PEO) poly(ethylene oxide-b-propylene oxide-b-ethylene oxide) and (PPO-b-PEO-b-PPO) poly (propylene oxide-b-ethylene oxide-b-propylene oxide) triblock coexisting polymers were produced (Chung, 2004). Adhesives prepared by dispersions unveiled better bond on least surface energy (PP—polypropylene) probe over the more surface energy probe made of steel. Furthermore, with the intention of enhancing the performance of adhesive made of WPUUs for (PSA-Pressure Sensitive Adhesive) applications, the designed optimized to bilayer system. The bilayer is made up of solid layer and as well liquid layer of two different types of WPUUs to attain the viscoelastic nature to generate a gradient with the property of viscoelasticity. This tactic directed to adhesives that unveiled fibrillation during the measurement of probe tack with the assistance of a steel probe, it is of PSA variant. At last, adhesive tapes prototype were developed by the system of bilayer (Lerf et al., 1998). The peel test was conducted over these adhesive tapes which unveil the performance in par with the other conventional adhesive tape, create the interest for its considerable application (Stankovich et al., 2006).

In general, to summary, the novel developed adhesives revealing the considerable performance compared over the steel and high surface energy. This outcome is partly accredited to higher surface energy, which lay path to better thermodynamic efficiency of adhesion, and triblock coexisting polymer-based WPUUs which diminishes the elastic modulus to accomplish the Dahl Quist criteria. Subsequently, these WPUU dispersions might be employed as a soft adhesive for attachment to least energy surfaces. Mixing these two WPUUs (in 2 dissimilar proportions) has not improved the performance adhesives, in comparison over the performance of adhesive with neat WPUUs. Quite reverse, the bilayer systems made up of the same 2 WPUUs employed in the combinations, liquid layer and solid layer in combination led to a plateau in the probe tack curves like steel, which result in fibrillation (Lomeda et al., 2008; Xu et al., 2009). Therefore, the built bilayer system is an efficacious tactic for the enhancement of the performance of adhesive made up of WPUUs to attain the performance like PSA, mean that can be employed as base of PSAs, led to formulate PSAs by amending the surface agents so as to tailor the various properties. The upper liquid layer provides tack, while the solid-based bottom layer donates to cohesion

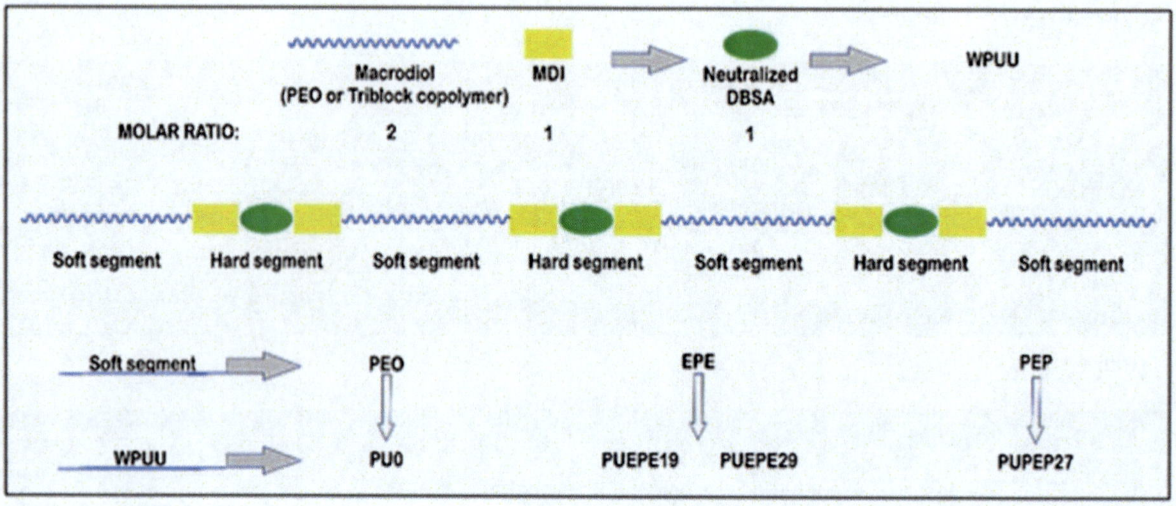

Fig. 5 Synthesis of the waterborne poly(urethane-urea)s (Díez-García et al., 2020)

and which aids fibrillation. Constructing on the performance of adhesive which developed by bilayer systems by selecting the best-performing bilayer of adhesives. The bilayer-based adhesive applied over the PP supporting the peel force is found to be in the range of domestic conventional tape. The peel force unveiled by the aluminum surface was found to be superior over the domestic tape. Nevertheless, the bilayer tape thickness is greater than that of the adhesive's thickness in conventional tape, and supplementary optimization is essential to attain satisfactory performance of thin layer thickness (Díez-García et al., 2020). Figure 5 illustrates the process of synthesis of WPUs based on triblock copolymers.

6 Water Polyurethanes Coatings

Metallic elements and particles were also used as additives and fillers in WPUs for coating them over the substrates. Some researchers tried to prepare a sturdy antistatic WPUs incorporated with ZrO_2 nano-particles by means of sol–gel technique (Bekyarova et al., 2009). In order to carry over this process, the coating surface resistivity and the performance of the resin has to be clearly analyzed. WPUs reinforced with 6% by weight of ZrO_2 nano-particles exhibited a surface resistivity of $9 \times 10^9\ \Omega\ sq^{-1}$ while it discharges the electrostatic charges effectively. Its strength of adhesion was found to be 31% higher than the unreinforced resin. Surface roughness of the above composite was measured to be 27 nm which impedes the penetration of dust into the surface of the composite and this was possible due to the homogenous distribution of the ZrO_2 nano-particles into WPUs (Bourlinos et al., 2003). Such sturdy WPU-ZrO_2 composite coatings can give way for the development of eco-friendly antistatic surfaces. The enhanced property of these coatings was primarily due to: (i) ZrO_2 nano-particles strongly adhere to the surface of WPU resin after undergoing cross-linking reaction, (ii) Surface of the coating is filled with uneven chemical structures, (iii) Homogenous distribution of the reinforcement is ensured by suitable adaptation of the resin. WPU- ZrO_2 antistatic coatings can be used in industrial antistatic surfaces for the aversion of dust over the surface and also for the purpose of electrical insulation (Hormaiztegui et al., 2018; Hormaiztegui et al., 2020; Madbouly et al., 2013; Pramoda et al., 2010).

Experiments were also carried out for the preparation of WPU coatings with fluorinated silicon (FSWPU) by using additives like 3-(2-Aminoethylamino) propyl trimethoxysilane (AEAPTES), PCL diol and 2,2,3,3-Tetrafluoro-1,4-butanediol. Particle size of FSWPUs were analyzed by dynamic light scattering, molecular weight by gel permeation chromatography (GPC), thermal properties by TGA and other characteristics of by FTIR and energy dispersive X-ray spectroscopy (EDS) (Larraza et al., 2020; Ma et al., 2020; Ramesh & Kumar 2020; Wijs, 1929; Wu et al., 2020). Its dynamic mechanical behavior, elongation, shear strength and tensile strength pre and post hydrolysis were investigated by a tensile testing machine. FSWPU emulsion coatings were applied over tiles for the measurement of surface tension, hydrophobicity and hydrophilicity through contact angle measuring apparatus (Aizpurua et al., 2020; Chang et al., 2020). Tape test was also carried out upon the tiles coated with FSWPUs for evaluation the coating adhesion over the substrate. From all the above test results it could be seen that AEAPTES groups enhanced the tensile strength, thermal stability and hydrophobicity while it minimized the surface energy and increased the coating adhesion with tiles

(Hsu et al., 2020; Sukhawipat et al., 2020; Yousefi et al., 2020). Hence, it could be concluded that FSWPUs can be used for better surface coating to improve the coating characteristics in various aspects.

7 Waterborne Polyurethane Nano-Composites

Solvent borne PUs have better hydrolytic strength, improved thermal stability and enhanced mechanical properties when compared with WPUs in humid conditions which is a note-worthy difference. Many researches underwent by performing experiments to enhance the properties of WPUs by all the way and it was determined that nano-particles could be incorporated into these materials to bridge the above-said lacuna. Blending of WPUs and silica nano-particles rendered a novel nano-particle dispersed WPUs (Serkis et al., 2016). This material blend comprises of two colloidal type commercial silica such as Ludox TMA and Ludox AS which were differentiated by their counter ions, shape and size. Results showcased the improvement in properties of WPUs when an appropriate content of silica was added to it and it was observed that when 5 wt.% of silica was added water resistance of the WPUs improved, when 32 wt.% was added functional properties improved and when 50 wt.% was added thermal stability was achieved. It was also noticed that TMA grade of silica performed better than AS grade owing to its strong cross-linking physical effect. Few attempts were also made to synthesize the WPUs using different proportions of cellulose nano-composites (CNC) to obtain WPU-CNC blends (Santamaria-Echart et al., 2016). Results showcased that soft segments crystallization was high when low proportions of CNC was used while the degree of crystallization reduced with the increase in content of CNC because of a strong PU chain and CNC interaction. Additionally few other characteristics such as hydrophilicity, thermal stability and mechanical properties were enhanced due to the incorporation of more amount of CNC to the WPUs. Few other researches attempted to fabricate WPUs with nano-silica composites by using preformed polymer (S1) and polytetrahydrofuran glycol (S2) dispersed silica individually (Cheng et al., 2012). From the results, it was observed that S1 rendered uniform dispersion of silica within them than S2 and alongside S1 rendered enhanced physical characteristics, thermal stability and water resistance to the WPUs than S2.

Few experiments focused in preparing aqueous PUs with functionally graded polyhedral oligomeric silsesquioxane (POSS) and between 0.3 and 4.6 wt.% of diols following prepolymer mixing technique (Honarkar et al., 2016a). An anionic emulsifier in the form of N,N-bis (2-hydroxyethyl-2-amino ethane sulfonic acid sodium salt was also used along with the above constituents. PU-POSS nano-composites were prepared based on the reaction between the precursor powders and the functional groups. Results of the experiments revealed that thermal stability, particle size of resulting PUs, tensile strength, viscosity, glass transition temperature (Tg) and modulus increased with the quantity of POSS. As a result of adding nano-particles into the PUs, their structural homogeneity and compatibility enhanced which could be noticed from the morphological analysis of PU-POSS composites (Honarkar et al., 2014, 2016b). In another experiment, WPU with transparent UV curable coating was prepared using silica nano-particles and they were characterized with good scratch resistance (Zhang et al., 2012). Results revealed that alkaline and acid silica had less dispersion within PU matrix while the nano-silica dispersed homogenously within it. It was also concluded that dispersion of silica nano-particles within PU matrix governed the transparency of the resulting thin films.

8 Castor Oil-Based WPU Nano-Composite Coatings

Traditional polymer matrix systems are getting replaced by nano-composite polymer materials as the nano-composites are characterized by improved properties and is of larger research importance these days. Nano-composites experimentally possess better properties than its organic counterparts due to the strong force of attraction at the interface of nano-sized grain boundaries at the material domain. Mechanical and thermal characteristics of the polymeric matrices could be enhanced by hybrid design and chemical treatment of the materials such as incorporation of organoclay or addition of heterocyclic group into the base matrix for thermal stability or using fillers to reduce the porosity of polymer network (Andjelkovic et al., 2005; Barikani et al., 2007; Cakić et al., 2013; Coutinho et al., 2001; Hourston et al., 1997; Li et al., 2014; Lu & Larock, 2010; Yang et al., 2002). Few experimenters prepared nano-composite prepolymer by substituting the CO by 15 wt.% into bio-polyol dispersing it in turn within Closite 30B (C30B) and diisocyanate where the catalyzed diol had undergone reaction with NCO prepolymer to result in a nano-composite PU (Alaa et al., 2015). Table 2 enlists various materials, processing techniques and applications of WPUs according to previous researchers.

Synergetic effect between the inorganic and organic constituents of the aforementioned materials enhance the characteristics significantly and induce a high level of coupling at the interface. Initially few works were carried out to prepare PU-based composite materials using functionally graded organoclay, polyol and organically modified silicate lamina by some researchers (Rahman et al., 2008). But

Table 2 Method of WPU synthesis and its applications

S. No.	Materials used for WPU synthesis	Processing method	Applications	References
1	Castor oil-based PUs with TDI and Closite 30B as nano-material	Prepolymer and blending by precursor melting	Advanced surface coating	Serkis et al., 2016)
2	Thermally cured castor oil-based Pus with hexamethylene diisocyanate and 3-mercaptopropyl trimethoxysilane as nano-material	Thiolene process and in situ sol–gel process	Mechanically modified PUs	Santamaria-Echart et al., 2016)
3	Castor oil/ pentaerythrito l triacrylate-based UV curable WPU with IPDI and 3-amino propyl trimethoxy silane and 3-glycidoxy propyl trimethoxy silaneas nanomaterial	Prepolymer method	Coating and related applications	Cheng et al., 2012)
4	Castor oil-based polyols and Hexamethoxyl methyl melamine for WPU with IPDI and Closite 30B as nano-material	Prepolymer method and ultrasonication process	Waterborne PU adhesive applications	Honarkar et al., 2016a)
5	Castor oil-based 2-package WPU with IPDI and 3-amino propyl trimethoxy silaneas nanomaterial	Two-step process	Wood and composite material coatings	Barikani et al., 2007)
6	UV curable WPU acrylate with IPDI and 3-amino propyl trimethoxy silaneas nanomaterial	Prepolymerizationtechnique	UV resistance coating and related applications	Rahman et al., 2008)
7	PU-prepolymer-based on castor oil with IPDI and 3-amino propyl trimethoxy silaneas nanomaterial	Siloxane-functionally graded CO and in situ sol–gel process	Substitutes for petrochemical polyols	Yang et al., 2006)
8	Polyurethane adhesives using castor oil derived polyols with IPDI and alkoxysilane castor oil as nanomaterial	Acetone process and transesterification reaction	Wood and steel panel coatings	Yu et al., 2004)
9	Aqueous castor oil-based PU/ polyamide sulfone copolymer dispersions with TDI and Closite 30 B as nano-material	Acetone process	Anticorrosive coatings	Jeon et al., 2007)
10	Epoxidatied CO on WPU with 1,4 TDI 80 and 3-mercaptopropyl trimethoxysilane as nanomaterial	Prepolymer method	Water resistance membrane in WPU dispersions	Deng et al., 2007)
11	Thermoset polyurethanes with TDI, HDI and IPDI	Mixing method	Coatings and mechanical applications	Kwon & Kim, 2005)

recently many materials such as multiwalled carbon nanotubes (MWCNT), silica, clay derivatives like laponite, bentonite, wollastonites, hectorite and montmorillonite, TiO_2 and ZnO were utilized for the manufacture of WPU nano-composites (Deng et al., 2007; Jeon et al., 2007; Kim et al., 2006; Kuan et al., 2005; Kwon & Kim, 2005; Lee & Lin, 2006; Liu et al., 2012; Meera et al., 2014; Mishra et al., 2010; Yang et al., 2006; Yeh et al., 2008; Yu et al., 2004). Clays are currently adopted as traditional thixotropic agents, among the other fillers, in WPU emulsions and coatings, but are recently analyzed as coating elements for some other applications and property improvements (Fu et al., 2015; Meera et al., 2014). Normally the properties of PU dispersions were controlled by many parameters like bonding between filler and host material, mean diameter of fillers and shape of the filler particle. In order to obtain the dispersion of nano-particles in high rates into the WPUs, the techniques used are: exfoliation method, nano-filler dispersion by interposing them within the layers of WPUs and mixing them with polymer solution (Chattopadhyay & Raju, 2007). Some researchers tried to manufacture bio-based WPU dispersions with CO/polyethylene glycol (PEG) and isophorone diisocyanate (IPDI) combined with a prepolymer of NCO/OH and HS in 1.4 and 50 wt.% respectively, in proportion with the solid constituent (Gao et al., 2012). These PU dispersions were converted into PU nano-composite film by intercalating the prepolymer layer with eucalyptus globulus CNC (ECN). When this material is subjected to mechanical characterization, results indicated that at low wt.% of ECN, the mechanical characteristics of the nano-composite films were higher. This improvement was correlated with the microphase segregation of HS from SS in the dispersion, homogenous dispersion of ECN within PU dispersions and the major contributor for the enhancement was reported to be the strong hydrogen bond between ECN and polymer matrix.

CO-based alkoxysilane and PU-based siloxane hybrid coating were prepared by the synthesis following the thiolene path by few researchers (Fu et al., 2014). It was observed from the results that PU-based Si coatings exhibited better mechanical and thermal characteristics when compared with the CO-based alkoxysilane. When the wt.% of Si varies from 0 to 5%, the contact angle of Si-PU systems increases which denotes the increase of hydrophobicity of the hybrid film and decrease of the free surface energy level. Few experimental trails focused on bio-based elastomeric PUs incorporated with hard segment content (HS) for assessing their properties and microstructure (Corcuera et al., 2010). Bulk polymerization method was used to synthesize the CO-based elatsomeric PUs in the form of SS and HDI hard domain without the application of any catalyst. From the results it was determined that due to the action of HS in the stress concentrated points of PUs at micro-level, the mechanical properties of the films did not show any significant variation with the increase in content of hard segment. A hybrid WPU based on CO and silica crosslinked material was developed by some experimenters (Fu et al., 2015) and was analyzed for various properties with respect to the content of silica. It was found that film transparency decreased with the increase in content of silica while other characteristics such as thermal stability, hydrophobicity and roughness of the films increased appreciably. Studies were also made focusing the analysis of effect of siloxane content on surface roughness and silica enrichment on the coating transmittance. It was noticed from the optical transmittance test results that the transmittance of the coatings decreased in ultraviolet region at a spectrum of 300–400 nm with the increase in silica content and phase separation rate which could be possibly due to optical loss during transmission from organic phase to the inorganic one. It was stated that when the hydrophobic coating material should possess better thermal stability, then core–shell structure fits better. Few other studies on the optical transmittance examination of organic PU-based coatings, same results were obtained owing to the scattering of the rays at PU and silica nano-particle interface (Meera et al., 2014).

Structural backbone network of CO was modified by incorporating novel hydrolyzable $-Si-OCH_3$ group into it and a smart hybrid coating was prepared by few researchers which potential and effective for many applications (Shaik et al., 2014) from which urea-PU/silica hybrid coatings were further processed. Results of experimental analysis revealed that these hybrid coatings exhibited enhanced thermal and mechanical characteristics when compared with the unmodified CO counterparts. Many experimental studies were performed to enhance various hybrid coating characteristics like surface properties, elasticity, thermal, toughness, mechanical, chemical resistance and rigidity using clay and other inorganic filler materials (Corcuera et al., 2010; Gao et al., 2012; Gurunathan et al., 2015; Shaik et al., 2014). Final application level characteristics for these hybrid coating systems were influenced by the cross-linking density in the three-dimensional network, filler used and the polymeric chain network. Many research group attempted to prepare CO-based WPU dispersions using C30 reinforcements and the effect of reinforcing C30 nano-filler into PU dispersion nano-composite was experimentally evaluated (Gurunathan et al., 2015). Results showcased that when the C30 nano-filler of 2 wt.% was incorporated into PU matrix, mechanical and thermal properties of the hybrid fil increased appreciably which was solely due to the matrix and filler interaction at the interface of C30 nano-filler and PUs. In few experiments 3 wt.% of C30B nano-filler rendered better tensile and other mechanical and coating characteristics for CO-based WPU systems when compared with their traditional counterparts (Panda et al., 2016). This could be due to the strong structural nano-filler network structure, enhanced matrix cross-linking with the nano-filler and better interfacial adhesion between filler and matrix.

Strength of the interfacial bonding was observed from the assessment of hydrogen bonding established between the PU matrix and the nano-filler. Due to the increase in NCO/OH ratio in PU nano-composites, many results concluded that the utilization of inorganioc content was high and so was the enhancement in mechanical characteristics of CO/PU-based nano-composites (Gurunathan et al., 2016; Yeganeh & Mehdizadeh, 2004). Trimethoxysilane groups undergo hydrolysis reaction and render functionally active silanols groups which renders improved mechanical properties for WPUs when potentially cross-linked with them. It was also studied by few authors that for enhancing mechanical characteristics and thermal stability of PU hybrid coatings, a small weight proportion of organoclay would suffice due to the intense formation of C=C double bonds within the inorganic and organic resin network. Few experiments were carried out to determine the adhesion, drying time, hardness, flexibility, chemical and impact resistance of mild steel substrate panels coated with bio-based WPU dispersion organic coatings (Gurunathan et al., 2015; Meera et al., 2014; Panda et al., , 2016; Patel et al., 2011) and it was found that the increase in inorganic constituents proportion increased the thermal stability of the WPU hybrid coatings. As stated above, few researchers tried to reinforce 3–4 wt.% of organoclay with WPUs and found that the values of strength and modulus improved when compared with pure PUs. It was also noticed that as the organoclay content increases mechanical, dynamic mechanical, modulus of elasticity and yield properties increased whereas the breaking stress and strain decreased considerably (Gurunathan et al., 2015; Yeganeh & Mehdizadeh, 2004). Many other researchers also studied the properties like operating temperature and heat resistance for the CO-based WPUs reinforced with organoclay and found that even those properties improved appreciably when compared with the unreinforced pure WPUs. It could be concluded from the above discussions that CO-based WPU dispersions when reinforced with nano-composites or nano-fillers may exhibit sound mechanical, thermal and other properties and could be potentially applied in various fields of coating and other applications.

9 Conclusion

This chapter was intended to provide an outline about the process of fabrication of WPUs and the materials associated with it. WPUs are promising class of materials with attributes like bio-based, ecofriendly, renewable and hydrophilic materials. Apart from its wide real-time applications, these materials have become the latest concern of academicians and industrialists. Due to its better performance, WPUs can be employed in manufacturing water-based inks. Particularly, the content of VOCs can be tailored for producing printing inks. This created a lot of interest among researchers for manufacturing novel WPUs. Addition of organic fluorides or natural elements may also increase the effectiveness of WPUs. Reduced mechanical property or poorly hydrophobic is the glitches concerned with the hydrophilic portions of WPUs, specifically by the ionic groups. This could be eradicated by hybridization of WPUs with composite emulsions, latex blending, latex interpenetrating network, copolymerization with emulsion and hybridizing with inorganic clay. From the above discussions, it could be concluded that, development and use of bio-based WPUs with fillers may contribute much for sustainable future and also paves way for the development of novel materials with enhanced performance. Current research progress on the processing and synthesis of WPUs is expected to reach newer heights for infusing novelty in it so that it can be used in various applications like water-based inks.

References

Aizpurua, J., Martin, L., Fernández, M., González, A., & Irusta, L. (2020). Recyclable, remendable and healing polyurethane/acrylic coatings from UV curable waterborne dispersions containing Diels-Alder moieties. *Progress in Organic Coatings, 139*, 105460.

Alaa, M. A., Yusoh, K., & Hasany, S. F. (2015). Synthesis and characterization of polyurethane–organoclay nanocomposites based on renewable castor oil polyols. *Polymer Bulletin, 72*(1), 1–17.

Andjelkovic, D. D., Valverde, M., Henna, P., Li, F., & Larock, R. C. (2005). Novel thermosets prepared by cationic copolymerization of various vegetable oils—synthesis and their structure–property relationships. *Polymer, 46*(23), 9674–9685.

Arnolds, R. (1990). In A. D. Wilson, J. W. Nicholson, & H. J. Prosser (Eds.), *Waterborne coatings, surface coatings* (Vol. 3, pp. 179–198). Elsevier Science.

Bai, C. Y., Zhang, X. Y., Dai, J. B., & Zhang, C. Y. (2007). Water resistance of the membranes for UV curable waterborne polyurethane dispersions. *Progress in Organic Coatings, 59*, 331–336.

Barikani, M., Ebrahimi, M. V., & Mohaghegh, S. S. (2007). Influence of diisocyanate structure on the synthesis and properties of ionic polyurethane dispersions. *Polymer-Plastics Technology and Engineering, 46*(11), 1087–1092.

Bekyarova, E., Itkis, M. E., Ramesh, P., Berger, C., Sprinkle, M., de Heer, W. A., & Haddon, R. C. (2009). Chemical modification of epitaxial graphene: Spontaneous grafting of aryl groups. *Journal of the American Chemical Society, 131*, 1336–1337.

Bloor, D., Graham, A., Williams, E. J., Laughlin, P. J., & Lussey, D. (2006). Metal–polymer composite with nanostructured filler particles and amplified physical properties. *Applied Physics Letters, 88*, 102103–102108.

Bourlinos, A. B., Gournis, D., Petridis, D., Szabo, T., Szeri, A., & Dekany, I. (2003). Graphite oxide: Chemical reduction to graphite and surface modification with primary aliphatic amines and amino acids. *Langmuir, 19*, 6050–6055.

Bozlar, M., He, D., Bai, J., Chalopin, Y., Mingo, N., & Volz, S. (2010). Carbon nanotube microarchitectures for enhanced thermal conduction at ultralow mass fraction in polymer composites. *Advanced Materials, 22*, 1654–1658.

Cakić, S. M., Špírková, M., Ristić, I. S., B-Simendić, J. K., Milena, M., & Poręba, R. (2013). The waterborne polyurethane dispersions based on polycarbonate diol: Effect of ionic content. *Materials Chemistry and Physics, 138*(1), 277–285.

Chattopadhyay, D. K., & Raju, K. V. S. N. (2007). Structural engineering of polyurethane coatings for high performance applications. *Progress in Polymer Science, 32*, 352–418.

Chang, J., Wang, X., Shao, J., Li, X., Xin, W., & Luo, Y. (2020). Synthesis and characterization of environmentally-friendly self-matting waterborne polyurethane coatings. *Coatings, 10*(5), 494.

Cheng, L., Zhang, X., Dai, J., & Liu, S. (2012). *Journal of Dispersion Science and Technology, 33*, 840–845.

Choi, J. T., Kim, D. H., Ryu, K. S., Lee, H. I., Jeong, H. M., Shin, C. M., Kim, J. H., & Kim, B. Q. (2011). Functionalized graphene sheet/polyurethane nanocomposites: Effect of particle size on physical properties. *Macromolecular Research, 19*, 809–814.

Chung, D. D. L. (2004). Electrical applications of carbon materials. *Journal of Materials Science, 39*, 2645–2661.

Corcuera, M. A., Rueda, L., d'Arlas, B. F., Arbelaiz, A., Marieta, C., Mondragon, I., & Eceiza, A. (2010). Microstructure and properties of polyurethanes derived from castor oil. *Polymer Degradation and Stability, 95*(11), 2175–2184.

Coutinho, F., Delpech, M. C., & Alves, L. S. (2001). Anionic waterborne polyurethane dispersions based on hydroxyl-terminated polybutadiene and poly (propylene glycol): Synthesis and characterization. *Journal of Applied Polymer Science, 80*(4), 566–572.

Deng, X., Liu, F., Luo, Y., Chen, Y., & Jia, D. (2007). Preparation, structure and properties of comb-branched waterborne polyurethane/OMMT nanocomposites. *Progress in Organic Coatings, 60*(1), 11–16.

Dieterich, D., Keberle, W., & Witt, H. (1970). *Angewandte Chemie International Edition, 9*, 40–50.

Díez-García, I., Keddie, J. L., Eceiza, A., & Tercjak, A. (2020). Optimization of adhesive performance of waterborne poly (urethane-urea)s for adhesion on high and low surface energy surfaces. *Progress in Organic Coatings, 140*, 105495.

Fan, W., Jin, Y., Shi, L., Zhou, R., & Du, W. (2020). Developing visible-light-induced dynamic aromatic Schiff base bonds for room-temperature self-healable and reprocessable waterborne polyurethanes with high mechanical properties. *Journal of Materials Chemistry A, 8*(14), 6757–6767.

Fu, C., Yang, Z., Zheng, Z., & Shen, L. (2014). Properties of alkoxysilane castor oil synthesized via thiol-ene and its polyurethane/siloxane hybrid coating films. *Progress in Organic Coatings, 77*(8), 1241–1248.

Fu, C., Hu, X., Yang, Z., Shen, L., & Zheng, Z. (2015). Preparation and properties of waterborne bio-based polyurethane/siloxane cross-linked films by an in situ sol–gel process. *Progress in Organic Coatings, 84*, 18–27.

Gao, Z., Peng, J., Zhong, T., Sun, J., Wang, X., & Yue, C. (2012). Biocompatible elastomer of waterborne polyurethane based on castor oil and polyethylene glycol with cellulose nanocrystals. *Carbohydrate Polymers, 87*(3), 2068–2075.

Garcia-Pacios, V., Jofre-Reche, J. A., Costa, V., Colera, M., & Martin-Martinez, J. M. (2013). Coatings prepared from waterborne polyurethane dispersions obtained with polycarbonates of 1,6-hexanediol of different molecular weights. *Progress in Organic Coatings, 76*, 1484–1493.

Gurunathan, T., Mohanty, S., & Nayak, S. K. (2015). Effect of reactive organoclay on physicochemical properties of vegetable oil-based waterborne polyurethane nanocomposites. *RSC Advances, 5*(15), 11524–11533.

Gurunathan, T., Arukula, R., Suk Chung, J., & Rao, C. R. (2016). Development of environmental friendly castor oil-based waterborne polyurethane dispersions with polyaniline. *Polymers for Advanced Technologies*. https://doi.org/10.1002/pat.3797

Han, Z., & Fina, A. (2011). Thermal conductivity of carbon nanotubes and their polymer nanocomposites: A review. *Progress in Polymer Science, 36*, 914–944.

He, X., Zhang, Y., He, J., & Liu, F. (2020). Synthesis and characterization of cathodic electrode position coatings based on octadecyl-modified cationic waterborne polyurethanes. *Journal of Coatings Technology and Research, 23*, 1–4.

Hepburn, C. (1992). *Polyurethane elastomers*. Applied Science. https://doi.org/10.1007/978-94-011-2924-4

Honarkar, H., Barmar, M., & Barikani, M. (2014). In *5th International Congress on Nanoscience & Nanotechnology (ICNN 2014)*, Tehran, October 22–24.

Honarkar, H., Barmar, M., & Barikani, M. (2016). In *6th International Congress on Nanoscience & Nanotechnology (ICNN 2016)*, Tehran, October 26–28.

Honarkar, H., Barmar, M., Barikani, M., & Shokrollahi, P. (2016). *Korean Journal of Chemical Engineering, 33*, 319–329.

Hormaiztegui, M. E. V., Aranguren, M. I., & Mucci, V. L. (2018). Synthesis and characterization of a waterborne polyurethane made from castor oil and tartaric acid. *European Polymer Journal, 102*, 151–160. https://doi.org/10.1016/j.eurpolymj.2018.03.020

Hormaiztegui, M. E., Daga, B., Aranguren, M. I., & Mucci, V. (2020). Bio-based waterborne polyurethanes reinforced with cellulose nanocrystals as coating films. *Progress in Organic Coatings, 144*, 105649.

Hourston, D. J., Williams, G., Satguru, R., Padget, J. D., & Pears, D. (1997). Structure–property study of polyurethane anionomers based on various polyols and diisocyanates. *Journal of Applied Polymer Science, 66*(10), 2035–2044.

Hsu, Y. T., Wang, W. H., & Hung, W. H. (2020). Evaluating the properties of a coating material with polycaprolactone-degradable fluorinated silicon-containing waterborne polyurethane. *Sustain, 12* (9), 3745.

Im, H., & Kim, J. (2012). Thermal conductivity of a graphene oxide–carbon nanotube hybrid/epoxy composite. *Carbon, 50*, 5429–5440.

Jeon, H. T., Jang, M. K., Kim, B. K., & Kim, K. H. (2007). Synthesis and characterizations of waterborne polyurethane–silica hybrids using sol–gel process. *Colloids and Surfaces a: Physicochemical and Engineering Aspects, 302*(1), 559–567.

Jeong, H. M., & Lee, S. H. (2003). Properties of waterborne polyurethane/PMMA/clay hybrid materials. *Journal of Macromolecular Science, Part B, 42*, 1153–1167.

Jeong, H. M., Jang, K. H., & Cho, K. (2003). Properties of waterborne polyurethanes based on polycarbonate diol reinforced with organophilic clay. *Journal of Macromolecular Science, Part B, 42*, 1249–1263.

Kim, B. K., Seo, J. W., & Jeong, H. M. (2003). Morphology and properties of waterborne polyurethane/clay nanocomposites. *European Polymer Journal, 39*, 85–91.

Kim, B. S., Park, S. H., & Kim, B. K. (2006). Nanosilica-reinforced UV-cured polyurethane dispersion. *Colloid and Polymer Science, 284*(9), 1067–1072.

Kuan, H. C., Ma, C. C. M., Chang, W. P., Yuen, S. M., Wu, H. H., & Lee, T. M. (2005). Synthesis, thermal, mechanical and rheological properties of multiwall carbon nanotube/waterborne polyurethane nanocomposite. *Composites Science and Technology, 65*(11), 1703–1710.

Kwon, J., & Kim, H. (2005). Comparison of the properties of waterborne polyurethane/multiwalled carbon nanotube and acid-treated multiwalled carbon nanotube composites prepared by in situ polymerization. *Journal of Polymer Science Part a: Polymer Chemistry, 43*(17), 3973–3985.

Larraza, I., Vadillo, J., Santamaria-Echart, A., Tejado, A., Azpeitia, M., Vesga, E., Orue, A., Saralegi, A., Arbelaiz, A., & Eceiza, A. (2020). The effect of the carboxylation degree on cellulose nanofibers and waterborne polyurethane/cellulose nanofiber nanocomposites properties. *Polymer Degradation and Stability, 173*, 109084.

Lee, H. T., & Lin, L. H. (2006). Waterborne polyurethane/clay nanocomposites: Novel effects of the clay and its interlayer ions on the morphology and physical and electrical properties. *Macromolecules, 39*(18), 6133–6141.

Lee, Y. R., Raghu, A. V., Jeong, H. M., & Kim, B. K. (2009). Properties of waterborne polyurethane/functionalized graphene sheet nanocomposites prepared by an in situ method. *Macromolecular Chemistry and Physics, 210*, 1247–1254.

Lee, D.-I., Kim, S.-H., & Lee, D.-S. (2019). Synthesis of self-healing waterborne polyurethane systems chain extended with chitosan. *Polymers, 11*(3), 503.

Lerf, A., He, H., Forster, M., & Klinowski, J. (1998). Structure of graphite oxide revisited. *The Journal of Physical Chemistry B, 102*, 4477–4482.

Li, J., Zheng, W., Zeng, W., Zhang, D., & Peng, X. (2014). Structure, properties and application of a novel low-glossed waterborne polyurethane. *Applied Surface Science, 307*, 255–262.

Liu, X. Q., Huang, W., Jiang, Y. H., Zhu, J. C. Z. Z., & Zhang, C. Z. (2012). Preparation of a bio-based epoxy with comparable properties to those of petroleum-based counterparts. *eXPRESS Polymer Letters, 6*(4), 293–298.

Liu, H., Li, C., & Sun, X. S. (2017). Soy-oil-based waterborne polyurethane improved wet strength of soy protein adhesives on wood. *International Journal of Adhesion and Adhesives, 73*, 66–74.

Lomeda, J. R., Doyle, C. D., Kosynkin, D. V., Hwang, W. F., & Tour, J. M. (2008). Diazonium functionalization of surfactant-wrapped chemically converted graphene sheets. *Journal of the American Chemical Society, 130*, 16201–16206.

Lu, Y., & Larock, R. C. (2010). Aqueous cationic polyurethane dispersions from vegetable oils. *Chemsuschem, 3*(3), 329–333.

Ma, Z., Zhang, X., Zhang, X., Ahmed, N., Fan, H., Wan, J., Bittencourt, C., & Li, B. G. (2020). Synthesis of CO_2-derived, siloxane-functionalized poly (ether carbonate) s and waterborne polyurethanes. *Industrial & Engineering Chemistry Research, 59* (7), 3044–3051.

Madbouly, S. A., Xia, Y., & Kessler, M. R. (2013). Rheological behavior of environmentally friendly castor oil-based waterborne polyurethane dispersions. *Macromolecules, 46*, 4606–4616. https://doi.org/10.1021/ma400200y

Maldovan, M. (2013). Sound and heat revolutions in phononics. *Nature, 503*, 209–214.

Meera, K. M. S., Sankar, R. M., Paul, J., Jaisankar, S. N., & Mandal, A. B. (2014). The influence of applied silica nanoparticles on a bio-renewable castor oil based polyurethane nanocomposite and its physicochemical properties. *Physical Chemistry Chemical Physics, 16*(20), 9276–9288.

Meng, Q. B., Lee, S. I., Nah, C., & Lee, Y. S. (2009). Preparation of waterborne polyurethanes using an amphiphilicdiol for breathable waterproof textile coatings. *Progress in Organic Coatings, 66*, 382–386.

Mishra, A. K., Mishra, R. S., Narayan, R., & Raju, K. V. S. N. (2010). Effect of nano ZnO on the phase mixing of polyurethane hybrid dispersions. *Progress in Organic Coatings, 67*(4), 405–413.

Panda, S. S., Panda, B. P., Mohanty, S., & Nayak, S. K. (2016). Synthesis and properties of castor oil based waterborne polyurethane Closite 30B nanocomposite coatings. *International Journal of Coating Research and Technology*. https://doi.org/10.1007/s11998-016-9855-8.

Patel, D. P., Nimavat, K. S., & Vyas, K. B. (2011). Surface coating studies of polyurethane derived from [(alkyd)-(epoxy resin treated castor oil)] isocyanate terminated castor oil mixture. *Advances in Applied Science Research, 2*, 558–566.

Poussard, L., Lazko, J., Mariage, J., Raquez, J. M., & Dubois, P. (2016). Biobased waterborne polyurethanes for coating applications: How fully biobasedpolyols may improve the coating properties. *Progress in Organic Coatings, 97*, 175–183.

Pramoda, K. P., Hussain, H., Koh, H. M., Tan, H. R., & He, C. B. (2010). Covalent bonded polymer–graphene nanocomposites. *Journal of Polymer Science Part A, 48*, 4262–4267.

Rahman, M. M., Kim, H. D., & Lee, W. K. (2008). Preparation and characterization of waterborne polyurethane/clay nanocomposite: Effect on water vapor permeability. *Journal of Applied Polymer Science, 110*(6), 3697–3705.

Ramesh, M., & Rajesh kumar, L. (2018). Wood flour filled thermoset composites. *Materials Research Foundations, 38*, 33–65. https://doi.org/10.21741/9781945291876-2.

Ramesh, M., & Kumar, L. R. (2020). Bioadhesives. In Inamuddin, R. Boddula, M. I. Ahamed, & A. M. Asiri (Eds.), *Green adhesives* (pp. 145–162). https://doi.org/10.1002/9781119655053.ch7.

Ramesh, M., Rajesh Kumar, L., Khan, A., & Asiri, A. M. (2020). Self-healing polymer composites and its chemistry. *Self-healing composite materials* (pp. 415–427). Woodhead Publishing. https://doi.org/10.1016/B978-0-12-817354-1.00022-3.

Riffat, S. B., & Ma, X. (2003). Thermoelectrics: A review of present and potential applications. *Applied Thermal Engineering, 23*, 913–935.

Rosthauser, J. W., & Nachtkamp, K. (1987). In K. C. Frisch, D& . Klempner (Eds.), *Advances in urethane science and technology* (pp. 121–162). Technomic.

Sanada, K., Tada, Y., & Shindo, Y. (2009). Thermal conductivity of polymer composites with close-packed structure of nano and micro fillers. *Composites Part a: Applied Science and Manufacturing, 40*, 724–730.

Santamaria-Echart, A., Ugarte, L., Garcia-Astrain, C., Arbelaiz, A., Corcuera, M. A., & Eceiza, A. (2016). *Carbohydrate Polymers, 151*, 1203–1209.

Saqr, K. M., Mansour, M. K., & Musa, M. N. (2008). Thermal design of automobile exhaust based thermoelectric generators: Objectives and challenges. *International Journal of Automotive Technology, 9*, 155–160.

Serkis, M., Spirkova, M., Hodan, J., & Kredatusova, J. (2016). *Progress in Organic Coatings, 101*, 342–349.

Shaik, A., Narayan, R., & Raju, K. V. S. N. (2014). Synthesis and properties of siloxane-crosslinked polyurethane-urea/silica hybrid films from castor oil. *Journal of Coatings Technology and Research, 11*(3), 397–407.

Sim, L. C., Ramanan, S. R., Ismail, H., Seetharamu, K. N., & Goh, T. J. (2005). Thermal characterization of Al_2O_3 and ZnO reinforced silicone rubber as thermal pads for heat dissipation purposes. *Thermochimica Acta, 430*, 155–165.

Stankovich, S., Piner, R. D., Nguyen, S. T., & Ruoff, R. S. (2006). Synthesis and exfoliation of isocyanate-treated graphene oxide nanoplatelets. *Carbon, 44*, 3342–3347.

Sukhawipat, N., Saetung, N., Pilard, J. F., Bistac, S., & Saetung, A. (2020). Effects of molecular weight of hydroxyl telechelic natural rubber on novel cationic waterborne polyurethane: A new approach to water-based adhesives for leather applications. *International Journal of Adhesion and Adhesives*, 102593.

Wang, D., Choi, D., Li, J., Yang, Z., Nie, Z., Kou, R., Hu, D., Wang, C., Saraf, L., Zhang, J., Aksay, L., & Liu, J. (2009). Self-assembled TiO_2–graphene hybrid nanostructures for enhanced Li-ion insertion. *ACS Nano, 3*, 907–914.

Wijs, J. J. A. (1929). The Wijs method as the standard for iodine absorption. *The Analyst, 54*, 12. https://doi.org/10.1039/an9295400012

Wu, G., Bian, J., Liu, G., Chen, J., Huo, S., Jin, C., & Kong, Z. (2020). Self-catalytic two-component waterborne polyurethanes with amino polyols from biomass based epoxy resin. *Journal of Polymers and the Environment, 28*(2), 713–724.

Xie, X. L., Mai, Y. W., & Zhou, X. P. (2005). Dispersion and alignment of carbon nanotubes in polymer matrix: A review. *Materials Science & Engineering, 49*, 89–112.

Xu, Y., Liu, Z., Zhang, X., Wang, Y., Tian, J., Huang, Y., Ma, Y., Zhang, X., & Chen, Y. (2009). A graphene hybrid material covalently functionalized with porphyrin: Synthesis and optical limiting property. *Advanced Materials, 21*, 1275–1279.

Yang, J. E., Kong, J. S., Park, S. W., Lee, D. J., & Kim, H. D. (2002). Preparation and properties of waterborne polyurethane–urea anionomers. I. The influence of the degree of neutralization and counter ion. *Journal of Applied Polymer Science, 86*(9), 2375–2383.

Yang, C. H., Liu, F. J., Liu, Y. P., & Liao, W. T. (2006). Hybrids of colloidal silica and waterborne polyurethane. *Journal of Colloid and Interface Science, 302*(1), 123–132.

Yeganeh, H., & Mehdizadeh, M. R. (2004). Synthesis and properties of isocyanate curable millable polyurethane elastomers based on castor oil as a renewable resources polyol. *European Polymer Journal, 40* (6), 1233–1238.

Yeh, J. M., Yao, C. T., Hsieh, C. F., Yang, H. C., & Wu, C. P. (2008). Preparation and properties of amino-terminated anionic waterborne-polyurethane–silica hybrid materials through a sol–gel process in the absence of an external catalyst. *European Polymer Journal, 44*(9), 2777–2783.

Yousefi E, Dolati A, Najafkhani H. (2020). Preparation of robust antistatic waterborne polyurethane coating. *Progress in Organic Coatings, 139*, 105450.

Yu, H., Yuan, Q., Wang, D., & Zhao, Y. (2004). Preparation of an ultraviolet curable water-borne poly (urethane acrylate)/silica dispersion and properties of its hybrid film. *Journal of Applied Polymer Science, 94*(4), 1347–1352.

Zhang, S., Yu, A., Liu, S., Zhao, J., Jiang, J., & Liu, X. (2012). *Polymer Bulletin, 68*, 1469–1482.

Zhang, Y., Zhang, W., Wang, X., Dong, Q., Zeng, X., Quirino, R. L., Lu, Q., Wang, Q., & Zhang, C. (2020). Waterborne polyurethanes from castor oil-based polyols for next generation of environmentally-friendly hair-styling agents. *Progress in Organic Coatings*, *142*, 105588.

Zhou, X., Li, Y., Fang, C., Li, S., Cheng, Y., Lei, W., & Meng, X. (2015a). Recent advances in synthesis of waterborne polyurethane and their application in water-based ink: A review. *Journal of Materials Science and Technology, 31*, 708–722.

Zhou, X., Li, Y., Fang, C., Li, S., Cheng, Y., Lei, W., & Meng, X. (2015b). Recent advances in synthesis of waterborne polyurethane and their application in water-based ink: A review. *Journal of Materials Science & Technology*. https://doi.org/10.1016/j.jmst.2015.03.002

Novel Research Areas of Applications for Water Borne Polyurethanes

Charles Oluwaseun Adetunji, Olugbemi Tope Olaniyan, Abel Inobeme, and Mohd Imran Ahamed

Abstract

Water borne polyurethanes have gained significant attention due to high-performance, excellent elasticity, no toxicity, abrasion resistance, adhesion functionality, flexibility, environmental friendliness, and broad substrate suitability. The physiochemical properties of water borne polyurethanes include low temperature, one component application, low viscosity, low solubility, and excellent coating performance. Therefore, this chapter intends to provide a detailed information sustainable production and applications of water borne polyurethanes. Different techniques such as thermal gravimetric analysis, Fourier transform infrared spectroscopy, mechanical measurements, and dynamic thermomechanical analysis used for the characterization of water borne polyurethanes were also highlighted. Moreover, detailed information on the application of water borne polyurethane plus polyurethane-urea dispersions for the production of novel eco-friendly nanomaterials were also highlighted.

Keywords

Water borne polyurethanes • Excellent elasticity • Non-toxic • Abrasion resistance • Adhesion • Flexibility • Environmentally friendly • Low viscosity • Eco-friendly • Low solvent

1 Introduction

The use of polyurethanes (PUs), which are solvent-borne as materials of remarkable performance, has been documented for long. The uniqueness of water borne polyurethane lies on the fact that the primary solvent employed for the synthesis is water. Water-based polyurethanes have wide applicability and have gained remarkable popularity due to their unique advantages over the organic solvents-based categories (Honarkar, 2017). Different approaches utilized for their production have also been widely employed. Atmospheric pollution, due to the emission of highly volatile organic compounds (VOCs), has been identified as one of the remarkable setbacks associated with the production of solvent polyurethanes since various organic solvents are used. Coatings produced from polyurethanes are known to possess some unique mechanical properties such as resistance to abrasion and toughness, which accounts for their durable nature. Apart from this, their improved esthetic effect also contributes to their high demand over time (Wen et al., 2019).

Various agencies and researchers around the globe have opined on the need for a novel technology that would help to minimize the exposure to various organic solvents associated with the processing and production of polyurethanes. Because of this, bold and practical measures have been put in place by such agencies with a view to reducing the release of various organic compounds that are volatile into the environment (Wen et al., 2019). The implementation of such guidelines has paved way for the birth of polymeric materials that are environmentally friendly. The emergence of the

C. O. Adetunji (✉)
Applied Microbiology, Biotechnology and Nanotechnology Laboratory, Department of Microbiology, Edo State University Uzairue, Iyahmo, Edo, Nigeria
e-mail: adetunji.charles@edouniversity.edu.ng

O. T. Olaniyan
Laboratory for Reproductive Biology and Developmental Programming, Department of Physiology, Edo State University Uzairue, Iyahmo, Edo, Nigeria

A. Inobeme
Department of Chemistry, Edo State University Uzairue, Iyahmo, Edo, Nigeria

M. I. Ahamed
Department of Chemistry, Faculty of Science, Aligarh Muslim University, Aligarh, Uttar Pradesh 202002, India

Inamuddin et al. (eds.), *Sustainable Production and Applications of Waterborne Polyurethanes*, Advances in Science, Technology & Innovation, https://doi.org/10.1007/978-3-030-72869-4_7

water borne polyurethane (WPU) has been recognized as the novel approach that is greener in nature. Water borne polyurethane has gained great recommendations because of its ease in processing, non-toxicity, and minimal energy utilization when compared to the conventional polyurethane (Mao et al., 2017). The extensive application of WPU in various industrial spheres cannot be overemphasized. One of such major industries is the coating sector where WPU is used in the coating of wood. They are also utilized in the finishing of leather, as biological materials, as well as materials used for damping (Yi et al., 2014).

However, the utilization of WPU, in which water is employed also, has few inherent limitations due to low resistance to water and oil. Addressing such setbacks is critical for industrial purposes, and this has brought about the modification of water borne polyurethane through the incorporation of groups such as silicon and fluorine (Bao et al., 2016). The formation of water borne polyurethane occurs when pre-polymer polyurethane that has functional group such as isocyanate is directly dispersed into water through emulsification step that involve the inversion of phase. The extension of polyurethanes after dispersion step is then carried out in water phase with the use of diamine. This implies that the primary dispersing solvent is water. This gives the WPU some extraordinary uniqueness such as eco-friendliness that is due to the decreased amount of the volatile organic compounds that are present (Wen et al., 2019). These alternative groups of polyurethane are also obtainable with varying and desired solid compositions and hardness of film. They are also known to lack residuals of the free isocyanate and have other advantageous features such as resistance to weather condition, zero content of volatile organic substances, isocyanate free, versatility and flexibility over a long period of time, resistance to ultra violet ray, and abrasion. Based on the standard requirements for coating materials, the polyurethane that is based on water is a fast growing component of the coating sector (Shin & Choi, 2018). Extensive studies have therefore been carried out on their synthesis, even though the positions of vital synthetic components have not been fully explored. More recently, water borne polyurethanes have been employed in the achievement of uncrosslinked thermoplastic coatings. Also in production processes where resistance to solvent and remarkable coating is vital, PUR is employed (Shan et al., 2020). The unique repelling potential of fabric materials has also been enhanced using water borne polyurethanes.

However, in other to obtain some other desired features such as antifouling properties, self-cleaning and higher repellants, and other additives such as silicon-based complexes, fluorine groups, and alkanes of long chain are introduced (Yi et al., 2014). In addition, more recently, WPU has been introduced in the production of different metallic primers, media for emulsion-based polymerization, fibers, additive in paint making, thickeners, defoaming agents, and in biomaterial production (Zhou et al., 2015). Novel ecologically friendly approaches permit the formulation of coatings that are of one component (1 K) as well as those of two components (2 K) giving rise to remarkable toughness, resistance to corrosion, unique substrate adhesion, and stain. The optimum overall features are found in the 2 K category of water borne polyurethanes (Shuaishuai Huang et al., 2018).

2 Specific Examples of Application for Water Borne Polyurethanes

In recent times, water borne polyurethanes have gained significant attention due to high-performance, excellent elasticity, non-toxic, abrasion resistance, adhesion, flexibility, environmentally friendly, and broad substrate suitability. The numerous and novel applications of water borne polyurethane are shown in Table 1.

3 Water Borne Polyurethanes as Adhesives

Water-based polyurethanes are widely used in adhesives. This type of adhesive is majorly a result of the physical force that exists between the layer of the adhesive and the adhesive to enhance a good bonding between the surfaces. This can be employed in fabric adhesives, woven, and non-woven cloths among others. It is, however, more suiting for materials with porous surfaces as well as those that are easily swollen by the bonding with solvents, low cost, and good bonding effect. Due to the absence of NCO groups in most of the water borne polyurethane adhesives, their curing is mainly as a result of the cohesive and adhesive forces that are generated by the polar functional groups present in the molecule (Shin & Choi, 2018).

Lei et al. (2015) studied the application of water borne polyurethane in adhesives and how to improve this property. Synthesis was carried out by varying the cross-linking densities in the various water borne polyurethanes produced. The trifunctional cross-linker used (trimethlolpropane) was varied to bring about products of different densities. The impacts of the effect of the cross-linker on the adhesive properties of the polyurethane were assessed with respect to strength of cohesion, interface adhesion strength between the substrate and the water borne polyurethane. It was concluded from the study that cross-linking plays a significant role in enhancing the adhesive property of WPUs.

Mohammad et al. (Rahman et al., 2016) utilized modified lignin amine during the polymer production process to generate sequence of water borne polyurethane. The authors discovered that the strength of adhesiveness increased with corresponding increase in modified lignin amine even at any

Table 1 Novel applications of water borne polyurethane

Applications	Properties	Authors
Production of leather	Mechanical properties and tensile strength	Yi et al. (2014), Kaiye Technology (2020), Perez-Liminana et al. (2005)
Coating material	Resistance to abrasion and UV	Shin and Choi (2018), Shan et al. (2020), Honarkar (2018), Howarth and Manock (1997), Humar et al. (2011)
Adhesive	Mechanical properties and water resistance	Lei et al. (2015)
Fiber production		Halima and Sharif (2018)
Ink binder	Pigment dispersion and surface tension	Fang et al. (2010), Zhihui et al. (2017)
Binder and adhesives	Mechanical strength	Ismoilov et al. (2019)

environmental condition. They, therefore, concluded that water borne polyurethane/ modified lignin amine adhesives are an excellent polymer with ability to withstand any harsh environmental condition. Strankowski et al. (2018) demonstrated that series of water borne polyurethanes could be synthesized using pre-polymer process from polycarbonatediol and isophorone diisocyanate polycaprolactone. Using different experimental techniques such as thermal gravimetric analysis, Fourier transform infrared spectroscopy, mechanical measurements, and dynamic thermomechanical analysis, they were able to reveal that isophorone diisocyanate and polycarbonatediol-based water borne polyurethanes contain the highest glass transition temperature.

Hengameh (2018) demonstrated that water borne polyurethanes could be used to produce films at room temperature, which has been seen to display good adhesion to many material surfaces. The author showed that this type of polymers is eco-friendly, nonflammable, hydrophobic, non-toxic, and does not exhibit wastewater production due to it insolubility in water. Thus, during the production processes, to disperse them in water solvent, addition of non-ionic plus ionic fragment to the polymers is encouraged.

Studies have revealed that water borne polyurethane plus polyurethane-urea dispersions can be utilized for the fabrication of novel environmental friendly materials: films and dispersions comprising of nanoparticle plus additives. In addition, electro spun nanocomposites can be generated from water borne polyurethane-cellulose nanocrystals dispersions utilizing ethylene oxide as form of polymer (Strankowski et al., 2018). The physiochemical properties of water borne polyurethanes include low temperature, one component application, low viscosity, eco-friendly, low solvent, and coating performance. Ismoilov et al. (2019) revealed that water borne polyurethane could reduce water resistance by self-emulsification on the chain of cationic water borne polyurethane. The authors showed that polyester polyurethane emulsion stability was excellent, but the water resistance or repellency plus mechanical property displayed a poor result.

Halima and Sharif (2018) studied polymeric materials particularly polyurethanes and discovered that they are one of the most resourceful materials generated using di or di or polyols plus polyisocyanates in a controlled polymerization reaction. They are generally used in the production of fibers, rubber, elastomers, paints, foams, films, gels, and coatings, with diverse industrial and commercial value. For several years, it application in biomedical science has been confirmed owing to the physiochemical properties such as high tensile strength, resistance to stress or fatigue, biodegradable, durability, and biocompatibility compared with other polymers. In cardiovascular physiology, polyurethanes are used in the fabrications of vascular grafts, blood coagulating devices, catheters, and prostheses plus heart valves. In addition, their role in tissue engineering as scaffolds, non-toxic implants, hydogels shape memory devices, diverse cardiovascular repair, and bone regenerations plus wound healing have received tremendous recognition.

4 Water Borne Polyurethane Ink Binder

Water-based ink binders as novel kinds of packaging materials have been produced from water borne polyurethanes. This is especially more suitable for food packaging, drinks, and drugs. In a related work, Fang et al. (2010) carried out a study in which they attempted to make a replacement for the usual binder resin in water base ink, by water borne substitute. They reported that gloss of aqueous-based ink if affected by the pigment dispersion, surface tension, and water content.

In attempts to boost the mechanical properties of polyurethanes so as to enhance their applications in binders and adhesives, Canak et al. (2013) carried out modification of the polyurethane by using a combination of 2-hydroxyethyl methacrylate and 3,5-bis(perfluorobenzyl)oxy benzyl alcohol. This also results to an improvement in the resistance of the PU to water and heat. Unlike the convectional materials used in ink binders that are made of acrylates alongside acrylamide as the cross-linkers thereby resulting to release of pollutants into the environment, the introduction of ink binders derived from WPUs has further resolved this challenges. The fabric in this case has a resulting higher elastic elongation, feels soft with higher quality and less contaminant. Various properties of the traditional ink binders are enhance through the use of calculated amount of water borne polyurethanes especially in cases where higher elasticity is needed (Shan et al., 2020).

Zhihue et al. (2017) studied the application of water borne polyurethane dispersion in ink binding by preparing WPU with a low viscosity and high solid content using two-step process of polymerization. The raw materials used during the synthesis include dimethylol propionic acid, poly glycol, and isophorne. They observed that the synthesized water borne polyurethane dispersed inside acetone concentration of 5–7% by weight could be used directly in the production of ink binder without the need for the removal of the acetone.

5 Water Borne Polyurethane Coatings

The formulation of water borne polyurethane is centered on the use of additives with water and emulsification of the binder. Formation of films is achieved through the use of emulsion coatings that are of single components. Agents commonly employed as binders include styrene, acrylic, and other co-polymers such as vinyl. For cosmetic top coating, water borne coatings of single components are used, and these are not good for immersion purposes. The application of such coatings is usually at 50–100 μm DFT (Shan et al., 2020).

Water borne polyurethanes are very useful in production of coatings and related materials in various industries. Apart from the unique advantage of being odorless and environmentally friendly due to the absence of free residues of isocyanates, they are also specifically unique in their coating potential. They are widely employed as eco-friendly materials in various industrial coatings and found their ways into a remarkable performance niche. Researchers have adduced the wide applications of water borne polyurethanes in coatings to be due to their ease of formulation as well as higher performance of their products. They are also capable of delivering results in varying areas (Zhihui et al., 2017).

Guo et al. (2012) targeted to enhance the water resistance and mechanical parameters of polyurethane by applying styrene and n-butyl acrylate to the PU. Due to the contribution effect of water resistance to adapting the specific use of some PU, Zhang et al. (2008) further boosted the water resisting properties of PU by using polydimethylsiloxane. Bondthane is a typical industrial use of water borne polyurethane. Some properties that have been utilized in this, thereby making it useful in production of coatings include high chemical resistance, low flexibility with respect to temperature, resistance to chemicals and abrasion (Honarkar, 2018).

Howarth and Manock (1997) reported that there were great predictions as long as early 1970s that there will be a great shift to water borne polyurethanes in the coating sector. At moment, various industries are complying with the legislation by making sure that the water borne coating is dominantly used. This occurrence, however, should not be seen as threat by the suppliers of resins but rather as a remarkable opportunity. Some resin suppliers that are of the medium and small categories who have the willingness to channel water borne resin such as hybrids and PUDs to specific demands of customers will have a secured future in this respect. There is a rapidly growing market on the use of water borne polyurethanes in coatings.

The use of dispersions type water-based polyurethane is one of the rapidly developing segments of the surface coating enterprise because of the advanced technology. Water-based polyurethanes dispersion (PUD) have become good replacement for the solvent counterpart especially in the coatings. They have several industrial applications. They can be more easily used in coatings when compared to the organic solvent types. Their formulation is like baked coatings and air dried making them suitable for rigid and flexible substrates (Xu et al., 2016).

Humar et al. (2011) carried out study on improvement of surface coatings. This was based on the application of two acrylic coatings that are water borne on spruce wood, with impregnation of two ethanolamine of copper having preservatives in them. In the studies, they observed a deeper penetration and as well as a lower contact angles compared to the wood that was not treated. They concluded an improvement of the adhesion coating.

It has been reported that the zeal from researchers to synthesize new water borne polyurethane is based on the fact that for several years, this material has been receiving consideration attention and recognition in the production of products relevant to diverse industrial sectors such as primers for metals, adhesives, emulsion polymerization media, associate thickeners, caulking materials, defoamers, textile dyes, paint additives, biomaterials, and plus pigment pastes. The authors noted that poly (lactic acid) plus poly (ε-caprolactone) are the foremost materials commonly utilized in the manufacture of high-performance water borne polyurethane using different polymerization processes to monitor the desired chain sequence (Fang et al., 2010).

Xu et al. (2016) studied water borne polyurethane pre-polymer by synthesizing cross-linked water borne polyurethane using trimethylolpropane as cross-linker. The authors analyzed the effects of various factors such as 1, 6-hexanediol, dimethylol propionic acid, poly (butylene itaconate) ester, plus trimethylolpropane. The structural integrity plus morphology was determined and characterized using different spectra analysis. From the results obtained, the authors concluded that the temperature resistance of cross-linked water borne polyurethane film was superior to water borne polyurethane film.

Kim and Kim (2005) revealed that the hydrolytic stability in water borne polyurethanes can be improved by utilizing polyester polyol containing alkyl side groups. They also showed that the physicochemical features of the dispersion cast films were excellent. Thus, they concluded that the combination of poly (2,4-diethyl-1, 5-pentamethylene adipate) glycol with poly (tetramethylene adipate) glycol used in the experiment gave the polymer the improved hydrolytic plus adhesion properties.

Doğan (2014) revealed that with the utilization of acetone procedure, a new polyurethanes made up of sodium 2-[(2-aminoethyl) amino] ethanesulphonate as its ionic group have been produced. The author went further to investigate the physiochemical properties of the novel structure such as hard segment content, ionic content, type of polyol, plus aqueous dispersions, and level of chain extension,. The author explained that incorporation of halloysite nanoparticles into the chemical structure of polyurethanes changed the thermomechanical properties.

Fatima et al. (2013) demonstrated that due to environmental and health concerns, various forms of water borne polyurethane dispersions have immerged as substitute to the other liquid-based polyurethane. The authors revealed that through polymerization reaction, water borne polyurethane with carboxylate anion as hydrophilic groups has been synthesized. They concluded that water borne polyurethane greater stability was observed with decrease in peel strength and increase in molecular weight.

Hsu et al. (2018) revealed that recently progress has been made in the development of an anti-adhesion material for individual having problems with joint motion and tendon gliding. The authors showed that water borne biodegradable polyurethane films are capable of producing biocompatible, non-cytotoxic, and good anti-adhesion material based on their mechanical plus physiochemical properties such as thickness change, gelatination, and plus water absorption. From their results, it was demonstrated using biochemical and histological assessment that water borne biodegradable polyurethane films showed immense potential as anti-adhesion property when compared with others. They, therefore, concluded that water borne biodegradable polyurethane films has low inflammatory response, appropriate degradation period, and better mechanical activity.

Yong et al. (2015) showed that the solvent-borne systems are the conventional polyurethane-ureas and polyurethane with hydrophobic nature; thus, in the synthesis of water borne polyurethane plus polyurethane-urea, the use of internal emulsifier which is environmental friendly due to less production of volatile organic molecules is encouraged. Studies have shown that polyurethanes are chemically synthesized co-polymers made up of urethane bonds in their dynamic chemical structure. In their structure, there are three monomers; polyol, chain extender, and diisocyanate facilitating the synthesis of large polyurethanes with great physicochemical properties dependent on the polyols conformation. The structural integrity of the polyols such as chain length, ether or ester groups, aromatic or aliphatic unit, and hydroxyl functionality determines the flexibility or rigidity of the polyurethanes. The utilization of polyurethanes in biomedical industries represents an important and dynamic area of tremendous advantage. Many studies have highlighted the importance or utilization of water borne polyurethanes as medical or biomedical implants in the time past; but in the recent years, the utilization of polyurethanes has extended tremendous to other biomedical areas such as wound dressing, tissue engineering, and drug delivery. In the area of wound dressing, medical and health workers find water borne polyurethane materials very highly compatible biologically.

6 Water Borne Polyurethane Leather

Water borne polyurethane are useful in fabrication of various kinds of leathers ranging from artificial leathers, leather resins use in shoes as well as finishing agents in leathers. They are used in belts, clothing, and shoes. In 2016, alone, more than two million metric tons of artificial leathers from water borne polyurethanes were produced. The manufacturing process of all involves the use of coagulation, however, the most recent advantageous shift that has benefited the process and environment at large is the use of water borne substitutes (Kaiye Technology, 2020).

The use of water borne polyurethane dispersal as material for coating has enhanced the leather production process. The water borne polyurethane leather is sometimes referred to as 'water-based leather.' They are also environmentally friendlier due to the use of water in their production. Zero emission of contaminant is associated with the production of these substances (Evonik, 2020). These products are used widely in luggage decoration, shoes, furniture, and vehicles and recognized widely in global market. The advantages of using water borne polyurethane leather also surpass that of the natural leather. This leather is also known to have outstanding styling effect and a bright surface (Kaiye Technology, 2020).

Also in footwear industries, water borne polyurethane has become a good replacement to the current solvent-based counterpart. Perez-Liminan et al. (2005) carried out a study in which different water polyurethane dispersing media was prepared through the use of mixing process of pre-polymer. There was an increase in resistance of the material to degradation by heat due to lowering of hard segment content.

7 As Novel Material for Wound Plastering and Other Clinical Usage

Polyurethanes are extensively utilized in wound and plaster industry due to its flexible, barrier protective, and antimicrobial plus soft properties. The chemistry of the frequently used polyurethane is perforated films, semi-permeable adhesive films, and foam plus hydrocolloids nature. Several protective properties show transparency and thin films with no exudate secretion during or post-operative wounds. They also reveal no permeability to pathogens and water but only to gases so that the wound can respite. In the synthesis of polyurethane wound dressing agents, different bioactive constituents are present like antiaging, supplements, antibacterial agents, glycerine, local anesthetics, zinc or collagen, absorbents, antifungal agents, gel-forming agents, and analgesic agents. Yari et al. (2012) have significantly reviewed the development of novel polyurethane membrane by synthesizing novel design using amine reaction from epoxy-polyurethane pre-polymer utilized for wound care with antimicrobial properties and wound exudates absorbent properties.

Heping et al. (2012) described the importance of polyurethane by demonstrating that water borne polyurethane is gaining tremendous attention for biomedical utilization owing to the excellent physiochemical properties and application in breast implants, vascular prostheses, pacemaker lead insulation, and bioadhesives plus heart valves. The authors revealed silent points on the surface properties of polyurethane, which is crucial to its performance; hence, medical devices developers should pay serious attention to the surface characteristics and different surface modifiers plus techniques. Many surface modifiers have been used in the past such as chemical, immobilization of biological compounds, salinization, and plus radiation grafting of monomers to improve the quality of polyurethane without altering the entire properties.

Gaddam et al. (2017) in an experiment used phosphorylated polyols produced from cottonseed oil to generate novel water borne polyurethane dispersions devoid of catalyst and dimethylol propanoic acid. The authors characterized the phospols synthesized from the epoxidized cottonseed oil using different spectra analysis. From the results obtained, they revealed that all the three water borne polyurethane displayed good storage stabilities, the films showed excellent thermal stability, anticorrosive activity plus good tensile strength, and hence, they concluded that the hydroxyl contents and degree of siloxane cross-linking gave water borne polyurethane dispersions films such an excellent physicochemical properties.

Kim et al. (1996) demonstrated that water borne polyurethane dispersions displayed volatile organic content with higher degree of adhesion when non-isocyanate is added. The authors revealed that the role of non-isocyanate is to serve as a pigment separating resin due to the existence of carboxyl, urethane plus functional groups showing good physicochemical properties. Increased molecular weight displayed by other dispersing agents makes them poor material, but non-isocyanate possess the ability to separate organic and inorganic pigments. They concluded that development of a more identical linkage system gives coatings with enhanced performance activities.

Tao et al. (2017) revealed that tinted water borne polyurethanes are used during different industrial activities; hence, the authors utilized black water borne polyurethanes to formulate step-growth polymerization. They observed that water borne polyurethanes emulsions generated displayed excellent physiochemical properties and characterized by different spectra analysis showing good color stronghold plus resistance to thermal effects.

Hercule et al. (2011) revealed that diverse cross-linkers could be used to improve the physicochemical activity of water borne polyurethane such as urea bridge introduced in between the structure has many advantages due to development of solid network between polyurethane structures showing increased strength of bands of urea plus urethane. The authors showed that cross-linkers in the urea bonds created intercellular hydrogen bonds that enhanced the structural integrity. Due to the catalyst and poly additive reaction used by the authors such as polyethylene glycol, dibutyltin dilaurate, and isophorone diisocyanate, high-performance water borne polyurethanes were created.

8 Conclusion and Future Recommendation to Knowledge

This chapter has provided a detailed information of the sustainable production and applications of water borne polyurethanes. The features of water borne biodegradable polyurethane films showed immense potential as anti-adhesion property when compared with others. They, therefore, concluded that water borne biodegradable polyurethane films have low inflammatory response, appropriate degradation period, and better mechanical activity. The utilization of polyurethanes in medical or biomedical areas such as wound dressing, drug delivery, and tissue engineering was highlighted. The biological compatibility of water borne polyurethane materials most especially in the area of wound dressing, medical, and health workers was also highlighted because of their highly compatible biologically nature.

References

Bao, L., Fan, H., Chen, Y., Yan, J., Yang, T., & Guo, Y. (2016). Effect of surface free energy and wettability on the adhesion property of water borne polyurethane adhesive. *RSC Advances, 6*(101), 99346–99352.

Canak, T. C., & Serhatli, I. E. (2013). Synthesis of fluorinated urethane acrylate based UV-curable composites. *Progress in Organic Coatings, 76*(2–3), 388–399.

Doğan, N. A. (2014). *Synthesis and characterization of novel water borne polyurethane dispersions*. Graduate School of Engineering and Natural Sciences. Master of Science thesis.

Evonik. (2020). Leading beyond chemistry.

Fang, C.O., Zhou, S.S., & Ren, P.G. (2010). Study on the Gloss of Polyurethane Water-Based Ink. Key Engineering Materials 428, 429:431–435 ·

Fatima, M., Zuber, M., Zia, K. M., Jamil, T., & Hussain, R. (2013). Synthesis and properties of aqueous polyurethane dispersions: Influence of molecular weight of polyethylene glycol Korean. *Journal of Chemical Engineering, 30*(12), 2259–2263. https://doi.org/10.1007/s11814-013-0166-9

Gaddam, S. K., Raju Kutcherlapati, S. N., & Palanisamy, A. (2017). Self-cross-linkable anionic water borne polyurethane−silanol dispersions from cottonseed-oil-based phosphorylated polyol as ionic soft segment. *ACS Sustainable Chemistry & Engineering, 5*, 6447–6455. https://doi.org/10.1021/acssuschemeng.7b00327

Guo, Y. H., Li, S. C., Wang, G. S., Ma, W., & Huang, Z. (2012). Water borne polyurethane/poly(n-butyl acrylate-styrene) hybrid emulsions: Particle formation, film properties, and application. *Progress in Organic Coatings, 74*(1), 248–256.

Halima, K., & Sharif, A. (2018). Polyurethane: A versatile scaffold for biomedical applications. *Significances of Bioengineering & Biosciences, 2*(3). SBB. 000536. https://doi.org/10.31031/SBB.2018.02.000536

Heping, X., Qiu, F., Wang, Y., Yang, D., Wenling, W., Chen, Z., & Zhu, J. (2012). Preparation, mechanical properties of water borne polyurethane and crosslinked polyurethane-acrylate composite. *Journal of Applied Polymer Science, 124*, 958–968.

Hercule, K. M., Yan, Z., & Christophe, M. M. (2011). Preparation and Characterization of Water borne Polyurethane Crosslinked by Urea Bridges. *International Journal of Chemistry, 3*(2), 88–96.

Honarkar, H. (2017). Water borne polyurethanes: A review. *Journal of Dispersion Science and Technology*. https://doi.org/10.1080/01932691.2017.1327818

Honarkar, H. (2018). Water borne polyurethanes: A review. *Journal of Dispersion Science and Technology, 39*(4), 507–516. https://doi.org/10.1080/01932691.2017.1327818

Howarth, G. A., & Manock, H. L. (1997). Water-borne polyurethane dispersions and their use in functional coatings. *Surface Coatings International, 80*, 324–328. https://doi.org/10.1007/BF02692680

Hsu, S.-H., Dai, L.-G., Hung, Y.-M., & Dai, N.-T. (2018). Evaluation and characterization of water borne biodegradable polyurethane films for the prevention of tendon postoperative adhesion. *International Journal of Nanomedicine, 13*, 5485–5497.

Humar, M., Pavlic, M., & Tomzic, D. (2011). Performance of water borne acrylic surface coatings on wood impregnated with Cu-ethanolamine preservatives. *Bulletin of Materials Science, 34* (1), 113–119. https://doi.org/10.1007/s12034-011-0052-1

Ismoilov, K., Akram, W., Chauhan, S., Ergasheva, K., Artikboeva, R., et al. (2019). Synthesis and evaluation of properties of a novel cationic wate rborne polyurethane finishing agent. *Journal of Chemical Engineering & Process Technology, 10*, 398. https://doi.org/10.35248/2157-7048.19.10.398

Kaiye Technology. (2020). Cloth Leather [Water-based PU Leather].

Kim, B. S., & Kim, B. K. (2005) Enhancement of hydrolytic stability and adhesion of water borne polyurethanes. *Journal of Applied Polymer Science*.

Kim, B. K., & Lee, J. C. (1996). *Journal of Polymer Science Polymer Chemistry Edition, 34*, 1095–1104.

Lei, L., Xia, Z., Ou, C., & Zhong, L. (2015). Effects of crosslinking on adhesion behavior of water borne polyurethane ink binder. *Progress in Organic Coatings, 88*, 155–163. https://doi.org/10.1016/j.porgcoat.2015.07.002

Mao, H., Qiang, S., Xu, Y., & Wang, C. (2017) Synthesis of polymeric dyes based on UV curable multifunctional water borne polyurethane for textile coating. *New Journal of Chemistry, 41*(2), 619627.

Perez-Liminana, M., Rorro-Palau, F. A., Orgiles-Barcelo, A. C., & Martin-Martinez, J. (2005). Characterization of water borne polyurethane adhesives containing different amounts of ionic groups. *International Journal of Adhesion and Adhesives, 25*(6), 507–515.

Rahman, M. M., Zahir, M. H., & Kim, H. D. (2016) Synthesis and properties of water borne polyurethane (WBPU)/modified lignin amine (MLA) adhesive: A promising adhesive material. *Polymers, 8*, 318. https://doi.org/10.3390/polym8090318

Shan, C., Ning, C., Lou, J., et al. (2020). Design and preparation of UV-curable water borne polyurethane based on novel fluorinated chain extender. *Polymer Bulletin*. https://doi.org/10.1007/s00289-020-03202-7

Shin, E. J, & Choi, S. M. (2018). Advances in water borne polyurethane-based biomaterials for biomedical applications. *Advances in Experimental Medicine and Biology, 2*. https://doi.org/10.1007/978-981-13-0947-2_14

Shuaishuai Huang, S., Liu, G., & Hu, H. (2018). Water-based anti-smudge NP-GLIDE polyurethane coatings. *Chemical Engineering Journal, 351*, 210–220.

Strankowski, M., Korzeniewski, P., Strankowska, J., & Thomas, S. (2018). Morphology, mechanical and thermal properties of thermoplastic polyurethane containing reduced graphene oxide and graphene nanoplatelets. *Materials (Basel), 11*, 82.

Wang, T., Sun, W., Zhang, X., Xu, H., Xu, F. (2017) Water borne polyurethane coatings with covalently linked black dye Sudan Black B. *Materials, 10*, 1247. https://doi.org/10.3390/ma10111247

Wen, J., Sun, Z., Xiang, J., Fan, H., Chen, Y., & Yan, J. (2019). Preparation and characteristics of water borne polyurethane with various lengths of fuorinated side chains. *Applied Surface Science, 494*, 610–618.

Xu, J., Li, T., Zhao, W., Li, P., & Wu, P. (2016). Synthesis and characterization of water borne polyurethane emulsions based on poly(butylene itaconate) ester. *Designed Monomers and Polymers, 19*(4), 309–318. https://doi.org/10.1080/15685551.2016.1152541

Yari, A., Yeganeh, H., & Bakhshi, H. (2012). Synthesis and evaluation of novel absorptive and antibacterial polyurethane membranes as wound dressing. *Journal of Materials Science: Materials in Medicine, 23*(9), 2187–2202. https://doi.org/10.1007/s10856-012-4683-6

Yi, L., Meng, X., Tian, X., Zhou, W., & Chen, R. (2014). Wettability of electrospun films of microphaseseparated block copolymers with 3,3,3-trifuoropropyl substituted siloxane segments. *Journal of Physical Chemistry C, 118*(46), 26671–26682.

Yong, Q., Nian, F., Liao, B., Huang, L., Wang, L., & Pang, H. (2015). Synthesis and characterization of solvent-free water borne polyurethane dispersion with both sulfonic and carboxylic hydrophilic chain-extending agents for matt coating applications. *RSC Advances, 5*, 107413–107420.

Zhang, C. Y., Zhang, X. Y., Dai, J. B., & Bai, C. Y. (2008). Synthesis and properties of PDMS modified water borne polyurethane-acrylichybrid emulsion by solvent-free method. *Progress in Organic Coatings, 63*(2), 238–244.

Zhou, X., Li, Y., Fang, C., & Li, S. (2015) Recent advances in synthesis of water borne polyurethane and their application in water-based ink: A review. *Journal of Materials Science and Technology –Shenyang, 31*(7) April 2015. https://doi.org/10.1016/j.jmst.2015.03.002

Zhihui, U., Guangfeng, W., & Huixuan, Z. (2017). Effects of the reagent molar ratio on the phase separation and properties of water borne polyurethane for application in a water-based ink binder. *Journal of Applied Polymer Science*. https://doi.org/10.1002/app.45406Citations:5

Applications of Polymeric Materials in Biomedical Engineering

Ahmad Mukhtar, Sidra Saqib, Sami Ullah, Muhammad Sagir, M. B. Tahir, Abid Mahmood, Abdullah G. Al-Sehemi, Muhammad Ali Assiri, Muhammad Ibrahim, and Amna Zulfiqar

Abstract

The class of multipurpose materials having countless properties for various applications in particular based on their relationship between structure and property is known as polyurethanes (PUs). Particular attention was given to the adaption of different applications of PUs with its unique physical, mechanical, chemical, and biological characteristics. Increasing the potential and efficiency of materials based on PU may be accomplished by modifying the production process or utilizing state-of-the-art characterization techniques. Changes to raw materials and production procedures by suitable methods will easily yield PUs fit for different unique applications. Over the last decade, significant experiments have been performed on the usage of polymers such as medication delivery devices, implants, tissue engineering scaffolds, etc. The discovery of a better synthesis method for polymer systems based on biologically dependent macromolecules may be encouraged by this information.

Keywords

Tissue engineering • Biomedical • Polymers • Polyurethanes • Polymerization

1 Introduction

In biomedical applications, designing polymers is a fast-growing field. Polyurethanes (PUs), and other polymers, are used for a variety of biomedical uses, e.g., tissue engineering, orthopedic devices, transdermal sections, catheters, and pharmaceutical carriers. Tiny molecules, enzymes, DNA, or peptides are supplied with the PU, helping to establish efficient medication supplies (Cheng et al., 2016; Cherng et al., 2013). It is because PUs can be designed for biocompatibility, biodegradability, ease of adjustment, high quality of product loading, and improved product stability (Bazban-Shotorbani et al., 2017). Any mechanism contributing to the advancement of these properties can be viewed in biomedical applications as an optimal alternative for usage. Of PUs, the introduction like additives has had a significant effect on their final properties in addition to becoming main raw materials like polyol and isocyanate. Concerning environmental problems and stability of systems, carbohydrates were recently explored as a PU additive (Desai et al., 2000; Lalwani & Desai, 2010). We also expected based on our work that there is a broad variety of carbohydrates in this area for combined PUs.

Polyurethanes (PUs) are a specific category of polymeric compounds, which vary in many respects from the other forms of plastic. They are ideal for several specific things including oils, liquid coatings, elastomers, insulators, sprinklers, mold, integrated skins, etc. Professor Otto Bayer (A German Professor) and his colleague's innovations, in

A. Mukhtar
Department of Chemical Engineering, Universiti Teknologi PETRONAS, 32610 Bandar, Seri Iskandar, Perak, Malaysia

S. Saqib
Department of Chemical Engineering, COMSATS University Islamabad, Lahore Campus, Lahore, 54000, Pakistan

S. Ullah · A. G. Al-Sehemi · M. A. Assiri
Department of Chemistry, College of Science, King Khalid University, P. O. Box 9004 Abha, 61413, Saudi Arabia

M. Sagir (✉)
Department of Chemical Engineering, University of Gujrat, Gujrat, 50700, Pakistan
e-mail: m.sagir@uog.edu.pk

M. B. Tahir
Department of Physics, University of Gujrat, Gujrat, 50700, Pakistan

A. Mahmood · M. Ibrahim
School of Environmental Sciences and Engineering, Government College University, Faisalabad, 38000, Pakistan

A. Zulfiqar
Institute of Soil and Environmental Sciences, University of Agriculture Faisalabad, Faisalabad, 38040, Pakistan

Inamuddin et al. (eds.), *Sustainable Production and Applications of Waterborne Polyurethanes*, Advances in Science, Technology & Innovation, https://doi.org/10.1007/978-3-030-72869-4_8

which PUs now exist, is a pure development (Bayer, 1946). In 1937, the formation of the technique of diisocyanate polyaddition contributed to World War II and pure polyurethane industry of PU created by a reaction between diisocyanate and polyester diol (Delebecq et al., 2013; Islam et al., 2014). German company TORRS developed PU as an alternative rubber. Aimed at beginners, PU coatings were used to impregnate paper and to produce clothing that avoided mustard gas and erosion (Mushtaq et al., 2014b).

PU has also been used as high-gloss coatings in the aircraft, as a chemical-resistant timber, stonework, and metals (Seymour & Kauffman, 1992). Besides, the production of inexpensive polyether polyols through the late '50s contributed too many automobile and taping applications which still are important to this day. PU-coating was widely used in viscoelastic polymers, coatings, polyurethane foam, and glues (Petrović & Ferguson, 1991; Sagir et al., 2014a). Currently, PUs is one of the most widely produced, flexible, and studied products in the world (Delebecq et al., 2013). To make it ideal in substitution, intermingled power and durability of metals were used through elastic properties of rubber (Sagir et al., 2014b).

Constantly developed manufacturing methods with additive forms and composition have been discussed in this chapter in a broad range of applications. It is thoroughly discussed how essential are the individual components of PUs, including the use of the synthesized PU for each part. Besides, this chapter explains and advances a broad variety of biomedical applications for embedded carbohydrate Pus (Sagir et al., 2016; Shahzad et al., 2018; Talebian et al., 2015).

2 Chemistry of Polyurethane

The chemistry of PUs refers to the aggregation of certain products, which are known as reaction polymers. PUs is synthesized through the reactions between isocyanate and polyol, either in a catalytic or in an ultraviolet activation (Soto et al., 2014). More than two isocyanate hydroxyl groups ($R'–(OH)_{n \geq 2}$) and ($R–(N=C=O)_{n \geq 2}$) are needed by isocyanate and polyol molecules, respectively. Such compounds involve phenolic compounds, unsaturated polyesters and sterilizers, and epoxies. The long-chain low cross-linked chains are used for the formation of stretch polymers, whereas shorter, strong cross-linked chains are used for producing hard polymers (Azam et al., 2014; Mushtaq et al., 2014a). In comparison, a combination of long chains with an average crosslink will create polymers appropriate for foam production (Ionescu, 2005). Due to the combination of PU, they strive to a combination of three-dimensional (3D) network aggregation with a molecular weight of infinite. That is why, a tiny amount of PUs may be considered a massive molecule and why standard PU's do not go so far as to melt or scatter when warm. The inclusion of various additive elements alongside isocyanates and polyols, along with the adjustment of their manufacturing conditions, enables a broad range of characteristics to be obtained, rendering them ideal for several applications (Mushtaq et al., 2014c; Pauzi et al., 2014; Sagir et al., 2014c). Various forms of polyols are available and can be produced using different methods in laboratories. The propylene and ethylene oxide which is copolymerized by a well-suited polyol precursor, for example, creates polyether polyols (Petrović, 2008), while the production of polyester polyols in a method is identical to that used in the preparation of polyester polymers. Polyethylene polyol, poly(tetramethylene ether), can also be developed for use in highly productive elastic applications by polymerizing tetrahydrofuran (Fox & Edmund, 2016). Rajendran and co-workers documented an illustration of preparing prepolymers with isocyanate terminal in poly tetrahydrofuran and its characterization studies (Rajendran et al., 1989). We have various amounts of –OH groups in their molecules. Therefore, given the nature of the isocyanate, industrial-grade diols have formulations that regulated cautiously to attain coherent material goods, which are required to manufacture POUs with specific properties. The formulations are therefore sufficient for average functionality polyols (Petrović, 2008). Besides, isocyanates are introduced in the PU synthesis utilizing a high-reactivity hydroxyl group-containing compound, although the reaction at room temperature is small (Sonnenschein, 2014). Their reactivity implies that they are weak and that they are produced by large molecular polyols (over ten thousand). To achieve a faster reaction rate, appropriate surfactants and catalysts are desired. As compared to aliphatic materials, aromatic isocyanates like hexamethylene diisocyanate, diphenylmethane diisocyanate (MDI), toluene diisocyanate, and isophorone diisocyanate are more reactive (Sagir et al., 2014d; Ullah et al., 2015).

Most polymers are dysfunctional (each molecule has two groups of isocyanates), only diphenylmethane and diisocyanate contain mixtures of two or more isocyanate groups (2.7 is the highest number of isocyanate group) and are functional to nature. The fluctuations in the unprocessed matter and blending of PU decides the characteristics of the drug (Mushtaq et al., 2015; Sagir et al., 2018; Talebian et al., 2018).

3 Waterborne Polyurethane

Water is used primarily for solvent coatings and adhesives, such as waterborne polyurethanes (WPU) (Chen & Chen, 1997). There are many articles of legislation limiting the volume of permissible organic solvents and other harmful air contaminants eliminated hooked on the surroundings. Thus, many profitable and industrialized uses rely on the polyurethane dispersions (PUD) or waterborne polyurethane dispersions (WPUD) (Fang et al., 2014; Fangcq & Zhous, 2011; Huber & Mecking, 2010). PUD's have a unique benefit since the dispersion's viscosity is not based on a polymer's molecular weight. The drying phase may therefore only prepare high-solid content WPUs (HSCWPUs). The scattered system contains the PU units besides the aqueous milieu in a two-stage colloidal system (Ullah et al., 2019b).

Many classes of pendent acid or tertiary nitrogen are neutralized within the PU chain to form salts that produce mainly water dispersibility centers. Specific properties of this dispersion are responsible for the polyol, isocyanate, ionomers, and chain extension forms and amounts used (Ullah et al., 2019a).

A new method aimed at synthesizing HSCWPUs (2-stage emulsion process) was established recently (Peng et al., 2015) where the bimodal particle size distribution was strictly regulated. It has been attributed that the greater distribution of fragments as a criterion for viscous purpose and solid substance interrelationship (Chu & Guyot, 2001). The spatial and time-output of the reactors and the time are taken for the development of the reactors have both been improved by this kind of high-solid substance material (Ai et al., 2010). There was a significant improvement in the organic solvent/water tolerance and thermal and mechanical characteristics of the processed material. In other significant WPUs, formerly a polycarbonate diol WPU which was strengthened with silicate, the power of the traction increased from 9 to 15 MPa, while the extensiveness decreased from 520 to 280%. It has been reported that the material produced has an acceptable coating application for flexible materials such as fabrics, paper, and leather, particularly for cases requiring high-abrasion resistance (Vermette et al., 2001). The process was cleared up to strictly conform to environmental protection requirements, and excellent hydrophobic and thermal properties were seen for the synthesized drug. It was also suggested for conventional oil-based products as an acceptable alternative (Fu et al., 2014).

4 Synthesis of Polyurethane

The best and most useful process can be used in the development of PUs through various routes (Ulrich, 1996). The most important and useful method is to create a polyol reaction (alcohol with 2 or many hydroxyl groupings in one molecule) and diisocyanate (Bayer, 1946; Ionescu, 2005; Lat et al., 2020). PUs can be formed by differing quantities and forms of polyol, insulates or additives in all styles, such as longevity and density.

4.1 Polyols

Polyols can be classified primarily into polyether or polyester polyols. The reaction of an epoxide to the active hydrogen-containing compound is the product of polyether polyols. The polymerization of epoxy monomers may also be generated by ring-opening (Bailosky et al., 2013; Brocas et al., 2013; Tahir et al., 2019). Polyesters polyols, which can also be derived by hydroxyl and carboxy compounds and multifunctional acids, are sometimes included in the group. Polyols may also be classified by their final application. Polyester polyols are derived from fresh pristine products and are formed by the direct method of polyester production by the processing of very pure diacids and glycols. An indicator is 1,4-butanediol and adipic acid. Polyester polyols are generally more viscous (Domanska & Boczkowska, 2014; Sagir & Talebian, 2020; Sagir et al., 2020), and more volatile than polyether polyols.

However, the fact that they manufacture PU with enhanced abrasion, solvent, and cuts resistance is also quite significant. Another polyester group is derived from reclaimed raw materials. The glycolysis, of recycling poly(-ethylene terephthalate) (PET) or dimethyl terephthalate (DMT) distillation bottles, such as diethylene glycol, is produced by transesterification treatment otherwise known as glycolysis. These aromatic polyols having low-molecular weight are utilized for the formation of rigid foam development, as they have reduced cost and a beneficial effect on polyisocyanurate (PIR) board foam and PU separation foams. Of sealants, elastomers, and adhesives which are of high-quality standards to stand up to chemical and environmental conditions, a group of polyols called specialty polyols is needed. Polycarbonate polyols, polysulfide polyols, polybutadiene polyols, and polycaprolactone polyols are some of these polyols.

4.2 Isocyanates and Non-isocyanates

For PU preparation, isocyanates are vital key elements. The difunctional or heterofunctional and aromatic or aliphatic form may be defined. Methylene diphenyl diisocyanate (MDI), toluene diisocyanate (TDI), and aliphatic diisocyanate are two of the more widely used alternatives available. MDI and TDI manufacturing groups are the most widely made of polymeric components, which are isomeric blends. It is typically used to produce flexible foams, such as molded foams for car seats or slabstock foams for the production of mattresses (Islam et al., 2014). Strong foams such as coolant insulation materials, and for elastomers (for shoe soils, etc.) may also be manufactured. The partial polyol reaction or the incorporation of certain materials can lead to isocyanate modification. The volatility of the isocyanates is reduced and toxicity invariably. The melting point should, therefore, be popular so that they can be more effectively handled, and the properties of the resultant polymers strengthened. Despite, the value of diisocyanate, ecology has led researchers to investigate strategies to minimize or even prevent their usage, primarily to mitigate environmental issues and other toxicity-related diisocyanate issues. The non-isocyanate route has followed the reaction with amines between cyclic carbonate and polyol could be prepared from lignin using an oxy-propylation method (Cateto et al., 2009; Rokicki & Piotrowska, 2002).

4.3 Catalysts

Catalysts found in PU can be divided into two main types: metal complexes and compounds of amines. Traditionally, amine catalysts consist of tertiary amines, such as dimethyl cyclohexylamine (DMCHA), dimethylethanolamine (DMEA), 1,4-diazabicyclo[2.2.2]octane (DABCO), and triethylenediamine (TEDA). The option of tertiary amine catalysts depends on how they will trigger reactions to the trimerization of urea, urethane, or isocyanates. Strong metal from bismuth, plum, copper, dye, and mercury derivatives can also be used as a catalyst for urethane. Specific effectiveness is very efficient in the manufacture of PU sealants, coats, and elastomers, and mercury carboxylates. This is because the polyol and isocyanate reactions are preferentially selected. They are, however, potentially poisonous, thereby contributing in recent times to the usage of zinc and bismuth carboxylates as replacements. Alkyl tin carboxylates, mercaptides, and oxides are also found in numerous applications. Tin commodities are usually applied to water bottle formulas, since tin carboxylate may be undesired in the tin through hydrolysis in the tin commodities.

In the formulation of different types of PUs, catalysts are mostly used for selective purposes. Throughout the form of minimizing CO production and other threats, for example, the latest $CuCo_2O_4$/graphitic carbon nitride nanohybrid has been used. The reactivity of such catalysts differs due to their design (Shi et al., 2015). Comparisons were made of the catalytic behavior of two catalysts, zirconium, and tin, for the production of waterborne polyurethanes dependent on isophorone diisocyanate (IPDI) (Sardon et al., 2009). The reactivity of isocyanate for the catalyst of tin was considered significant whereas the reactivity of zirconium is the same for the zirconium catalyst.

4.4 Chain Extenders and Cross-Linkers

The chain extenders ($f = 2$) and cross-linkers ($f = 3$ or more) are also a category of compounds that play important roles in PU polymer morphology. These compounds are typically finished with small molecular weights of amine and hydroxyl. The elastomeric properties of these composites are extracted from so-called rough and hard segments of polymer copolymers interfaces. These compounds are extremely useful when improving the morphology of PU adhesives, elastomers, biscuits, and other important microcellular and skin foam. As such, the urethane harsh segment dominates the domain of amorphous polyesters (or polyether) so that these domain areas are related. This interfacial distinction occurs because of the incompatibility with the hard segments (high melting) and the immiscibility (while all phases are amorphous). There is, therefore, no impact on phase separation in crystallization. The hard segments produced with isocyanates and chain extender are usually stable and rigid, whereas the so-called "filter segments," which are made of polyols (high-molecular weight), can pass freely and oasis appear in foil shapes. Binding the strong and then the binding contributes to the reduction of plastic oscillations of polymer chains and thereby to elastomeric durability.

The mechanical deformation of these compounds causes certain portions of the stressed segment to be uncoiled and the tough segments to be aligned with the stress. The adjustment of the rigid segments combined with the subsequent strong hydrogen binding provides high resistance to tensile, tear, and good stretching characteristics (Blackwell & Gardner, 1979; Musselman et al., 1998). A good choice of chain extenders may also have chemical resistance, heat, and flexural characteristics. Some of the chain extension devices most commonly used are 1,4-butanediol (BDO), cyclohexane dimethanol, ethylene glycol, bis(2-hydroxyethyl)ether and 1,6-hexanediol. These glycols may be used for thermoplastic PU production. They are also well-ordered domains of the hard group, which will differentiate and can be handled in the molten state. The main example is ethylene glycol which is extracted from bis-phenyl in the case of unnecessary deterioration if the hard segment amount becomes too high (Howard, 2002; Romaškevič et al., 2006).

4.5 Surfactants

Surfactants are utilized to strengthen foam properties and non-foam PU polymers. They resemble block polymers of polydimethylsiloxane–polyoxyalkylene, nonylphenol ethoxylates, silicone oils, and some other organic compounds. These are identical to polymers. For the emulsification of liquid components, the control of cell-sizes, and the stabilization of cell structures for the control of collapse and against vacuum at the subsurface, they are applied to applications involving foams. These are used as anti-foaming and air release agents and weathering agents for non-foam applications. These are often used to eliminate imperfections in the paint, such as sink traces, orange peeling, and pinholes. Different forms of surfactants, including nonionic and cationic surfactants, are essential for the preparation of PU products (Motamedi et al., 2014; Zheng et al., 2010). The usage of nonionic surfactants has demonstrated outstanding surface behavior without a crucial micelle concentration. The usage of cationic surfactants for corrosion resistivity was, on the other side, stronger. The surfactants utilization to produce PU has several drawbacks. Like, modern surfactants with low-molecular weight cause delamination and corrosion (Jin et al., 2006). They may also quickly move to the surface of PU materials often. Therefore, another researcher proposed a surfactant-free PU (Lu et al., 2011).

5 Carbohydrates Incorporated Polyurethane

5.1 Potential of Carbohydrates in Polyurethane

The collection of carbohydrates as components of the drug delivery structures is focused on several important factors. Reproducibility from natural sources such as animals and plants animals are one of the most attractive factors. Polysaccharides for example starch, cellulose, hemicellulose, alginate, hyaluronic, guar gums, etc., are produced from seeds, while chitin and chondroitin are extracted from livestock and aquatic waste. Usage of carbohydrates not only promotes organic growth but also promotes the optimization of existing raw materials that are well-defined by recent progress in the application and modification of the carbohydrate techniques (Galanakis, 2016; Li et al., 2017; Pontis, 2017). Biocompatibility can also be considered before deciding on medication admission applications. The chemical structure that matches well-known macromolecules in the human body characterizes carbohydrates. Polysaccharides are natural polymers and, compared with synthetic polymers, typically are biocompatible. The capacity for degradation in synthetic polymers with residual initiators/additives that can be harmful to cell production. This is the correct characteristic explaining its usage as components in oral medicinal items (Leclerc et al., 2004; Liu et al., 2017; Mattu et al., 2015). Carbohydrate derivative such as hydroxypropyl cellulose is decreased by platelet adhesion, suggesting it for usage as a biocompatible substance nominee (Filip et al., 2013; Macocinschi et al., 2012, 2013).

However, the degradable components of degradable polymers of carbohydrates in soil and water are attractive to specific microorganisms (Tamburini et al., 2017). Certain desirable characteristics of polysaccharides are simplicity of use and cost-efficiency. While polysaccharides can be extra costly in the purifying phase, quantity and ease of supply may minimize the costs of transport in conjunction with other synthetic alternatives. Sub-products of fishing operations or agricultural practices may contain large quantities of carbohydrates, thus providing a waste accumulation solution (Abdou et al., 2008; Blanco-Fernandez et al., 2011; Wang et al., 2011).

Carbohydrate structural properties allow them easy to alter, derivatize, and integrate them in polymer matrices. Cross-links, which form a three-dimensional network for the entanglement and release of drugs, can translate carbohydrates into hydrogels (Nieuwenhove et al., 2017). Many carbon derivatives, such as carboxymethyl cellulose (Dev et al., 2010; Paramakrishnan et al., 2016; Sencadas et al., 2012) are used in several biomedical applications. Carbohydrates may also make a major input to the creation of smart drug delivery systems as an important special function (Liu et al., 2017). This is because of their adaptability and the ability to bind with other reacting materials that release medications depending on external stimuli including temperature, pH, or magnet field (Cui et al., 2017; Poorgholy et al., 2017; Zhao et al., 2013, 2017). Taking into consideration the extraordinary properties of PUs and carbohydrates, a device of natural and synthetic materials may have a mutual benefit to maintain all the ingredients' foundations. The carbohydrates are available as fillers or cross-linkers in the PU network (Solanki et al., 2015; Valodkar & Thakore, 2011; Wu et al., 2014).

5.2 Role of Carbohydrates in Polyurethane

Including (1): composites/fillers in the PU network and (2): covalent linkages in isocyanates for the formation of the link are two important approaches that report carbohydrates as embedded in the PU network.

5.3 Role of Carbohydrates in Polyurethane as Filler

Another approach is the use of carbohydrates as fillers that can illustrate the synthesis of embedded PU carbohydrates. Including nanoparticles of cellulose (Aranguren et al., 2013), starch nanoparticles, and chitosan nanoparticles (Wu et al., 2007, 2014) are reported for improved mechanical strength and enhanced properties of PUs for the inclusion in PU network of nano-formats of carbohydrates such as cellular nanowhiskers. A hydrogen bonding with the resin macromolecular chains may also be formed in the active carbohydrates in addition to the covalent bonding. The use of hydrogen may have a significant effect on rheological properties, including the viscosity of PU (Wu et al., 2014). Specific reasons for upgraded dispersion, enhancement, and interfacial adhesion among the PU matrix chains are allegedly hydrogen bonding (Rueda et al., 2013; Rueda-Larraz et al., 2009). The explanation is that the contact between the hydrophobic polymer matrix and hydrophilic carbohydrates may be improved because of the hydrogen bonding. Carbohydrate contact with PU matrix often influences the thermal properties of PUs through hydrogen bonding. Improved activity, between storms and PU matrix (Barikani & Mohammadi, 2007), was responsible for growing the PU's glass transfer temperature. The attraction of hydrogen among the starch and the PU matrix is reasoned to influence PU channel mobility.

5.4 Role of Carbohydrates in Polyurethane as Cross-Linkers

The carbohydrates have specific properties for the polymer due to the relationship between carbohydrates and the PU matrix/components. In the PU network, the carbohydrates could be utilized as interconnecting agents because of the existence of many hydroxyl groups that can react to isocyanate groups. The developed interconnected network can be used in the PU matrix for product processing. To order to encapsulate product particles in the PU network, macromolecules such as cyclodextrins may also be included in the polyurethanes structure. This may contribute to improved medication effectiveness. A deterioration of the PU matrix can result in molar amounts of carbon hydrates incorporated into PUs, contributing to a regulation of the drug release profiles of PUs. Cross-linkers may be added and the final properties of PUs can be strengthened. This is because shifts in the number of cross-linkers have been induced to alter the degree of cross-linkage and the molecular weight between PU network crosslinks (Desai et al., 2000; Lalwani & Desai, 2010).

6 Characterization of Carbohydrates Incorporated Polyurethane

Different analytical techniques describe the effective integration of carbohydrates with the urethane network. E.g., FTIRs of polymers of this sort display a maximum pitch of 1669 cm^{-1} that could be allocated to the urethane band C=O (NH=COO). The relation between polyurethane and carbohydrate process is thus verified. The absence of the 2285–2250 cm^{-1} N=C=O band would guarantee that the groups of isocyanates in polyol or hydroxyl groups of carbohydrates were used in full during a cross-connect phase. Besides, there can be evidence that carbon hydrates in the polymer cross-linking are retained with the existence of stretching vibrations of C–H bonds (3500–3200 cm^{-1}), C–O (1320–1000 cm^{-1}) and O–H (2905 cm^{-1}) expanding vibrations of polymers, particularly polysaccharides (Okoli et al., 2014). In another report, NMR exhibited urethane proton (NHCOO) at 7.10–7.91 ppm, however, the peaks were detected at 3.77–4.88 ppm and 2.63–3.03 ppm were assignable to protons of the CH_2N, CH_2, CH_2OCO and CH_2O supporting the development of starch-modified polyurethane. Moreover, as proof of carbon involvement in the chemical reaction, the Pu's thermal characteristics were determined by TGA/DSC, are known. This is due to an increase in carbon content, which results in a large cross-linking intensity of polyurethane (Barikani & Mohammadi, 2007; Solanki & Thakore, 2014). This is the result of the increase in glass exchange temperature. Carbohydrate incorporation can have a significant impact on polymer hydrophilicity. The contact angle for PU polyurethane with starch graft was reported to increase from 0° for starch to 107° (Barikani & Mohammadi, 2007). The surface characteristics of the PU network can be used to demonstrate homogeneity, adhesion, and distribution in the SEM (Solanki et al., 2015). A "sea-island structure" which can be due to the energy dissipation cycle at the interfaces of the carbon and PU matrix is an unusual morphological characteristic of carbohydrate built-in PU. This implies the carbohydrate filler and matrix are spread evenly, scattered, and stable (Gao et al., 2012).

7 Applications of Polyurethane in Biomedical Science

In a wide range of biomedical applications, chitin and chitosan PU's are utilized, such as stent coatings, injury repair, tissue engineering, separation membranes, electrodes, sutures, and a drug delivery system. Remember that the bioactive properties of chitin, polyurethane, $NHCOCH_3$, NH_2, and NHCOO groups, are important for biomedical applications. The NHCOO-group in PU can be deduced from the interaction of protein-based peptides (NHCO). The use of chitin or chitosan improves biocompatibility and the cytotoxicity of the relevant material (Usman et al., 2016).

Using starch as one of the polyols, superparamagnetic polyurethane microspheres were made. The investigators investigated the usage of theophylline as a standard treatment in a controlled drug delivery system. The starch improved the microspheres' explosion effect significantly. Starch was believed to have successful connections with the PU polymer causing kinetic drug release (Mahmoudi & Laurent, 2010). The microsphere of PU was formed using starch and polycaprolactone (PCL) as a polyol in another analysis, which analyzed drug release profiles according to the starch concentration. The rate of release was observed to be 24% lower by utilizing starch in conjunction with PCL and 2000 g/mol molecular weight. Interestingly, there was no impact of the introduction of starch in conjunction with PCL with a molecular weight of 530 g/mol on opioid release levels with PUs. It indicates the starch is often consistent with certain ingredients and implies the pace of release of drugs.

Cellulose cross-linked PUs have been reported to be smart polymers intended for use in pH reactive delivery of drugs (Solanki et al., 2015). Cellulose acetate and PU have been combined in another analysis to generate nanofibrous groats. The medication loading in PU scaffolds has been conducted using an electrospinning method for particular applications in the fire, recurrent, and diabetic injury (Lee & Kim, 2012; Unnithan et al., 2014). The loading method is focused on the possible use in intravaginal opioid abuse has been shown by cellulose acetate phthalate core–shell and PU-hybrid micro/nanofibers (Hua et al., 2016). Cellulose acetate phthalate was the pH reactivity of these nanofibers. To prepare bimodal foam nano-composites for application in tissue technique, polyester PU is filled with cellulose nanowhiskers (CNWs) (Shahrousvand et al., 2017). The designed scaffolds possibly will provide effective cell proliferation as well as osteogenic differentiation in the three-dimensional synthetic extracellular matrix of human mesenchymal stem cells.

8 Conclusions

This chapter examined several methods for the design, characterization, and biomedical applications of various embedded carbohydrate polyurethanes. The evidence gathered here reveals that carbohydrates can be integrated into their structural structure and hence improved suitability to biomedical applications. Concerning the final properties of polyurethane, the discussion has been done on the influence of stoichiometry of ingredients and carbohydrates. The carbohydrate addition in the form of the structural elements of PU could improve mechanical properties, biocompatibility, and biodegradability in thermal stability without much compromise. The review offers guidance for the selection of the appropriate synthetic procedure and type of carbs. Their low solubility in most solvents and weak homogeneity and complex processability are problematic for the use of carbohydrates. Formamide dimethyl and sulfoxide dimethyl's are widely used solvents to degrade this issue. Besides, a further disadvantage is the poor thermal stability of carbohydrates. Nevertheless, it is possible to react at lower temperatures by using catalysts and by the ideal reactive existence of materials used in PU chemistry. It is worth noting that many of the reports examined are evidence-of-concept studies and require progress for the actual application. This is a fairly new and effective concept of polymer synthesis. However, it is emerging, several problems need to be tackled, including biocompatibility in vitro, biodegradation, and the impact of in vivo degradations. The in vivo properties of these polyurethanes along with the clinical trials should be regarded in future efforts.

Acknowledgements The authors gratefully acknowledge the Departments at their respective universities for providing state of the art research facilities.

References

Abdou, E. S., Nagy, K. S., & Elsabee, M. Z. (2008). Extraction and characterization of chitin and chitosan from local sources. *Bioresource Technology, 99*, 1359–1367.

Ai, Z., Deng, R., Zhou, Q., Liao, S., & Zhang, H. (2010). High solid content latex: Preparation methods and application. *Advances in Colloid and Interface Science, 159*, 45–59.

Aranguren, M. I., Marcovich, N. E., Salgueiro, W., & Somoza, A. (2013). Effect of the nano-cellulose content on the properties of reinforced polyurethanes. A study using mechanical tests and positron anihilation spectroscopy. *Polymer Testing, 32*, 115–122.

Azam, M. R., Tan, I. M., Ismail, L., Mushtaq, M., Nadeem, M., & Sagir, M. (2014). Kinetics and equilibria of synthesized anionic surfactant onto berea sandstone. *Journal of Dispersion Science and Technology, 35*(2), 223–230.

Bailosky, L. C., Bender, L. M., Bode, D., Choudhery, R. A., Craun, G. P., Gardner, K. J., et al. (2013). Synthesis of polyether polyols with epoxidized soy bean oil. *Progress in Organic Coatings, 76*, 1712–1719.

Barikani, M., & Mohammadi, M. (2007). Synthesis and characterization of starch-modified polyurethane. *Carbohydrate Polymers, 68*, 773–780.

Bayer, O. (1946). *Polyurethanes*. HM Stationery Office.

Bazban-Shotorbani, S., Hasani-Sadrabadi, M. M., Karkhaneh, A., Serpooshan, V., Jacob, K. I., Moshaverinia, A., et al. (2017). Revisiting structure-property relationship of pH-responsive polymers for drug delivery applications. *Journal of Controlled Release, 253*, 46–63.

Blackwell, J., & Gardner, K. H. (1979). Structure of the hard segments in polyurethane elastomers. *Polymer, 20*, 13–17.

Blanco-Fernandez, B., Lopez-Viota, M., Concheiro, A., & Alvarez-Lorenzo, C. (2011). Synergistic performance of cyclodextrin–agar hydrogels for ciprofloxacin delivery and antimicrobial effect. *Carbohydrate Polymers, 85*, 765–774.

Brocas, A.-L., Mantzaridis, C., Tunc, D., & Carlotti, S. (2013). Polyether synthesis: From activated or metal-free anionic ring-opening polymerization of epoxides to functionalization. *Progress in Polymer Science, 38*, 845–873.

Cateto, C. A., Barreiro, M. F., Rodrigues, A. E., & Belgacem, M. N. (2009). Optimization study of lignin oxypropylation in view of the preparation of polyurethane rigid foams. *Industrial & Engineering Chemistry Research, 48*, 2583–2589.

Chen, G. N., & Chen, K. N. (1997). Self-curing behaviors of single pack aqueous-based polyurethane system. *Journal of Applied Polymer Science, 63*, 1609–1623.

Cheng, X., Jin, Y., Sun, T., Qi, R., Li, H., & Fan, W. (2016). An injectable, dual pH and oxidation-responsive supramolecular hydrogel for controlled dual drug delivery. *Colloids and Surfaces B: Biointerfaces, 141*, 44–52.

Cherng, J. Y., Hou, T. Y., Shih, M. F., Talsma, H., & Hennink, W. E. (2013). Polyurethane-based drug delivery systems. *International Journal of Pharmaceutics, 450*, 145–162.

Chu, F., & Guyot, A. (2001). High solids content latexes with low viscosity. *Colloid and Polymer Science, 279*, 361–367.

Cui, X., Guan, X., Zhong, S., Chen, J., Zhu, H., Li, Z., et al. (2017). Multi-stimuli responsive smart chitosan-based microcapsules for targeted drug delivery and triggered drug release. *Ultrasonics Sonochemistry, 38*, 145–153.

Delebecq, E., Pascault, J.-P., Boutevin, B., & Ganachaud, F. (2013). On the versatility of urethane/urea bonds: Reversibility, blocked isocyanate, and non-isocyanate polyurethane. *Chemical Reviews, 113*, 80–118.

Desai, S., Thakore, I., Sarawade, B., & Devi, S. (2000). Effect of polyols and diisocyanates on thermo-mechanical and morphological properties of polyurethanes. *European Polymer Journal, 36*, 711–725.

Dev, A., Mohan, J. C., Sreeja, V., Tamura, H., Patzke, G. R., Hussain, F., et al. (2010). Novel carboxymethyl chitin nanoparticles for cancer drug delivery applications. *Carbohydrate Polymers, 79*, 1073–1079.

Domanska, A., & Boczkowska, A. (2014). Biodegradable polyurethanes from crystalline prepolymers. *Polymer Degradation and Stability, 108*, 175–181.

Fang, C., Zhou, X., Yu, Q., Liu, S., Guo, D., Yu, R., et al. (2014). Synthesis and characterization of low crystalline waterborne polyurethane for potential application in water-based ink binder. *Progress in Organic Coatings, 77*, 61–71.

Fangcq, Y., & Zhous, S. (2011). The effect of additives to the polyurethane water-based ink. *Research Journal of Chemistry and Environment, 15*, 377–379.

Filip, D., Macocinschi, D., & Gradinaru, L. (2013). Thermal and surface characteristics of some β-cyclodextrin-based side-chain azo amphiphilic polyurethanes. *Polymer Degradation and Stability, 98*, 951–957.

Fox, R. B., & Edmund, B. (2016). Mechanically frothed gel elastomers and methods of making and using them. Google Patents.

Fu, C., Zheng, Z., Yang, Z., Chen, Y., & Shen, L. (2014). A fully bio-based waterborne polyurethane dispersion from vegetable oils: From synthesis of precursors by thiolene reaction to study of final material. *Progress in Organic Coatings, 77*, 53–60.

Galanakis, C. M. (2016). *Nutraceutical and functional food components: Effects of innovative processing techniques*. Academic Press.

Gao, Z., Peng, J., Zhong, T., Sun, J., Wang, X., & Yue, C. (2012). Biocompatible elastomer of waterborne polyurethane based on castor oil and polyethylene glycol with cellulose nanocrystals. *Carbohydrate Polymers, 87*, 2068–2075.

Howard, G. T. (2002). Biodegradation of polyurethane: A review. *International Biodeterioration & Biodegradation, 49*, 245–252.

Hua, D., Liu, Z., Wang, F., Gao, B., Chen, F., Zhang, Q., et al. (2016). pH responsive polyurethane (core) and cellulose acetate phthalate (shell) electrospun fibers for intravaginal drug delivery. *Carbohydrate Polymers, 151*, 1240–1244.

Huber, J., & Mecking, S. (2010). Aqueous poly (arylacetylene) dispersions. *Macromolecules, 43*, 8718–8723.

Ionescu, M. (2005). *Chemistry and technology of polyols for polyurethanes*. iSmithers Rapra Publishing.

Islam, M. R., Beg, M. D. H., & Jamari, S. S. (2014). Development of vegetable-oil-based polymers. *Journal of Applied Polymer Science, 131*.

Jin, L., Liu, Z., Xu, Q., & Li, Y. (2006). Preparation of soap-free cationic emulsion using polymerizable surfactant. *Journal of Applied Polymer Science, 99*, 1111–1116.

Lalwani, R., & Desai, S. (2010). Sorption behavior of biodegradable polyurethanes with carbohydrate crosslinkers. *Journal of Applied Polymer Science, 115*, 1296–1305.

Lat, D. C., Ali, N., Jais, I. B. M., Yunus, N. Z. M., Razali, R., & Talip, A. R. A. (2020). A review of polyurethane as a ground improvement method. *Malaysian Journal of Fundamental and Applied Sciences, 16*, 70–74.

Leclerc, E., Furukawa, K., Miyata, F., Sakai, Y., Ushida, T., & Fujii, T. (2004). Fabrication of microstructures in photosensitive biodegradable polymers for tissue engineering applications. *Biomaterials, 25*, 4683–4690.

Lee, S. J., & Kim, B. K. (2012). Covalent incorporation of starch derivative into waterborne polyurethane for biodegradability. *Carbohydrate Polymers, 87*, 1803–1809.

Li, H., Yang, J., Li, P., Lan, T., & Peng, L. (2017). A facile method for preparation superhydrophobic paper with enhanced physical strength and moisture-proofing property. *Carbohydrate Polymers, 160*, 9–17.

Liu, G., Gu, Z., Hong, Y., Cheng, L., & Li, C. (2017). Electrospun starch nanofibers: Recent advances, challenges, and strategies for potential pharmaceutical applications. *Journal of Controlled Release, 252*, 95–107.

Lu, Y., Xia, Y., & Larock, R. C. (2011). Surfactant-free core–shell hybrid latexes from soybean oil-based waterborne polyurethanes and poly (styrene-butyl acrylate). *Progress in Organic Coatings, 71*, 336–342.

Macocinschi, D., Filip, D., Vlad, S., Butnaru, M., & Knieling, L. (2013). Evaluation of polyurethane based on cellulose derivative-ketoprofen biosystem for implant biomedical devices. *International Journal of Biological Macromolecules, 52*, 32–37.

Macocinschi, D., Filip, D., Vlad, S., Oprea, A. M., & Gafitanu, C. A. (2012). Characterization of a poly (ether urethane)-based controlled release membrane system for delivery of ketoprofen. *Applied Surface Science, 259*, 416–423.

Mahmoudi, M., & Laurent, S. (2010). Controlling the burst effect of a drug by introducing starch in the structure of magnetic polyurethane microspheres containing super paramagnetic iron oxide nanoparticles.

Mattu, C., Wang, T., Siri, A., Sartori, S., & Ciardelli, G. (2015). Ionic cross-linking of water-soluble polyurethane improves protein encapsulation and release. *Engineering in Life Sciences, 15*, 448–455.

Motamedi, M., Tehrani-Bagha, A., & Mahdavian, M. (2014). The effect of cationic surfactants in acid cleaning solutions on protective performance and adhesion strength of the subsequent polyurethane coating. *Progress in Organic Coatings, 77*, 712–718.

Mushtaq, M., Tan, I. M., Ismail, L., Lee, S. Y. C., Nadeem, M., & Sagir, M. (2014a). Oleate ester-derived nonionic surfactants: Synthesis and cloud point behavior studies. *Journal of Dispersion Science and Technology, 35*(3), 322–328.

Mushtaq, M., Tan, I. M., Ismail, L., Nadeem, M., Sagir, M., Azam, R., & Hashmet, R. (2014b). Influence of PZC (point of zero charge) on the static adsorption of anionic surfactants on a Malaysian sandstone. *Journal of Dispersion Science and Technology, 35*(3), 343–349.

Mushtaq, M., Tan, I. M., Nadeem, M., Devi, C., Lee, S. Y. C., & Sagir, M. (2014c). A convenient route for the alkoxylation of biodiesel and its influence on cold flow properties. *International Journal of Green Energy, 11*(3), 267–279.

Mushtaq, M., Tan, I. M., Rashid, U., Sagir, M., & Mumtaz, M. (2015). Effect of pH on the static adsorption of foaming surfactants on Malaysian sandstone. *Arabian Journal of Geosciences, 8*(10), 8539–8548.

Musselman, S., Santosusso, T., & Sperling, L. (1998). Structure versus performance properties of cast elastomers. In *Polyurethanes' 98 Conference Proceedings*.

Okoli, C. P., Adewuyi, G. O., Zhang, Q., Diagboya, P. N., & Guo, Q. (2014). Mechanism of dialkyl phthalates removal from aqueous solution using γ-cyclodextrin and starch based polyurethane polymer adsorbents. *Carbohydrate Polymers, 114*, 440–449.

Paramakrishnan, N., Jha, S., & Kumar, K. J. (2016). Effect of carboxymethylation on physicochemical, micromeritics and release characteristics of Kyllinga nemoralis starch. *International Journal of Biological Macromolecules, 92*, 543–549.

Pauzi, N. N. P. N., Majid, R. A., Dzulkifli, M. H., & Yahya, M. Y. (2014). Development of rigid bio-based polyurethane foam reinforced with nanoclay. *Composites Part B: Engineering, 67*, 521–526.

Peng, S.-J., Jin, Y., Cheng, X.-F., Sun, T.-B., Qi, R., & Fan, B.-Z. (2015). A new method to synthesize high solid content waterborne polyurethanes by strict control of bimodal particle size distribution. *Progress in Organic Coatings, 86*, 1–10.

Petrović, Z. S. (2008). Polyurethanes from vegetable oils. *Polymer Reviews, 48*, 109–155.

Petrović, Z. S., & Ferguson, J. (1991). Polyurethane elastomers. *Progress in Polymer Science, 16*, 695–836.

Pontis, H. (2017). Protein and carbohydrate separation and purification. In *Methods for analysis of carbohydrate metabolism in photosynthetic organisms* (pp. 45–63). Academic Press.

Poorgholy, N., Massoumi, B., & Jaymand, M. (2017). A novel starch-based stimuli-responsive nanosystem for theranostic applications. *International Journal of Biological Macromolecules, 97*, 654–661.

Rajendran, G., Mahadevan, V., & Srinivasan, M. (1989). Synthesis of some low glass transition temperature polytetrahydrofuran polymers. *European Polymer Journal, 25*, 461–463.

Rokicki, G., & Piotrowska, A. (2002). A new route to polyurethanes from ethylene carbonate, diamines and diols. *Polymer, 43*, 2927–2935.

Romaškevič, T., Budrienė, S., Pielichowski, K., & Pielichowski, J. (2006). Application of polyurethane-based materials for immobilization of enzymes and cells: A review. *Chemija, 17*, 74–89.

Rueda, L., Saralegui, A., d'Arlas, B. F., Zhou, Q., Berglund, L. A., Corcuera, M., et al. (2013). Cellulose nanocrystals/polyurethane nanocomposites. Study from the viewpoint of microphase separated structure. *Carbohydrate Polymers, 92*, 751–757.

Rueda-Larraz, L., d'Arlas, B. F., Tercjak, A., Ribes, A., Mondragon, I., & Eceiza, A. (2009). Synthesis and microstructure–mechanical property relationships of segmented polyurethanes based on a PCL–PTHF–PCL block copolymer as soft segment. *European Polymer Journal, 45*, 2096–2109.

Sagir, M., Mushtaq, M., Tahir, M. S., Tahir, M. B., & Shaik, A. R. (2020). CO_2 philic surfactants, switchable amine-based surfactants and wettability alteration for EOR applications. *Surfactants for Enhanced Oil Recovery Applications*, 89–102.

Sagir, M., Mushtaq, M., Tahir, M. B., Tahir, M. S., Ullah, S., Shahzad, K., & Rashid, U. (2018). CO_2 foam for enhanced oil recovery (EOR) applications using low adsorption surfactant structure. *Arabian Journal of Geosciences, 11*(24), 789.

Sagir, M., & Talebian, S. H. (2020). Screening of CO_2-philic surfactants morphology for high temperature-pressure sandstone reservoir conditions. *Journal of Petroleum Science and Engineering, 186*, 106789.

Sagir, M., Tan, I. M., Mushtaq, M., Ismail, L., Nadeem, M., & Azam, M. R. (2014a). Synthesis of a new CO_2 philic surfactant for enhanced oil recovery applications. *Journal of Dispersion Science and Technology, 35*(5), 647–654.

Sagir, M., Tan, I. M., Mushtaq, M., Ismail, L., Nadeem, M., Azam, M. R., & Hashmet, M. R. (2014b). Novel surfactant for the reduction of CO_2/brine interfacial tension. *Journal of Dispersion Science and Technology, 35*(3), 463–470.

Sagir, M., Tan, I. M., Mushtaq, M., & Nadeem, M. (2014c). CO_2 mobility and CO_2/brine interfacial tension reduction by using a new surfactant for EOR applications. *Journal of Dispersion Science and Technology, 35*(11), 1512–1519.

Sagir, M., Tan, I. M., Mushtaq, M., Pervaiz, M., Tahir, M. S., & Shahzad, K. (2016). CO_2 mobility control using CO_2 philic surfactant for enhanced oil recovery. *Journal of Petroleum Exploration and Production Technology, 6*(3), 401–407.

Sagir, M., Tan, I. M., Mushtaq, M., & Talebian, S. H. (2014d). FAWAG using CO_2 philic surfactants for CO_2 mobility control for enhanced oil recovery applications. In *SPE Saudi Arabia Section Technical Symposium and Exhibition 2014*. Society of Petroleum Engineers.

Sardon, H., Irusta, L., & Fernández-Berridi, M. (2009). Synthesis of isophorone diisocyanate (IPDI) based waterborne polyurethanes: Comparison between zirconium and tin catalysts in the polymerization process. *Progress in Organic Coatings, 66*, 291–295.

Sencadas, V., Correia, D. M., Areias, A., Botelho, G., Fonseca, A., Neves, I., et al. (2012). Determination of the parameters affecting electrospun chitosan fiber size distribution and morphology. *Carbohydrate Polymers, 87*, 1295–1301.

Seymour, R. B., & Kauffman, G. B. (1992). Polyurethanes: A class of modern versatile materials. *Journal of Chemical Education, 69*, 909.

Shahrousvand, E., Shahrousvand, M., Ghollasi, M., Seyedjafari, E., Jouibari, I. S., & Salimi, A. (2017). Preparation and evaluation of polyurethane/cellulose nanowhisker bimodal foam nanocomposites for osteogenic differentiation of hMSCs. *Carbohydrate Polymers, 171*, 281–291.

Shahzad, K., Čuček, L., Sagir, M., Ali, N., Rashid, M. I., Nazir, R., Nizami, A. S., Al-Turaif, H. A., & Ismail, I. M. I. (2018). An ecological feasibility study for developing sustainable street lighting system. *Journal of Cleaner Production, 175*, 683–695.

Shi, Y., Yu, B., Zhou, K., Yuen, R. K., Gui, Z., Hu, Y., et al. (2015). Novel CuCo2O4/graphitic carbon nitride nanohybrids: Highly effective catalysts for reducing CO generation and fire hazards of thermoplastic polyurethane nanocomposites. *Journal of Hazardous Materials, 293*, 87–96.

Solanki, A. R., Kamath, B. V., & Thakore, S. (2015). Carbohydrate crosslinked biocompatible polyurethanes: Synthesis, characterization, and drug delivery studies. *Journal of Applied Polymer Science, 132*.

Solanki, J. M., & Thakore, S. (2014). Structure–property relationships and biocompatibility of carbohydrate crosslinked polyurethanes. *Carbohydrate Polymers, 110*, 338–344.

Sonnenschein, M. F. (2014). *Polyurethanes: Science, technology, markets, and trends* (Vol. 11). Wiley.

Soto, M., Sebastian, R. M., & Marquet, J. (2014). Photochemical activation of extremely weak nucleophiles: Highly fluorinated urethanes and polyurethanes from polyfluoro alcohols. *The Journal of Organic Chemistry, 79*, 5019–5027.

Tahir, M. B., Sagir, M., & Abas, N. (2019). Enhanced photocatalytic performance of CdO-WO3 composite for hydrogen production. *International Journal of Hydrogen Energy, 44*(45), 24690–24697.

Talebian, S. H., Sagir, M., & Mumtaz, M. (2018). An integrated property–performance analysis for CO_2-philic foam-assisted CO_2-enhanced oil recovery. *Energy & Fuels, 32*(7), 7773–7785.

Talebian, S. H., Tan, I. M., Sagir, M., Muhammad, M. (2015). Static and dynamic foam/oil interactions: Potential of CO_2-philic surfactants as mobility control agents. *Journal of Petroleum Science and Engineering, 135*, 118–126.

Tamburini, D., Łucejko, J. J., Zborowska, M., Modugno, F., Cantisani, E., Mamoňová, M., et al. (2017). The short-term degradation of cellulosic pulp in lake water and peat soil: A multi-analytical study from the micro to the molecular level. *International Biodeterioration & Biodegradation, 116*, 243–259.

Ullah, S., Bustam, M. A., Ahmad, F., Nadeem, M., Naz, M. Y., Sagir, M., & Shariff, A. M. (2015). Synthesis and characterization of melamine formaldehyde resins for decorative paper applications. *Journal of the Chinese Chemical Society, 62*(2), 182–190.

Ullah, S., Bustam, M. A., Assiri, M. A., Al-Sehemi, A. G., Sagir, M., Kareem, F. A. A., Elkhalifah, A. E. I., Mukhtar, A., & Gonfa, G. (2019a). Synthesis, and characterization of metal-organic frameworks-177 for static and dynamic adsorption behavior of CO_2 and CH_4. *Microporous and Mesoporous Materials, 288*, 109569.

Ullah, S., Suleman, H., Tahir, M. S., Sagir, M., Muhammad, S., Al-Sehemi, A. G., Zafar, M., Kareem, F. A. A., Maulud, A. S., & Bustam, M. A. (2019b). Reactive kinetics of carbon dioxide loaded aqueous blend of 2-amino-2-ethyl-1, 3-propanediol and piperazine using a pressure drop method. *International Journal of Chemical Kinetics, 51*(4), 291–298.

Ulrich, H. (1996). *Chemistry and technology of isocyanates*. Wiley.

Unnithan, R., Gnanasekaran, G., Sathishkumar, Y., Lee, Y. S., & Kim, C. S. (2014). Electrospun antibacterial polyurethane–cellulose acetate–zein composite mats for wound dressing. *Carbohydrate Polymers, 102*, 884–892.

Usman, K. M. Z., Zuber, M., Tabasum, S., Rehman, S., & Zia, F. (2016). Chitin and chitosan based polyurethanes: A review of recent advances and prospective biomedical applications. *International Journal of Biological Macromolecules, 86*, 630–645.

Valodkar, M., & Thakore, S. (2011). Organically modified nanosized starch derivatives as excellent reinforcing agents for bionanocomposites. *Carbohydrate Polymers, 86*, 1244–1251.

Van Nieuwenhove, I., Salamon, A., Adam, S., Dubruel, P., Van Vlierberghe, S., & Peters, K. (2017). Gelatin-and starch-based hydrogels. Part B: In vitro mesenchymal stem cell behavior on the hydrogels. *Carbohydrate Polymers, 161*, 295–305.

Vermette, P., Griesser, H. J., Laroche, G., & Guidoin, R. (2001). *Biomedical applications of polyurethanes* (Vol. 6). Landes Bioscience Georgetown.

Wang, W., Xu, J., & Wang, A. (2011). A pH-, salt-and solvent-responsive carboxymethylcellulose-g-poly (sodium acrylate)/medical stone superabsorbent composite with enhanced swelling and responsive properties. *Express Polymer Letters, 5*.

Wu, G.-M., Chen, J., Huo, S.-P., Liu, G.-F., & Kong, Z.-W. (2014). Thermoset nanocomposites from two-component waterborne polyurethanes and cellulose whiskers. *Carbohydrate Polymers, 105*, 207–213.

Wu, Q., Henriksson, M., Liu, X., & Berglund, L. A. (2007). A high strength nanocomposite based on microcrystalline cellulose and polyurethane. *Biomacromolecules, 8*, 3687–3692.

Zhao, J., Zheng, K., Nan, J., Tang, C., Chen, Y., & Hu, Y. (2017). Synthesis and characterization of lignosulfonate-graft-poly (acrylic acid)/hydroxyethyl cellulose semi-interpenetrating hydrogels. *Reactive and Functional Polymers, 115*, 28–35.

Zhao, Q., Sun, G., Yan, K., Zhou, A., & Chen, Y. (2013). Novel bio-antifelting agent based on waterborne polyurethane and cellulose nanocrystals. *Carbohydrate Polymers, 91*, 169–174.

Zheng, J., Luo, J., Zhou, D., Shen, T., Li, H., Liang, L., et al. (2010). Preparation and properties of non-ionic polyurethane surfactants. *Colloids and Surfaces A: Physicochemical and Engineering Aspects, 363*, 16–21.

Applications of Waterborne Polyurethanes Foams

K. V. Narayana Saibaba

Abstract

Polyurethane foams are one of the versatile commercial materials ranging from solvent-based polyurethane foams to waterborne polyurethane foams (WPU). Waterborne polyurethanes are eco-friendly and exhibit excellent mechanical characteristics such as strength, flexibility, ease of moulding into required shapes, durability, lightweight and resistance to heat and moisture. Waterborne polyurethane foams have become an essential part of our daily life; we cannot even imagine the future without these foams. These foams are used in almost all industries viz. construction, automobile, footwear, leather, textile, pharmaceutical and furniture, etc. Foams are also used in packaging industries for safeguarding delicate and fragile valuable goods during storage and transportation in addition to their use in applications in filtration, insulation and fashion designing. Waterborne polyurethanes also contribute to sustainable energy through their excellent heat and water insulating characteristics. Recent advancements in technology allowing the preparation of special-purpose WPU foams provide tailor-made solutions to all applications. New eco-friendly composite materials being developed in this area have ample opportunities for the growth of the WPU market in the near future.

Keywords

Polyurethane foams • Waterborne polyurethane foam • WPU foams • PU foams • Polyurethanes

K. V. Narayana Saibaba (✉)
Department of Biotechnology, GITAM Institute of Technology, GITAM University, Visakhapatnam, Andhra Pradesh 530045, India
e-mail: skvn@gitam.edu

1 Introduction

Polyurethanes are one of the most useful multipurpose materials available in the market today. They have excellent properties like toughness, flexibility, durability and resistance to water, oils and most of the chemicals. These unique properties make them suitable for almost all applications; some important applications are footwear, insulation, elastic fibre, foams and medical devices. Polyurethanes offer elasticity together with toughness and are hence widely used in footwear industries. In the coating industry, polyurethanes are widely used since materials coated with polyurethanes are durable and cost of production is also affordable. Excellent bonding capacities of polyurethanes find their applications as adhesives in the adhesive industry. This versatile feature makes polyurethanes application in foam, wood, plastic and furniture industries. Polyurethane elastomers can be moulded easily into any shape and size and can be used to fill the gaps. They are lightweight, cheap but offer excellent resistance to environmental physical conditions and, hence, widely used in packaging and transportation industries. Polyurethanes are used as excellent insulation material in building and construction and refrigeration industries (Romaškevič et al., 2006).

The polymer foams are flexible, rigid, lightweight and easy to be moulded into desired shape and size. They are used for a wide range of applications such as construction, automobile, electrical, wood and furniture, leather and textile, packaging, etc. These foams are produced predominantly from polymers such as polyurethane (PU), polystyrene (PS), polyethylene (PE) and to some extent from other polymers such as polyvinyl chloride and polypropylene, etc. However, worlds foam production is mainly dominated by polyurethanes. The global polymer foam sales were more than 100 billion dollars in 2015 with consumption of 22 million tonnes, and it is expected to reach more than 25 million tonnes by the end of 2019 (Eaves, 2004; Gama et al., 2018; Lee & Ramesh, 2004; Rapra, 2018, 2019;

Inamuddin et al. (eds.), *Sustainable Production and Applications of Waterborne Polyurethanes*, Advances in Science, Technology & Innovation, https://doi.org/10.1007/978-3-030-72869-4_9

Titow, 2001). Polyurethanes are used in almost all applications of human life. It is estimated that consumption of PU alone was 60.5 billion USD in 2017 and is expected to be over 79 billion USD by the end of 2021 (Plasticsinsight.com, 2018). Table 1 lists the important developments and various applications of polyurethane foams.

2 Production of Polyurethanes

Polyurethane production was started with the use of the diisocyanate polyaddition process invented by Prof. Otto Bayer, who is considered the father of the polyurethanes industry. Polyurethanes were first synthesised as a replacement material for rubber during World War II. During World War II, PUs were used for the impregnation of paper, manufacturing of mustard gas resistant fabrics, high-gloss aeroplane finishes and as coatings for wood, metal and masonry. By the end of 1950, polyurethane foam production was started to replace the existing foams. These polyurethanes later substituted the other polymer materials in almost all fields due to their versatile properties. Polyurethanes have excellent elastomeric and thermoplastic properties. Polyurethanes are polymeric materials that exist in many forms (liquid state, solid-phase and ionomers in aqueous dispersions). PU

Table 1 Research developments and applications of polyurethanes

S.No.	Nature of work	Reference
1	Stability of waterborne polyurethanes for applications in coating industry	Lin et al. (2015)
2	Evaluation and characterization of waterborne polyurethane for tendon anti-adhesive applications	Hsu et al. (2018)
3	Colored waterborne polyurethane syntheses	Wang et al. (2017)
4	Study of expanded graphite-waterborne polyurethane foams of oil-water filtration applications	Vásquez et al. (2019)
5	Fabrication and characterization of waterborne polyurethane/silver nanocomposite foams	Zhao et al. (2019)
6	Biomedical applications of low density biodegradable shape memory foams	Singhal et al. (2014)
7	Production of green polyurethanes from bio based renewable sources	Konieczny and Loos (2019)
8	Synthesis of palm-oil based rigid polyurethane foams	Chuayjuljit et al. (2010)
9	Application of thermoset waterborne polyurethanes	Wu et al. (2014)
10	Studies on Waterborne polyurethane prepared from Waterborne polyurethane wood coatings based on rapeseed fatty acid methyl esters in wood coatings applications	Philipp and Eschig (2012)
11	Applications of polyurethane foams as biointegrable matrices for the preparation of scaffolds for bone reconstruction	Tanzi et al. (2003)
12	Modelling studies of polyurethane foams for automobile industry applications	Deng et al. (2003)
13	Studies on applicability of PUF for thermal insulation of buildings	Serrano et al. (2016)
14	Study of temperature bearing characteristics of PUF in flooring applications	Garrido et al. (2016)
15	Studies on applicability of waterborne polyurethanes in drug delivery systems	Omrani et al. (2017)
16	Application of waterborne polyurethanes in textile industry	Xinrong et al. (2014)
17	Study of biodegradable waterborne polyurethane foam as in sole in foot wear industry applications	Mukherjee et al. (2019)
18	Synthesis of castor oil based polyurethane adhesives for wood-to-wood bonding	Somani et al. (2003)
19	Study of influence of crosslinking agent on the properties of waterborne polyurethane adhesives	Yang et al. (2011)
20	Study on thermal characteristics of polyurethanes derived from bio based method	Zhang and Kessler (2015)

materials can be synthesised as either rigid or flexible. PUs are available in many forms and find a broad range of applications such as chairs, desks, folders, footwear, cushions, insulation materials, etc.

Production of PU is carried out in two steps. First, an intermediate polymer, having medium molecular weight, is produced from the reaction between isocyanate and polyol. This intermediate polymer is called a prepolymer, which is in high viscosity liquid state. Later, it reacted with a diol chain extender to produce the final polyurethane having high molecular weight. Different types of polyurethanes have been prepared in suitable physical conditions by adding additives and catalysts (Honarkar, 2018).

Polyurethane is a polymer composed of organic units joined together by urethane (–NH–CO–O) links. The reaction of alcohol with diols or polyols and isocyanates (di-isocyanate or polyisocyanate) produces polyurethanes. Petrochemical-based diols such as ethylene glycol and propylene oxide are generally used for this purpose (Tu et al., 2007), but vegetable oils such as soybean and castor oils can also be used for the production of polyurethanes. Urethane link in polyurethanes imparts better hardness, weather and chemical resistance, scratch resistance, protection of material against moisture, biological, chemical, radiation deterioration and glossy properties. Due to these excellent wide ranges of properties, polyurethanes are being used in a wide variety of applications (Chattopadhyay & Raju, 2007; Miao et al., 2014; Panda et al., 2018; Vincent & Natarajan, 2014), such as automobile seating, insulations, packaging, construction, foot ware, plastic furniture, military and automotive industries, etc. The consumption of polyurethanes is increasing rapidly and is ranked 5th in the production of plastic in terms of volume (Tu et al., 2007). Polyurethanes are also used in the biomedical sector for applications such as nasogastric catheters, peritoneal dialysis, infusion pumps, implanted pacemaker part, etc. (Panda et al., 2018).

3 Polyurethane Foams (PUF)

Polyurethanes are versatile engineering materials that are available in many forms such as foams, coatings, elastomers, hard plastic, flexible plastic, cast and injection moulding components of agriculture, etc. Among these, foam is a widely used form of polyurethane, exceeding over 75 per cent of total polyurethane use (Zhou et al., 2015). Polyurethane foams (PUF) are prepared by reacting polyisocyanates with a polyol in the presence of a blowing agent, surfactant and a catalyst without external heating. Polyurethane formation and gas generation occur simultaneously in PU foam preparation.

3.1 Solvent-Based Polyurethane Foams and Waterborne Polyurethane Foams

Polyurethane foams are generally either solvent-based or water-based; but bio-based polyurethanes and other types are also available these days. In general, solvent-based polyurethanes are produced from petroleum-based feedstocks, and these are used widely because of their flexibility, abrasion resistance and stability against water, certain chemicals and common solvents. However, their disposal creates a lot of environmental problems due to the release of volatile toxic organic compounds. In some cases, they also release a smallamount of free isocyanate into the environment. Due to this reason, applications of water-based dispersion polyurethanes are gaining popularity. Waterborne polyurethanes (WPU) are environmentally friendly, and hence, usage of these waterborne materials is increasing throughout the globe in textile, adhesive, metal and wood industries (El-Wahab et al., 2015; Ljubic et al., 2015; Massalha et al., 2015; Qu et al., 2014; Sheikhy et al., 2013; Wan & Chen, 2017). Waterborne polyurethanes are made of up poly ϵ caprolactone and polylactic acid. Waterborne polyurethane is extensively used in various fibres, adhesives, emulsions, metal primers, additives of paint, defoaming agents, colour pigments such as inks and textile dyes, etc.

4 Applications of Waterborne Polyurethane Foams

Waterborne polyurethane foams (WPUFs) have excellent characteristics such as lightweight, low odour and resiliency, etc. They provide excellent cushioning characteristics and can be easily moulded to fit the shape of the required application. They have excellent applications in packaging due to their cushioning properties. WPU foams are used in the packaging industry mainly for protecting delicate and highly fragile objects such as glass, ceramic during transportation and storage. WPU foams are used in medical applications where high resistance to abrasion is required.

The waterborne polyurethane materials show good resistance to temperature, flexibility and mechanical strength. However, they are prone to microcrack development due to physical damage; later, these micro-cracks extent to larger areas resulting in a decrease in the lifespan of materials (Koh et al., 2014). Self-healing of waterborne polyurethanes overcomes this disadvantage and shows good dispersion stability, thermal healing stability and good lifespan (Wan & Chen, 2017). Wide varieties of waterborne polyurethane foams are available for different applications such as construction, footwear, leather, refrigeration, leather, fibre and garment, adhesive and coatings, etc.

4.1 Reticulated Waterborne Polyurethanes (RWPU) Applications

Filtration foams made of waterborne polyurethane foams are called reticulated waterborne polyurethane (RWPU) filters. Waterborne polyurethane foams used in water filtration applications are lightweight, have low odour and are highly resistant to mildew. These foams have good chemical resistance and, hence, are widely used for filtration applications and other applications such as lawnmower engines and wiping machines used in houses and offices. RWPU foams have enormous applications due to low density, flexibility, porous, high tensile, elongation and tear characteristics. They are used as filters, sound absorbers in acoustic chambers, microphones and facemasks and wiping pads, etc. RWPU filters are made in such a way that they allow only fluid and restrict the passage of dust or any solid particles. RWPU foams can be used for both air and liquid filtration. They are available in different densities and pour sizes ranging from 10 to 100 pores per inch. In general, RWPU foams consist of uniform cells in a dodecahedron shape; cell walls are entirely made up of waterborne polyurethane material. These cells obstruct the passage of dust particles due to their high porosity and large surface area. Filter foams are depth loading type filters in which particles are concentrated inside the cells. They are homogeneous and have uniform structural porosity, uniform and can withstand high pressures (Fig. 1). They are widely used in furnaces, air conditioners, automobiles, air cleaners and other applications. RWPU foams are compressible and are used as ink rollers, blood filters, etc. Some of the examples of filter foams are pictorially shown in Fig. 1.

Filtration foams are also used for cleaning water ponds so that clean water can be obtained after the removal of odour and other debris. These filters can be used to sieve out the silt and particles, dirt and other waste materials present in water. Filtration foams are classified into three categories: mechanical, biological and chemical foam media depending upon their use. Mechanical foams are used for general water cleaning for the removal of dirt, silt and other solid particles. Commonly used commercial general-purpose cleaning foams are shown in Fig. 2. They have high abrasion characteristics and are available as sponges and pads. Biological foams are used as media for the growth of useful bacteria. In contrast, chemical foams are designed for the removal of chemicals or toxic components from the water. The polyurethane foam filter combined with activated carbon during the polyurethane foam-forming process, i.e. reaction of isocyanates and polyesters in the presence of activated carbon exhibits a very high adsorption capacity. These filters are used in various applications such as ion-exchange, adsorption of odour-causing gases and water purification, etc.

4.2 Construction Applications

Polyurethane foams are important thermoplastics, and their market demand has been increased primarily in the construction, automotive and packaging industries. Polyurethane foams are globally widely used in building and construction purposes because of their excellent insulation properties, strength, energy-saving characteristics, aesthetic appearance and ambience improvement to the construction. The USA is one of the major polyurethane foam consumption countries, and market demand for these foams has been increased from 1214.11 million USD in the year 2012 to 1493 million USD in 2016 (Plasticsinsight.com, 2018). Environmental concerns activated the shift from the usage of solvent-based polyurethane to water-based polyurethane in all applications. Waterborne polyurethane foams are used in the construction field for a wide range of applications such as filling and sealing, thermal insulation, acoustical insulation, etc. These

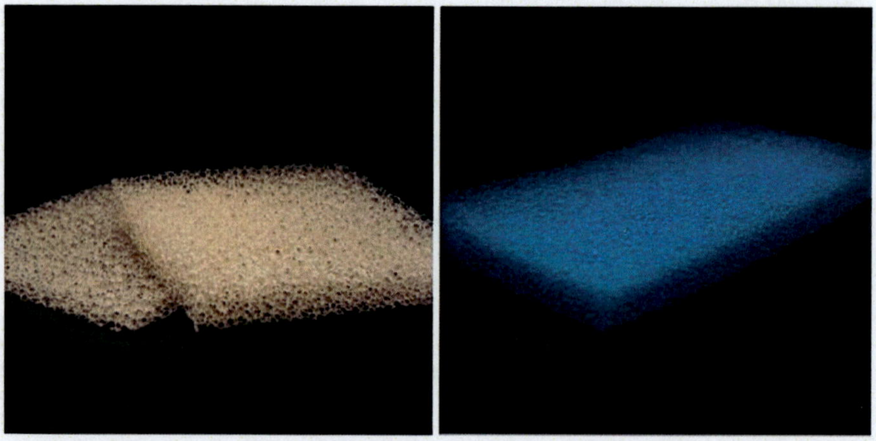

Fig. 1 Foams used for filtration applications

Fig. 2 Foams used in general cleaning purposes

Fig. 3 Foams applied in wall and roof insulation

are environmentally friendly, low cost and does not contain any toxic organic compounds. Figure 3 presents the waterborne polyurethane foam used for wall and roof insulation. Special formulations are used for making these foams for special use in construction applications; foam expands and cures by moisture in the air. These foams can be used to fill the gaps between windows, frames and doors during installation. These are used for bonding and mounting in building construction, used for fire retardant applications. They possess quick setting formulations, flexible and do not crack or dry out; hence, widely used for filling, sealing gaps, joints, cracks and other openings. Nowadays, polyurethane spray foam is gaining extended applications in the thermal insulation of buildings and constructions. Polyurethane spray foam provides high reflectivity, thereby improves the thermal insulation. These foams provide excellent adhesive properties to any type of roof because of being thermally stable over a wide range of temperatures, resistant to detergents, oils and other domestic chemicals.

4.3 Waterproofing Applications

Hydrophobic water-based polyurethane foams are used for waterproofing applications in building construction. These types of foams have large expansion capacity, high density and quick setting properties. They are flexible, uniform and foaming can be trimmed or cut into the required shape easily. They do not absorb moisture because of their hydrophobic nature and can be used for all waterproofing applications. Waterproof and fireproof WPU foams are available at an economical prices and are also used as insulating materials for electrical and water pipes used in air conditioners. Roof surface sprayed with waterborne polyurethane foams along with urethane, acrylic, or silicone rubber coatings improves the quality of insulation against heat and cold, water and also improves the life of a roof.

Spray polyurethane foam (SPF) is widely used for waterproofing applications of the roof. SPF acts as both waterproofing material as well as insulation material. In closed-cell SPF cells, pores are closed and not connected to each other; hence, roofs covered with SPF remain dry when exposed to water. During the spraying process, two liquids present in the sprayer get chemically cross-linked and react instantaneously. When SPF liquid hits the roof, the liquid expands to 20–30 times its original volume in few seconds and dries immediately. When sprayed on the roof, it forms a single monolithic barrier that seals the entire roof surface from edge to edge with no seam or joints. The overall thickness of SPF can be around one to two-inch thickness based on the roof type.

4.4 Textile/Fabric/Carpet Industry Applications

The textile industry is one of the major markets for water-based polyurethane foam consumption. Waterborne polyurethane foams find their applications in the textile industry for textile lamination, integral carpet manufacturing and rebounded foam manufacturing. WPU foams are combined with textile or garments for higher durability, flexibility and lamination against water and fire. Textile lamination is achieved either by solution adhesives or by the application of heat. WPU foams are glued to the textile with water-based polyurethane adhesives, which on evaporation from the adherent surfaces after application and become hard to give bonding. Fabrics made with water-based polyurethane foam adhesives, or thermoplastic polyurethane adhesives are very soft and have good bond strength, water resistance and easy drying. Textile garments layered with waterborne polyurethane foam show good permeability of moisture, conduction of heat, adequate resistance to the stick and abrasion during washing of clothes. These are widely used for a seamless zipper, seamless pockets, tops, innerwear of men and women garments, etc.

On the application of heat, polyurethane foam melts and is nipped between two textile roles to get the textile lamination. Low-density waterborne polyurethane foams are used as pre-coat in carpet making. The amount of foam in coating depends on the type of application and can be adjusted to get the uniform coating. WPU foam is used in a carpet for imparting abrasion resistance, improving load-bearing characteristics and improving walking comport. Integral carpet making involves the binding of WPU foam with nylon, wool, or polypropylene tufts through urethane adhesives to get proper cushioning of the carpet. Rebonded foam is made with scrap or recycled WPU foams and is used as carpet underlay (Szycher, 1999).

4.5 Leather Industry Applications

In recent years, demand for comfort accessories has increased, which created the rapid growth in the usage of artificial leather prepared from WPU foams. Artificial leather coated with a water-based polyurethane foam layer is used widely in automobile seat manufacturing. Artificial leather is cheaper than natural leather; however, the porosity of natural leather is less and not as comfortable as artificial leather. Artificial leather layered with the WPU foam has increased porosity, better cushioning, improved heat dissipation and enhanced moisture permeability of automobile seats lowering the temperature and preventing wrinkle formation. WPU foam provides a pleasant and comfortable driving experience and aesthetic environment in automobiles. These foams are added to the leather cloths for additional protection against water and heat. WPU coated leather textiles are used in fashion clothing, shoes, ladies handbags, T-shirts and leather fabrics, tarpaulins and other fabrics used for industrial purposes. WPU foams provide an unlimited range of opportunities in the design of fashion and footwear industries. Global use of WPU foam coated fabric and synthetic leather has increased significantly in the form of bags, jackets, footwear, and other fashionable outwear.

4.6 Foot Wear Industry Applications

Polyurethane foams are comfortable, affordable and long-lasting in good quality footwear. Foot soles made of polyurethane foam are lightweight and have excellent mechanical properties. WPU foams are extensively used in trekking and sport's shoes (Fig. 4). The application of polyurethanes in footwear is increasing due to long-lasting performance and cost-effectiveness. Mainly three types of polyurethane foams (i.e. soft foam, rigid foam and microcellular foams) are used in the footwear industry. Footwear consists of different parts and components such as upper and lining, sole, insole, in-sock, adhesives, etc. In-sock is generally placed above the inside sole to provide cushioning to the foot, to absorb sweat during usage. In-sock is generally fixed in the footwear either permanently with adhesive or replaceable manner, and it is generally made of either leather, fabric, or water-based polyurethane foam coated fabric. Waterborne polyurethane gives flexibility, compressibility and elastic properties to the sole (Mishra, 2019). Waterborne polyurethane foams are extensively used in the upper part of the shoe for different designs. Textile uppers are also being widely used in footwear for lightweight and good appearance in

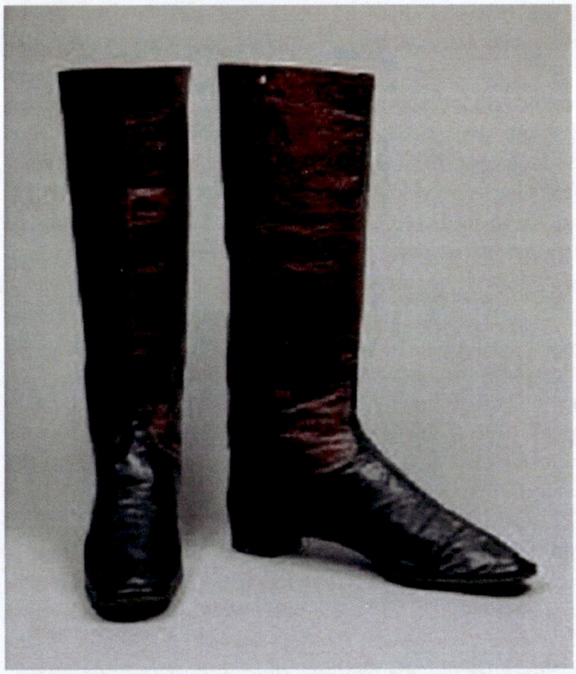

Fig. 4 Foam applied in fashion footwear

children shoes, canvas shoes and sports shoes. Textile membranes glued with WPU foam are being soft and resemble leather, used for this purpose. WPU foam laminated linen is used for providing a high degree of water resistance to shoes. In general, these materials have micropores in the range 1–2 μm and obstruct the entry of water molecules to pass through. However, these pores allow the permeation of water vapours and hence offer wet resistance.

4.7 Automobile Industrial Applications

Flexible waterborne polyurethane foams have a significant global market. Automobile industries are one of the major consumers of WPU foams due to their good cushioning properties and mechanical strength and they are flexible to mould into different shapes easily, provides good comfort and protection, and energy conservation. WPU foams are lightweight; therefore, fuel consumption is greatly reduced. Semi-flexible foams are used in the construction of the car dashboard, armrests, headrests and door liners, etc. Figure 5 shows some of the applications of WPU foams in the automobile industry. WPU foams are used in automobile seat making because of their high permeability towards moisture and excellent cushioning properties. Special formulations of WPU foams are used for the insulation of buses to protect passengers from noise pollution. These foams form a seamless airtight sealer layer inside the bus compartment and reduce heat loss, noise and vibration to comfort passengers during travelling.

Fig. 5 Foam applied in automobile applications

4.8 Furniture Industry Applications

Waterborne polyurethane foams are slowly replacing solvent-based polyurethane foams in the manufacturing of mattresses, pillows and cushions in furniture industries due to environmental concerns. These are widely used in furniture items due to their compressibility, flexibility, appearance, lightness, easy moulding, good resistance towards moisture and chemicals as well as appropriate mechanical strength (Biron, 2004). Flexible foams are used as filling components in wood openings, decorative furnisher products such as beads, feathers, etc. WPU foams are available in different types, and this versatility helps the designer for creating novel products. These foams can be moulded in accordance with body contours to give extra comfort. Memory foams developed so far adapts (Fig. 6) to the sleeping person's body to provide good sleep. WPU foams used in designing hospital beds help to prevent bedsores in patients who are bound to the bed for a longer period of time.

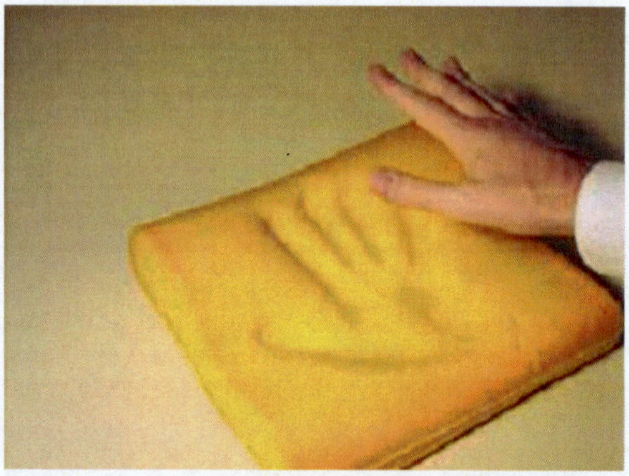

Fig. 6 Memory foam used for comport sleeping

4.9 Refrigeration Applications

Polyurethanes are one of the widely used materials in home and industrial appliances. The most common use of PUF in the cooling industry is as insulation material in refrigerators and freezers. Rigid PUF are cost-effective and have very good insulating properties, thereby used to reduce energy consumption in refrigerators. In modern refrigerators, the PUF layer is placed between the outer metal sheet casing and inner polystyrene liner to impart structural strength and insulation properties. Being a good adhesive, PUF fit properly between inner and outer walls and help to prevent heat transfer between the system and surroundings.

Polyurethane foam is one of the major contributors to eco-friendly sustainability through less energy consumption in cooling appliances such as refrigerators, freezers and air conditioners, etc. Polyurethane foams are the most widely used insulating material in the world for refrigerators. Rigid waterborne polyurethane foams used as the insulating material in refrigeration appliances provide high mechanical strength and can be moulded into the required shape as per the design. The low thermal conductivity of these foams provides good insulation for heat. Insulation is the key factor for the energy consumption of appliances for maintaining the required temperature in refrigerating and air conditioner systems. It is estimated that approximately 50% of the food would have been rotten without the use of insulating materials in refrigerating cooling systems. In this way, WPU foams are contributing to the universe in maintaining sustainable energy and food resources for the future generation.

4.10 Separation Process Applications

The use of polyurethane foams as a sorbent in the solid-phase extraction process has been in use since the 1970s. PUF has been in use for the separation of organic pollutants from water since 1972s. Later due to their high adsorption and regenerative properties, they have been used for the removal of many pollutants such as metal ions, dyes, phenols, etc. Because of concerns about environmental pollution, waterborne polyurethane (WPU) has been replacing the conventional binders in preparing active polyurethane foams for adsorption applications. Studies on adsorption proved that waterborne polyurethane foams modified with various components give higher removal efficiency due to higher carbon content, enhanced porosity and improved hydrophilicity of foam. These observations have been further supported by scanning electron microscopy and wettability tests (Krupadam et al., 2010; Mao et al., 2013; Trang & Zenitova, 2019). Dyes containing effluents discharged into the water bodies harm marine animals and human beings and disturb the entire ecosystem.

Treatment of dye-containing effluents is generally carried out using the adsorption technique. Activated carbon is widely used as an adsorbent for dye removal; however, its high procurement and regenerative costs restrict its usage in developing countries. Studies showed that polyurethane foams coated with chitosan have been very effective in the removal of dyes. These adsorbents are available at a low cost; however, the adsorption capacity is high. These adsorbents also have a good recycling capacity. Modified polyurethane foams have been widely used in the removal of various pollutants from effluent using the adsorption technique. They have been used for the separation of mercury, dyes, metal ions, etc. (Machado et al., 2017).

Waterborne polyurethane foams have been used for the separation of oil from oil-in-water emulsions. WPU foams can be combined with graphite granules or calcium carbonate for preparing highly porous interconnected networks with significant improvement in pore sizes. These foams can be used as 3D filters for the removal of surfactant-free oil-in-water mixtures through gravity. These foams show significant oil rejection ratios and can be used for several filtration techniques with low cost and higher efficiency in oil–water systems and other emulsions (Vásquez et al., 2019).

4.11 Medical and Biotechnological Applications

Polyurethanes have been in use in the medical field as breast implants for the past 60 years due to their unique properties such as biocompatibility, biostability, biodegradability and good mechanical strength. Due to these properties, they find various applications in the medical field. These properties can be achieved by the controlled addition of appropriate monomers during the polyurethane syntheses. Polyurethane foam-based meniscal scaffold is used to treat patients suffering from partial meniscal loss. Arthroscopic polyurethane meniscal scaffold implantation is a safe and effective procedure for treating patients suffering from partial meniscal loss (Filardo et al., 2012). Studies reported that polyurethane foam-based materials are still a preferred option in the plastic surgery industry for the reconstruction of the breast. Polyurethanes are also used as membranes for the wound dressing application process. These membranes were prepared by adding antimicrobial compounds to the epoxy-terminated polyurethane prepolymers. Membranes prepared with polyurethanes have good hydrophilicity and water vapour transition rate and thereby prevent the accumulation of exudates and minimises the inflammation in the wounded area (Yari et al., 2012).

Polyurethane materials can be used in safe and effective nanotherapeutic systems. Biocompatible polyurethane prepared from polyoxyethylene castor oil derivative surfactants is in use as cell carriers for endovascular applications (Morral-Ruíz et al., 2014). Nanoparticles synthesised from polyurethane foams show good mechanical properties and biocompatibility and are used in drug delivery systems. Polyurethane composite materials are used as a bone replacement due to higher viscosity and good lubricating properties compared to other polymers. One of the important applications of polyurethanes is in cardiovascular applications; in fact, polyurethanes were used in the first artificial heart. Polyurethanes are used for various cardiovascular applications due to their good physicochemical, mechanical and biocompatibility properties. Polyurethanes have good durability that is the primary feature required for cardiovascular applications such as the development of prosthetic heart valves. Biodegradable polyurethanes are used in cardiac tissue engineering (Boffito et al., 2016).

Polyurethanes are versatile components and have wide applications in biochemical and biotechnological fields (Romaškevič et al., 2006). These are highly stable, and the biocompatibility of polyurethane foam made them the first choice for various biosensor applications, cell and enzyme immobilisations. Enzymes are called biocatalysts since they activate the rate of biochemical reactions. They are widely used in biochemical, biomedical, pharmaceutical and food industries due to their catalytic efficiency, non-toxic nature and high specificity. WPU foams are considered suitable carriers for enzyme immobilisation due to their easy control of pore size and low price. Waterborne polyurethane foams have been used as cell carriers because they are inert and have strong mechanical characteristics. Reticulated polyurethane foam filters have high porosity and freely suspended cells and hence act as very good cell growth carriers for mycelia or flocculent microorganisms, plant and animal cell cultures.

4.12 Foam Adhesive Applications

Water-based polyurethane adhesives have good eco-friendly properties such as low odour and can be easily cleaned so that they are very safe for indoor applications. These foams are flexible, having low shrinkage, are non-toxic and create a very strong bond. All these advantages make these adhesives very important for building applications. Waterborne polyurethane foam adhesives are used for many applications such as structural applications, mechanical and electrical installations, carpentry, metalwork, etc. WPU foam adhesives are used for steel reinforcement in new and repair works in structural work. These are also used for threaded rods, bolts and other fastening systems. WPU foam adhesives are used as anchoring supports for ducting and other equipment in mechanical and electrical installations. These are used for fixing metal railing, windows and door frames and other supports used in metalwork. These foams are non-corrosive and do not require making any holes for nailing or screwing purposes. WPU foam adhesives are used in concrete, ceramic, brick and stone reinforcement works, plasterboards and woodworks.

4.13 Other Applications

Waterborne polyurethanes are primarily used as thermal, water and sound insulating materials due to their high flexibility, compressibility and good insulating properties. Because of excellent mechanical properties, stability and biocompatibility, they are also used widely in biomedical applications. WPU foams are also used in electrical applications as electrical conductors as well as electrical insulators. WPU foam shows good dielectric strength under different degrees of humidity and temperatures. A wide variety of carbon nanotubes are mixed with WPU foams for the improvement of the physical, thermal and electrical properties of WPU foams. A WPU/Fe_3O_4 nanocomposite prepared with oleic acid modified nanoparticles showed uniform dispersion in the WPU and improved thermal, electric and magnetic properties. It could be used for potential applications in microwave absorption (Zhang et al., 2013). Pipes insulated with rigid polyurethane foams prepared from the special blends of polyols are used for city heating and cooling requirements. Foams used for this purpose have low thermal conductivity so that heat losses can be minimised. The life of these foams is as long as 30 years, and they are highly stable and have high-temperature resistance and withstand up to temperatures of 140 °C. They are good heat insulators, therefore, used for heat insulation. Fragile and delicate materials are wrapped with WPU foams during packaging, transportation and storage. Other applications of WPU foams include sponges making, interior decorative items, insulation boards, mobile phone cases and ID card lamination, etc. WPU coatings are used to paint home appliances to achieve an aesthetic appearance to customers and protect from rust and heat.

5 Strategies Towards Sustainable Production and Applications

Waterborne polyurethane foam (WPUF) manufacturing technology is well established and widely used in every field. However, to get sustainable production and applications, the search for new raw materials and application areas need to be continued to overcome unforeseen problems in the future. A new technology such as 3D printing has emerged as a novel technique, and it may replace the market if tailored-made products and applications are not developed in WPUF technology. Though polyurethane foams have excellent structural and mechanical properties, researches are directed towards the invention of new composite materials for the improvement of mechanical properties for tailored-made applications. Different types of fillers, such as cellulose, date palm particles, eggshell wastes, etc., can be used to improve the structural and mechanical properties of waterborne polyurethane foams (Gandini & Belgacem, 2002; Gandini et al., 2010; Lammers et al., 1993).

5.1 Production of Polyols Through Renewable Resources

Most of the polyols used for the production of polyurethane foams are petroleum-based; due to environmental concerns, new polyols derived from renewable resources such as bio-related materials are gaining importance. The use of renewable resources provides the sustainable production of polyurethane foams as a result of this green biotechnology market is expected to grow in future. Renewable polyols are the obvious alternative materials for the sustainability of polyurethane production and applications. Already companies have started the production of polyols through the oxypropylation or the acid liquefaction of biomass feedstocks in the polyurethane foam industry. They could be able to produce foams with good density, morphology and thermal characteristics (Aniceto et al., 2012; Niu et al., 2011). Vegetable oils are abundantly available and can be used for the production of polyols. Vegetable oils are soluble in most of the industrial solvents, biodegradable and eco-friendly and capable of producing highly flexible foams with good structural and mechanical properties. Various types of vegetable oils such as soybean oil, castor oil, palm oil, rapeseed oil, etc., have been used for the production of polyurethane foams (Badri, 2012; Fridrihsone et al., 2013; Guo et al., 1999; Petrovi'c & Cvetkovi'c, 2012; Tamami et al., 2004).

5.2 Production of Isocyanates Through Renewable Resources

Isocyanates being petroleum-based, are toxic to the environment. Therefore, studies have been carried out for production of isocyanates from the non-petroleum-based feedstocks such as amino acids. Non-isocyanate polyurethanes are being produced; they possess good mechanical and thermal properties, significant resistance towards chemicals, high porosity and low permeability. They have been used in foam, coating and sealant applications (Guan et al., 2011).

5.3 Recycling of Foams

The widespread use of polyurethanes has increased their production on one side, and this has led to the increase of polyurethane foam wastes on another side. Polyurethane foams are non-biodegradable and are generally disposed of by landfilling or by incineration processes. However, these methods may pose an environmental threat to the natural ecosystem, and hence, proper recycling is required for the sustainable production of WPU foams. In addition to environmental benefits, recycling also offers economic benefits by reducing the cost of production of WPU foams. In general, physical and chemical recycling methods are used for this purpose. In physical recycling, waste scraps are recycled by using adhesives, binders and regrinding, moulding methods, whereas chemical treatment is used for the modification of waste foams in the chemical recycling method. Through this method, polyurethane foams are depolymerised into oligomers and even smaller molecules. These smaller components can be further used to produce WPU foams.

6 Summary

Despite the excellent properties of solvent-based polyurethanes, environmental concerns regarding the release of volatile toxic components into the environment motivated the use of waterborne polyurethanes in place of solvent-based polyurethanes. To maintain the sustainability of waterborne polyurethane (WPU) production and applications, it is required to produce different ranges of WPU foams that can be used directly to satisfy the needs of the consumer. It is required to come up with new strategies and tailor-made solutions for the wide applications of WPU foams. The production of WPU foams through renewable resources should be increased to the extent of petroleum-based feedstocks to maintain the eco-friendly system. New waterborne polyurethanes need to be developed and produced by novel renewable methods to take advantage of potential new opportunities available in future.

References

Aniceto, J. P. S., Portuga, I., & Silva, C. M. (2012). Biomass-based polyols through oxypropylation reaction. *ChemSusChem, 5*(8), 1358–1368.

Badri, K. H. (2012). *Biobased polyurethane from palm kernel oil-based polyol*. Intech.

Biron, M. (2004). The plastics industry: Economic overview. *Thermosets and Composites*.

Boffito, M., Sartori, S., Mattu, C., & Ciardelli, G. (2016). Polyurethanes for cardiac applications. *Advances in Polyurethane Biomaterials* 387–416.

Chattopadhyay, D. K., & Raju, K. V. S. N. (2007). Structural engineering of polyurethane coatings for high performance applications. *Progress in Polymer Science, 32*, 352–418.

Chuayjuljit, S., Maungchareon, A., & Saravari, O. (2010). Preparation and properties of palm oil-based rigid polyurethane nanocomposite foams. *Journal of Reinforced Plastics and Composites, 29*, 218–225. https://doi.org/10.1177/0731684408096949

Deng, R., Davies, P., & Bajaj, A. (2003). Flexible polyurethane foam modelling and identification of viscoelastic parameters for automotive seating applications. *Journal of Sound and Vibration, 262*, 391–417.

Eaves, D. (2004). *Handbook of Polymer Foams* (p. 289). Rapra Technology Ltd.

El-Wahab, H. A., El-Fattah, M. A., Ayman, H. A., Elhenawy, A. A., & Alian, N. A. (2015). Synthesis and characterisation of some arylhydrazone ligand and its metal complexes and their potential application as flame retardant and antimicrobial additives in polyurethane for surface coating. *Journal of Organometallic Chemistry, 791*, 99–106.

Filardo, G., Zaffagnini, S., Di Martino, A., Di Matteo, B., Muccioli, G. M. M., Busacca, M., & Marcacci, M. (2012). Biodegradable polyurethane meniscal scaffold for isolated partial lesions or as combined procedure for knees with multiple comorbidities: Clinical results at 2 years. *Knee Surgery, Sport Traumatology Arthroscopy, 22*, 128–134. https://doi.org/10.1007/s00167-012-2328-4

Fridrihsone, A., Stirna, U., Lazdin, B., Misane, M., & Vilsone, D. (2013). Characterisation of polyurethane networks structure and properties based on rapeseed oil derived polyol. *European Polymer Journal, 49*, 1204–1214.

Gama, N. V., Ferreira, A., & Barros-Timmons, A. (2018). Polyurethane foams: Past, present, and future materials *11*, 1841.

Gandini, A., Pinto, C., Costa, J. J., & Pascoal, N. C. (2010). Process for the production of liquid polios of renewable origin by the liquefaction of agro-forestry and agro-food biomass. WO Patent 2010020903 A1, February 25.

Gandini, A., & Belgacem, M. N. (2002). Recent contributions to the preparation of polymers derived from renewable resources. *Journal of Polymers and the Environment, 10*, 105–114.

Garrido, M., Correia, J. R., & Keller, T. (2016). Effect of service temperature on the shear creep response of rigid polyurethane foam used in composite sandwich floor panels. *Construction Building and Materials, 118*, 235–244.

Guan, J., Song, Y., Lin, Y., Yin, X., Zuo, M., Zhao, Y., Tao, X., & Zheng, Q. (2011). Progress in study of non-isocyanate polyurethane. *Industrial and Engineering Chemistry Research, 50*, 6517–6527.

Guo, A., Javni, I., & Petrovic, Z. (1999). Rigid polyurethane foams based on soybean oil. *Journal of Applied Polymer Science, 77*, 467–473.

Honarkar, H. (2018). Waterborne polyurethanes: A review. *Journal of Dispersion Science and Technology, 39*(4), 507–516.

Hsu, S. H., Dai, L. G., Hung, Y. M., & Dai, N. T. (2018). Evaluation and characterisation of waterborne biodegradable polyurethane films for the prevention of tendon postoperative adhesion. *International Journal of Nanomedicine, 13*, 5485–5497. https://doi.org/10.2147/IJN.S169825

Koh, E., Kim, N. K., Shin, J., & Kim, Y. M. (2014). Polyurethane microcapsules for self-healing paint coatings. *RSC Advances, 4*, 16214–16223.

Konieczny, J., & Loos, K. (2019). Green polyurethanes from renewable isocyanates and biobased white dextrins. *Polymers (basel), 11*(2), 256. https://doi.org/10.3390/polym11020256

Krupadam, R. J., Khan, M. S., & Das, S. (2010). Adsorption of fluoride from water by surface-functionalised polyurethane foam. *Water Science and Technology, 62*(4), 759–765.

Lammers, G., Stamhuis, E. J., & Beenackers, A. A. C. M. (1993). Kinetics of the hydroxypropylation of potato starch in aqueous solution. *Industrial and Engineering Chemistry Research, 32*, 835–842.

Lee, S. T., & Ramesh, N. S. (2004). *Polymeric foams: Mechanisms and materials*. CRC Press.

Lin, Y. Y., Hung, K. C., & Hsu, S. H. (2015). Stability of biodegradable waterborne polyurethane films in buffered saline solutions. *Biointerphases, 10*(3), 031006. https://doi.org/10.1116/1.4929357

Ljubic, D., Srinivasan, M., Szoszkiewicz, R., Javni, I., & Petrovic, Z. S. (2015). Surface modified graphene/single-pase polyurethane elastomers with improved thermo-mechanical and dielectric properties. *European Polymer Journal, 70*, 55–65.

Machado, G. S. N., Centenaro, B. R., AlexsandraValério, A. A., Souza, U., Silva, A., Oliveira, J. V., & Déborade, . (2017). Oliveira application of polyurethane foam chitosan-coated as a low-cost adsorbent in the effluent treatment. *Journal of Water Process Engineering, 20*, 201–206.

Mao, N., Zhou, L., Ye, Z., Zheng, W., Peng, L., & Li, Y. (2013). Preparation of waterborne polyurethane foam with active carbon and its adsorption for phenol in aqueous solution. *Journal of Environmental Engineering, 139*(8), 1070–1079.

Massalha, N., Brenner, A., Sheindorf, C., Haimov, Y., & Sabban, I. (2015). Enriching composite hydrophilic polyurethane foams with PAC to enhance adsorption of phenol from aqueous solutions. *Chemical Engineering Journal, 280*, 283–292.

Miao, S., Wang, P., Su, Z., & Zhang, S. (2014). Vegetable-oil-based polymers as future polymeric biomaterials. *Acta Biomaterialia, 10* (4), 1692–1704.

Mishra, M. (2019). *Encyclopedia of polymer applications*, (Vol. 3). CRC Press.

Morral-Ruíz, G., Melgar-Lesmes, P., García, M. L., Solans, C., & Garcia-Celma, M. L. (2014). Polyurethane and polyurea nanoparticles based on polyoxyethylene castor oil derivative surfactant suitable for endovascular applications. *International Journal of Pharmaceutics, 461*, 1–13. https://doi.org/10.1016/j.ijpharm.2013.11.026

Mukherjee, M., Gurusamy-Thangavelu, S. A., Chelike, D. K., Alagumalai, A., Das, B. N., & Mandal, A. B. (2019). Biodegradable polyurethane foam as shoe insole to reduce footwear waste: Optimisation by morphological physicochemical and mechanical properties. *Applied Surface Science, 499*, 143966. https://doi.org/10.1016/j.apsusc.2019.143966

Niu, M., Zhao, G., & Alma, M. H. (2011). Polycondensation reaction and its mechanism during lignocellulosic liquefaction by an acid catalyst: A review. *Forestry Studies in China, 13*, 71–79.

Omrani, I., Babanejad, N., Shendi, H. K., & Nabid, M. R. (2017). Fully glutathione degradable waterborne polyurethane nanocarriers: Preparation, redox-sensitivity, and triggered intracellular drug release. *Materials Science and Engineering, C: Materials for Biological Applications, 70*(1), 607–616. https://doi.org/10.1016/j.msec.2016.09.036

Panda, S. S., Panda, B. P., Nayak, S. K., & Mohanty, S. (2018). A review on waterborne thermosetting polyurethane coatings based on castor oil: Synthesis, characterisation, and application 500–522.

Petrovi'c, Z. S., & Cvetkovi'c, I. (2012). Vegetable oil-based hyperbranchedpolyols in flexible foams. *Contemporary Materials, 1*, 63–71.

Philipp, C., & Eschig, S. (2012). Waterborne polyurethane wood coatings based on rapeseed fatty acid methyl esters. *Progress in Organic Coatings, 74*, 705–711. https://doi.org/10.1016/j.porgcoat.2011.09.028

Plasticsinsight.com. (2018). *Polyurethane production, pricing and market demand*. Available online: https://www.plasticsinsight.com/resin-intelligence/resin-prices/polyurethane/. Accessed on September 13, 2018.

Qu, R. J., Gao, J. J., Tang, B., Ma, Q. L., Qu, B. H., & Sun, C. M. (2014). Preparation and property of polyurethane/nanosilver complex fibers. *Applied Surface Science, 294*, 81–88.

Rapra, S. (2018). *High-performance polymer foams to 2021—Market reports*. SmitherRapra.

Rapra, S. (2019). *Polymer foams market forecast to 2019*. SmithersRapra.

Romaškevič, T., Budrienė, S., Pielichowski, K., & Pielichowski, J. (2006). Application of polyurethane-based materials for immobilisation of enzymes and cells: a review. *CHEMIJA, 17*, 74–89.

Serrano, A. M., Borreguero, I., Garrido, J. F., & Rodríguez and M. Carmona, . (2016). Reducing heat loss through the building envelope by using polyurethane foams containing thermoregulating microcapsules. *Applied Thermal Engineering, 103*, 226–232.

Sheikhy, H., Shahidzadeh, M., Ramezanzadeh, B., & Noroozi, F. (2013). Studying the effects of chain extenders chemical structures on the adhesion and mechanical properties of a polyurethane adhesive. *Journal of Industrial and Engineering Chemistry, 19*, 1949–1955.

Singhal, P., Small, W., Cosgriff-Hernandez, E., Maitland, D. J., & Wilson, T. S. (2014). Low density biodegradable shape memory polyurethane foams for embolic biomedical applications. *Acta Biomaterialia, 10*(1), 67–76. https://doi.org/10.1016/j.actbio.2013.09.027

Somani, K. K., Patel, S., Rakshit, N., & Animesh, . (2003). Castor oil based polyurethane adhesives for wood-to-wood bonding. *International Journal of Adhesion and Adhesives, 23*, 269–275. https://doi.org/10.1016/S0143-7496(03)00044-7

Szycher, M. (1999). *Handbook of polymers* (1st edn, pp. 13–34). CRC Press.

Tamami, B., Sohn, S., & Wilkes, G. L. (2004). Incorporation of carbon dioxide into soybean oil and subsequent preparation and studies of nonisocyanate polyurethane networks. *Journal of Applied Polymer Science, 92*, 883–891.

Tanzi, M. C., Farè, S., Petrini, P., Tanini, A., Piscitelli, E., Zecchi Orlandini, S., & Brandi, M. L. (2003). Cytocompatibility of polyurethane foams as biointegrable matrices for the preparation of scaffolds for bone reconstruction. *Journal of Applied Biomaterials and Biomechanics, 1*(1), 58–66.

Titow, W. V. (2001). *PVC Technology* (pp. 146). Rapra Technology Ltd. ISBN 1859572405.

Trang, D. T., & Zenitova, L. A. (2019). Study on the sorption capacity of the adsorbent based on polyurethane and chitin to remove oil spills Y. *Earth and Environmental Science, 337*, 012008.

Tu, Y.-C., Kiatsimkul, P., Suppes, G., & Hsieh, F.-H. (2007). Physical properties of water-blown rigid polyurethane foams from vegetable oil-based polyols. *Journal of Applied Polymer Science, 105*, 453–459.

Vásquez, L., Campagnolo, L., Athanassiou, A., & Fragouli, D. (2019). *ACS Applied Materials and Interfaces, 11*(33), 30207–30217. https://doi.org/10.1021/acsami.9b07907

Vincent, B. J., & Natarajan, B. (2014). Waterborne polyurethane from polycaprolactone and tetramethylxylene diisocyanate: Synthesis by varying NCO/OH ratio and its characterisation as wood coatings. *Open Journal of Organic Polymer Materials, 4*, 37–42.

Wan, T., & Chen, D. (2017). Synthesis and properties of self-healing waterborne polyurethanes containing disulfide bonds in the main chain. *Journal of Material Science, 52*, 197–207.

Wang, T., Sun, W., Zhang, X., Xu, H., & Xu, F. (2017). Waterborne polyurethane coatings with covalently linked black dye Sudan Black B. *Materials (basel), 10*(11), 1247. https://doi.org/10.3390/ma10111247

Wu, G. M., Chen, J., Huo, S. P., Liu, G. F., & Kong, Z. W. (2014). Thermoset nanocomposites from two-component waterborne polyurethanes and cellulose whiskers. *Carbohydrate Polymers, 105*, 207–213. https://doi.org/10.1016/j.carbpol.2014.01.095

Xinrong, S., Nanfang, W., Kunyang, S., Sha, D., & Zhen, C. (2014). Synthesis and characterisation of waterborne polyurethane containing UV absorption group for finishing of cotton fabrics. *Journal of Industrial and Engineering Chemistry, 20*, 3228–3233.

Yang, D., Han, L., & Zhang, H. (2011). Monocomponent waterborne polyurethane adhesives: Influence of crosslinking agent on their properties. *Journal of Macromolecular Science, Part A Pure and Applied Chemistry, 48*, 277–283.

Yari, A., Yeganeh, H., & Bakhshi, H. (2012). Synthesis and evaluation of novel absorptive and antibacterial polyurethane membranes as wound dressing. *Journal of Materials Science. Materials in Medicine, 23*, 2187–2202. https://doi.org/10.1007/s10856-012-4683-6

Zhang, C., & Kessler, M. R. (2015). Bio-based polyurethane foam made from compatible blends of vegetable-oil-based polyol and petroleum-based polyol. *ACS Sustainable Chemical Engineering, 3* (4), 743–749. https://doi.org/10.1021/acssuschemeng.5b00049

Zhang, S., Li, Y., Peng, L., Li, Q., Chen, S., & Hou, K. (2013). Synthesis and characterisation of novel waterborne polyurethane nanocomposites with magnetic and electrical properties. *Composites. Part A, Applied Science and Manufacturing, 55*, 94–101.

Zhao, B., Qian, Y., Qian, X., Fan, J., & Feng, Y. (2019). Fabrication and characterization of waterborne polyurethane/silver nanocomposite foams *40*(4), 1492–1498.

Zhou, X., Fang, C., Li, S., Cheng, Y., & Lei, W. (2015). Recent advances in synthesis of waterborne polyurethane and their application in water-based ink: A review. *Journal of Materials Science & Technology, 31*(7), 708–722.

Water-borne Polyurethane-Metal Oxide Nanocomposite Applications

Prasun Banerjee, Adolfo Franco, K. Chandra Babu Naidu, and N. Suresh Kumar

Abstract

The biodegradable property of the water-borne polyurethanes (WPU) potentially replaces the use of conventional polyurethanes (PU). Some unique properties of the water-borne polyurethanes (WPU) include retardation capability of flames, non-toxic nature, and cheap production cost. Water-borne polyurethanes (WPU) can easily be synthesized in a medium of water with an organic synthesis process. It is also possible to remove the hydrophobic groups with new derivative polymers while synthesis. The main constituent of the water-borne polyurethanes (WPU) includes ionic moieties and ionomers. For a broad range of industrial applications, nanofillers need to add inside the matrix of the water-borne polyurethanes (WPU). The role of these nanofillers is in general to reinforce the water-borne polyurethanes (WPU) matrix with enhanced properties. For instance, iron oxide nanoparticles can be added with the water-borne polyurethanes (WPU) as a material for shape memory (SMM) applications. The addition of the iron oxide can reduce the recovery time as well as the hyperthermia behavior of the nanoparticles helps in induction heating. Hence, the Fe_3O_4—water-borne polyurethanes (WPU) material can be used inside the injured tissues for potential treatments as stents, removal of the blood clots, teeth/jaws malpositioning fixing, control drug delivery, etc. Similarly, another metal oxide like TiO_2 reinforcement in water-borne polyurethanes (WPU) can act as a biomimetic material for stem cell application. ZnO-water-borne polyurethanes (WPU) matrix was also successfully utilized as EMI shielding material. Some reports suggested SnO_2-water-borne polyurethanes (WPU) material as thermal insulators and SiO_2-water-borne polyurethanes (WPU) material as an adhesive for thin films. Hence, the overall behavior of the water-borne polyurethanes (WPU) enhances many folds due to the strong ionic interaction of the matrix of the water-borne polyurethanes (WPU) and metal oxides nanoparticles.

1 Introduction

Presently water-borne polyurethanes (WPU) are drawing a lot of attention to replace the polyurethanes (PU) due to their superior properties (Cao, Chang, & Huneault, 2008; Lee & Kim, 2012; Lee, Kwon, & Kim, 2014; Tian et al., 2010; Wang, Zhang, & Gu, 2005). Their ability to retard flames, non-toxic nature, and cheap production cost are the few superior properties (Cao et al., 2003; Gao et al., 2012; Lee, Bae, & Kim, 2018; Ma et al., 2012; Zhou et al., 2015). Water-borne polyurethanes (WPU) can be synthesized in a medium of water using organic synthesis techniques (Chen et al., 2008; Lee, Hur, & Kim, 2017; Wang & Zhang, 2008; Wu & Zhang, 2001; Zou et al., 2011). It is possible to produce a range of new polymers with the reaction between the prepolymer PUs and water molecules (Gogoi & Karak, 2014; Liu et al., 2013; Xu et al., 2011). Here the most important step is to complete removal of the hydrophobic groups generally present in the polyurethanes (PU) (Wang & Zhang, 2005; Yeh et al., 2008; Zhang et al., 2012a). The role of the water molecules is to attract the hydrophilic part and separate itself as pre-polymeric water-borne polyurethanes (WPU) in the synthesis (Benhamou et al., 2015; Miskolczi et al., 2018; Wang et al., 2010). So the water molecules help to emulsify as well as helps to dissolve. Ionic moiety and ionomer is, in general, the main constituent of the

P. Banerjee (✉) · K. Chandra Babu Naidu
Department of Physics, Gandhi Institute of Technology and Management (GITAM) University, Bengaluru, 561203, India
e-mail: pbanerje@gitam.edu

A. Franco
Instituto de Física, Universidade Federal de Goiás, Goiânia, Brazil

N. Suresh Kumar
Department of Physics, JNTUA, Anantapuram, Andhra Pradesh, India

Inamuddin et al. (eds.), *Sustainable Production and Applications of Waterborne Polyurethanes*, Advances in Science, Technology & Innovation, https://doi.org/10.1007/978-3-030-72869-4_10

water-borne polyurethanes (WPU) (Huang et al., 2011; Lai et al., 2011; Shin & Choi, 2018). So the hydrophilic role of the water-borne polyurethanes (WPU) mainly played by these ions normally situated at the side of the chain of the polymers (Guo et al., 2009; Huang et al., 2006; Wu et al., 2014).

The development of the research on polyurethane (PU) starts with the important contribution led by the famous German Prof Otto Bayer (Gogoi et al., 2015; Lee et al., 2018; Yu et al., 2006). Nearly 83 years ago groundbreaking research on polyurethane (PU) started the foundation of the polyurethane (PU) research. After the end of the 2nd World War gradually different industries around the world start showing interest in the further development of polyurethane (PU) (Bahadur et al., 2018; Shang et al., 2014; Wang, Lue, & Zhang, 2009; Zia et al., 2015). In its earlier days of research, the higher cost of the synthetic raw materials made it nearly impossible to apply for industrial applications (Liu et al., 2017a; Nikje & Tehrani, 2010; Omrani et al., 2017; Xiaojuan et al., 2010). But efforts to develop new and innovative synthesis techniques gradually help to reduce the cost of the synthesis (Li et al., 2014a, 2016; Ma et al., 2006; Naz et al., 2018; Wan & Chen, 2018). In these regards, the invention of the water-borne polyurethanes (WPU) can be considered the landmark (Cao et al., 2007; Chang et al., 2009; Lei et al., 2019; Li et al., 2012; Song et al., 2016; Wang et al., 2018a; Zhong & Cai, 2009). Due to the new developments, various countries around the world start growing the technology for further industrial applications. The production reached a peak level of 4 million tons by the end of 1980s (Fu et al., 2013; Liang et al., 2018; Liu et al., 2017b; She et al., 2013). Further use of the polyurethane (PU) helped to reach double production within 10 years thereafter. The present-day consumption of water-borne polyurethanes (WPU) has already reached many folds than that of the demand just a decade ago (Lee et al., 2006; Tsou et al., 2017; Zhang et al., 2006).

Reports suggested various superior properties like optical, thermal, mechanical, etc. by these water-borne polyurethanes (WPU) (Zhang, Zhang, & Guo, 2012b). A broad range of industrial applications can be observed for the water-borne polyurethanes (WPU) (Zeng Zhang, & Zhou, 2004). In these purposes, a range of nanofillers is in general incorporated inside the matrix of the water-borne polyurethanes (WPU) (Saeedi et al., 2019; Zeng et al., 2003). These nanofillers can be simple metal to different types of metal oxide powders as well (Banerjee & Franco, 2016, 2017, 2019; Banerjee et al., 2007, 2008, 2009, 2010, 2011; Bhattarchya et al., 2010; De et al., 2010; Franco et al., 2018a, 2018b; Pessoni et al., 2019; Ratkovski et al., 2018; Valentin et al., 2019). The role of these nanofillers is in general to reinforce the water-borne polyurethanes (WPU) matrix (Bahadur et al., 2019; Cui et al., 2008; Liu et al., 2016; Ren et al., 2019; Wang et al., 2004, 2020). The interaction between the metal oxides and the water-borne polyurethanes (WPU) matrix leads to the better overall behavior of the compounds (Li, Sun, & Zhang, 2014b; Rad, Sharifan, & Asadi, 2017; Saetung et al., 2016; Sun et al., 2012; Zhao et al., 2017). So the overall behavior of the water-borne polyurethanes (WPU) enhances many folds due to the strong ionic interaction of the matrix of the water-borne polyurethanes (WPU) and metal oxides throughout the body of the polymers (Dong, Jin, & Zhang, 2011; Ji et al., 2013; Zhang et al., 2018).

Hence, in this book chapter, the present advancement and application of the research of the reinforcement of the water-borne polyurethanes (WPU) with metal oxide powders have been highlighted in a holistic way.

2 Different Types of Reinforcement of Water-Borne Polyurethanes (WPU)s Using Metal Oxide Nanomaterials

2.1 Fe_3O_4-water-borne Polyurethanes (WPU) for Tissue Engineering and Control Drug Delivery Applications

Figure 1a shows the chemical structure of the water-borne polyurethanes (WPU) material (Ou et al., 2014). The nanoparticle structure of the water-borne polyurethanes (WPU) has been presented in Fig. 1b (Ou et al., 2014). It can be observed that the NPs are spherical in nature. The size of the nanoparticles is dependent on the temperature with a relatively loose structure. The biodegradable nature of the water-borne polyurethanes (WPU) can be used as a material for shape memory (SMM) shown in Fig. 2a (Wang, Jeng, & Hsu, 2018b). Here the water-borne polyurethanes (WPU) material attained a temporary angle due to the application of the stimulus externally. But the material regained its original shape within the recovery angle when such stimulus was removed. The state of the recovery is quite faster when water is present as a medium is shown in Fig. 2b (Wang et al., 2018b). There is a huge demand for shape-memory materials (SMM) in the healthcare industry for potential applications in tissue engineering. For instance, the shape memory materials (SMM) can be used inside the injured tissues for potential treatments as stents, removal of the blood clots, teeth/jaws malpositioning, etc. For these purposes, one can easily use CAD software to design the required 3D shapes. Thereafter 3D printing technologies can be utilized to print such required 3D scaffolds shown in Fig. 2c (Wang et al., 2018b). Here the main advantage of the use of the water-borne polyurethanes (WPU) is biodegradable as well as flexible in nature in comparison to the conventional tricalcium phosphate material. But the use of metal

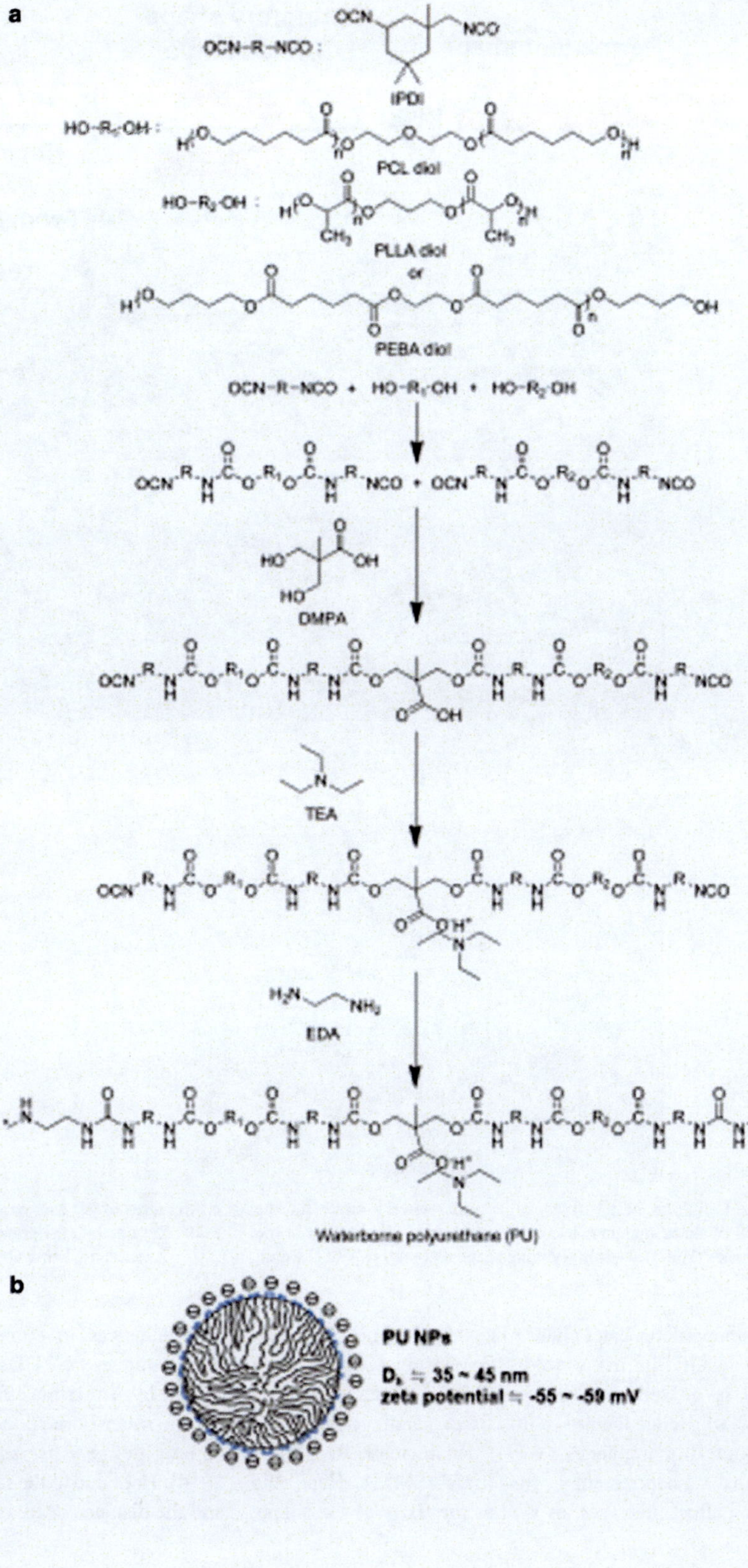

Fig. 1 **a** Chemical structures of water-borne polyurethanes (WPU) and **b** the sketch of PU NPs. Reprint with the permission from Ou et al. (2014). Copyright 2014, American Chemical Society

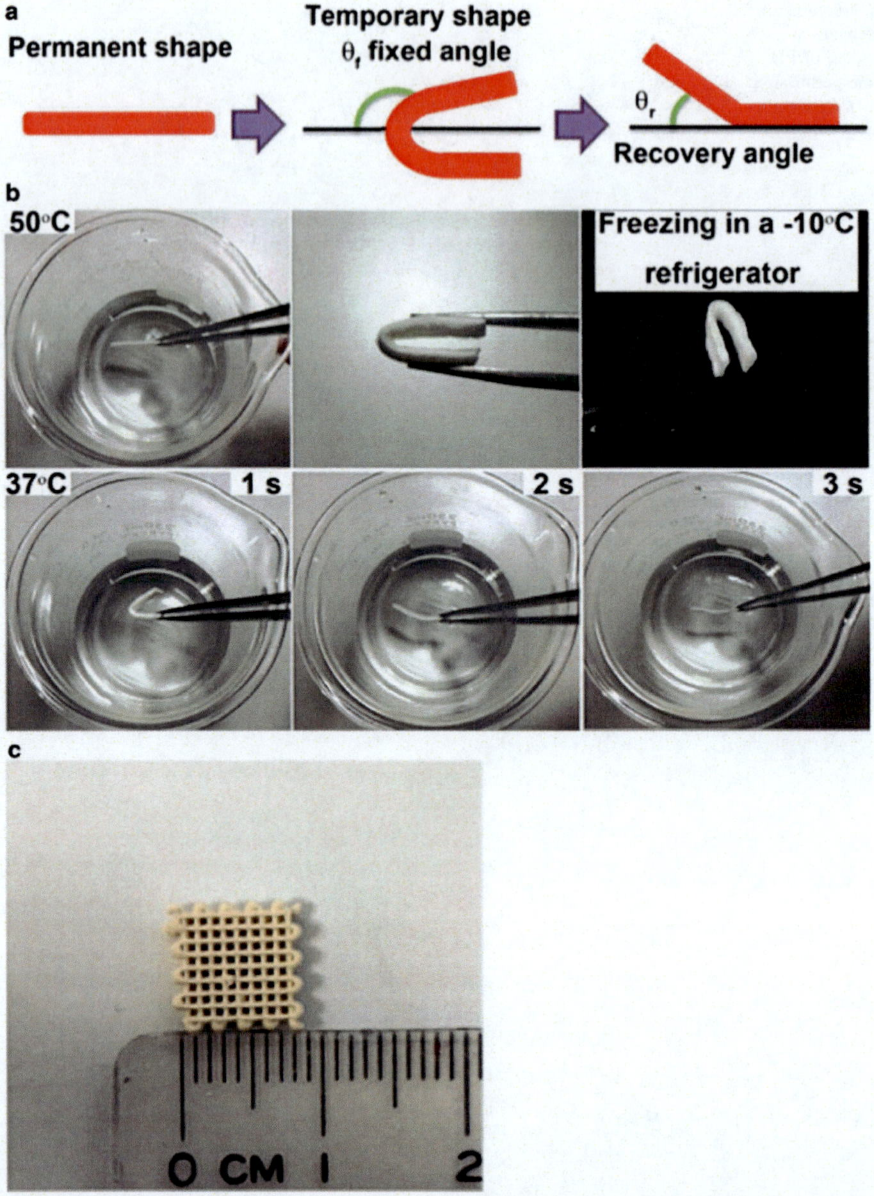

Fig. 2 **a** Definition of the fixed angle and recovery angle for the evaluation of shape memory, **b** continuous images of the water-borne polyurethanes (WPU) scaffold showing shape recovery in 37 °C water, **c** Structures of 3D Iron oxide–water-borne polyurethanes (WPU) scaffolds. Reprint with the permission from Wang et al. (2018b). Copyright 2018, American Chemical Society

oxide nanoparticles can enhance the shape memory material (SMM) quality in many folds. For instance, iron oxide (Fe_3O_4) is a well-known hyperthermia material (Pereira et al., 2020). Hence the use of iron oxide nanoparticles in the water-borne polyurethanes (WPU) can increase its performance as a shape memory materials (SMM). Here, the induction effect increases, as well as the fixity of the shape, also enhances in comparison to the original water-borne polyurethanes (WPU).

The hyperthermia effect of the Fe_3O_4 can also be utilized to use the water-borne polyurethanes (WPU) NPs for the control of drug delivery use as well as shown in Fig. 3 (Ou et al., 2014). Here due to the induction effect the NPs size expanded and the distance decreases between them considerably.

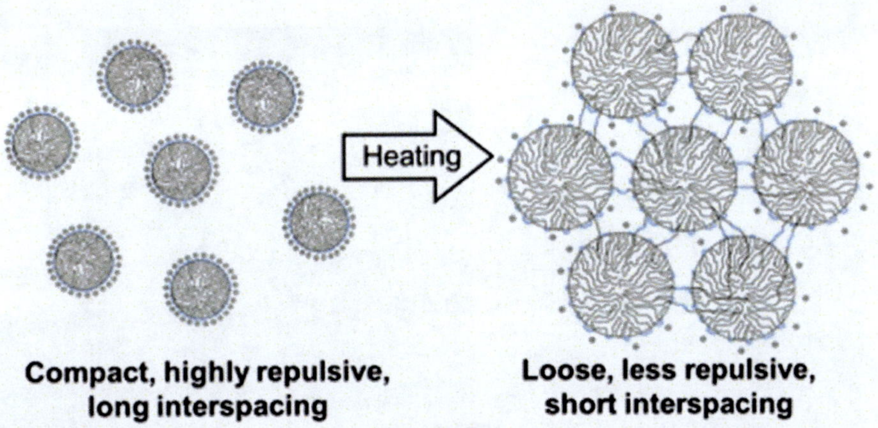

Fig. 3 Possible mechanisms for the temperature-induced self assembly and swelling. Reprint with the permission from Ou et al. (2014). Copyright 2014, American Chemical Society

2.2 ZnO-Water-Borne Polyurethanes (WPU) as an EMI Shielding Material

Electromagnetic interference shielding (EMI) materials are a very important class of material due to their superior property to protect electronic types of equipment effectively. There is a range of fillers, as well as polymers, are available for these purposes. But good mechanical strength, environment-friendly nature as well as excellent reforming film quality makes the use of water-borne polyurethanes (WPU) as one of the best choices for these purposes. On the other hand, graphene is also very popular as a filler in the water-borne polyurethanes (WPU) matrix as an electromagnetic interference shielding (EMI) material. But the effectiveness of the electromagnetic interference shielding (EMI) requires more amount of filler inside the water-borne polyurethanes (WPU). These indeed cause a great disadvantage for graphene as a filler material. Because higher content of the graphene material increases the thickness of the film which decreases its applicability. But the use of a very small amount of metal oxide filler can easily replace the shortcoming of the use of graphene material. The use of the ZnO needle-like whisker filler not only helped to design excellent electromagnetic interference shielding (EMI) material but also increases its flexibility property many folds, shown in Fig. 4. (I) (Xu et al., 2019).

The use of original water-borne polyurethanes (WPU) for electromagnetic interference shielding (EMI) purposes can result in poor performance as there is always a chance to penetrate the material due to the absence of the filler shown in Fig. 4 (II) (a) (Xu et al., 2019). On the other hand, the presence of the needle-like metal oxide NPs as a filler inside the water-borne polyurethanes (WPU) matrix blocks the microwave frequency more efficiently 4 (II) (b) (Xu et al., 2019).

2.3 TiO_2-water-borne Polyurethanes (WPU) as a Biomimetic Material for Stem Cell Applications

Biomimetic materials are well known as biomaterials with the capability of imitative biological properties. Due to that, they find wide application in various medical fields. Here the 3D scaffold made with water-borne polyurethanes (WPU) can easily replace conventional polycaprolactone. Here to avoid compromise with the toughness microsphere of the TiO_2 nanoparticle can be added inside the matrix of the water-borne polyurethanes (WPU). Here the biomineralization is a very important process for the biocompatibility of the 3D scaffolds which stimulates the regeneration process, shown in Fig. 5a (Zhu et al., 2018). Here it can be observed that the role of titanium oxide is solely to accelerate the biomineralization process.

The role of bone marrow mesenchymal stem cells (MSC) in cell growth is a well-known phenomenon. But the addition of 3D scaffolds of TiO_2-water-borne polyurethanes (WPU) along with it shows excellent stem cell regeneration process within 7 days of the tissue culture shown in Fig. 5b (Zhu et al., 2018). Here the biodegradable, as well as the hydrophilic nature as well as the presence of the titanium oxide nanoparticles helps to accelerate the process.

2.4 SiO_2-water-borne Polyurethanes (WPU) for Adhesive Applications

Polyolefin films (PF) finds a lot of popular applications, especially in the packaging industry. But adhesion is a very important property for the packaging of foods. Hence, the nonpolar nature of the PF makes it sometimes very difficult to work with it for these purposes. One of the effective

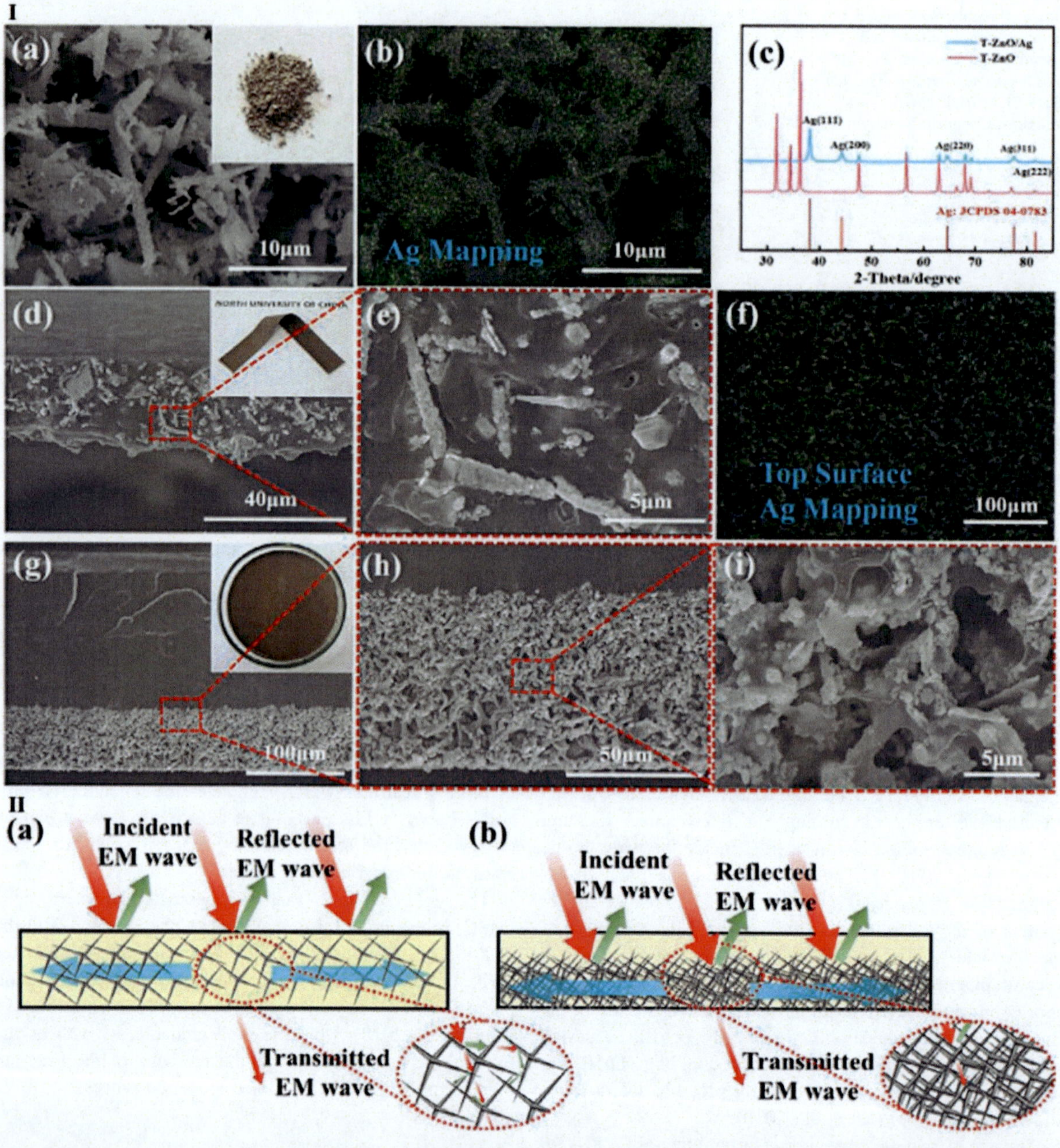

Fig. 4 **I** SEM images of (a) morphology and (b) elemental distribution of the T-ZnO/Ag nanofiller. (c) XRD patterns of the T-ZnO and T-ZnO/Ag nanofiller. SEM images of the T-ZnO/Ag/water-borne polyurethanes (WPU) composite film with uniform structure at 17.3 vol% T-ZnO/Ag: (d, e) cross-section images, (f) elemental distribution image of the top surface. (g–i) cross-sectional SEM images of the T-ZnO/Ag/water-borne polyurethanes (WPU) composite film with deposited layer structure at 11.1 vol% T-ZnO/Ag. **II** Schematic representation of the shielding mechanism for the T-ZnO/Ag/water-borne polyurethanes (WPU) composite films with different structures: (a) uniform structure, (b) deposited layer structure. Reprint with the permission from Xu et al. (2019). Copyright 2019, Elsevier

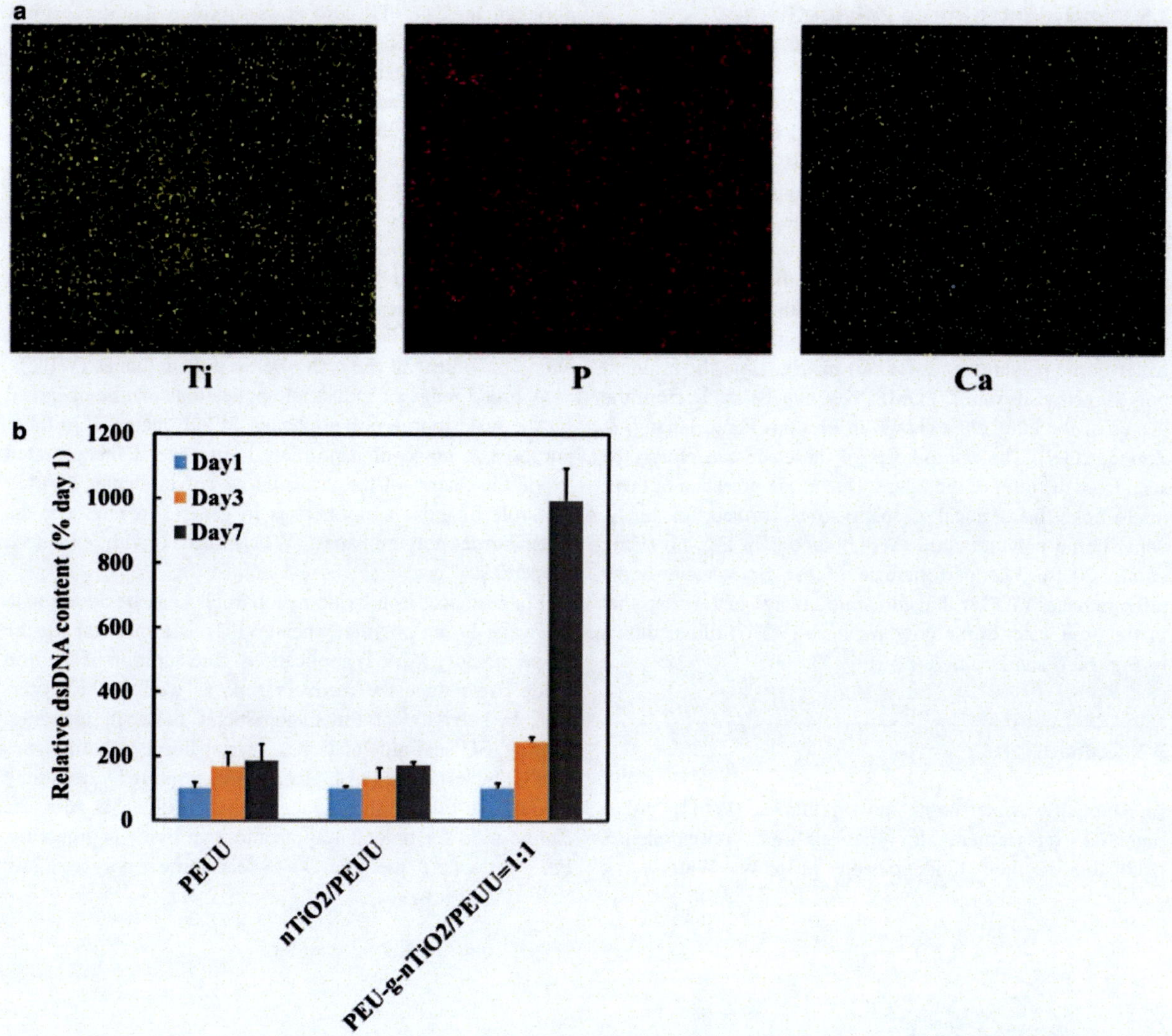

Fig. 5 **a** EDX characterization of Ca and P deposition in the scaffold PEU-g-nTiO_2/PEUU = 1:1. **b** Mesenchymal stem cell growth on the scaffolds with and without reinforcement with PEU-gnTiO_2. Reprint with the permission from Zhu et al. (2018). Copyright 2018, Elsevier

methods for proper adhesion is the discharge method. But due to the generation of high temperature in this case toxic byproduct elements generated. These toxic materials can cause harmful effects on the human body when the food is consumed. Hence the use of water-borne polyurethanes (WPU) can easily eliminate the problem along with the use of silicon oxide nanoparticles which acts as an excellent adhesive, shown in Fig. 6a (Fu et al., 2014). The AFM characterization is shown in Fig. 6b clearly demonstrated the relatively smooth surface of the original water-borne polyurethanes (WPU) (Fu et al., 2014). The presence of silicon moieties further increases the surface roughness of the SiO_2-water-borne polyurethanes (WPU) matrix. The excellent adhesive property of the SiO_2-water-borne polyurethanes (WPU) matrix demonstrated in Fig. 6c even at temperatures as high as 120 ^{0}C (Fu et al., 2014).

2.5 SnO_2-Water-Borne Polyurethanes (WPU) for Thermal Insulation Applications

SnO_2 doped with antimony is a well-known material for heat insulation applications. When the metal oxide mixed together with polymers and dried subsequently in thin-film form, can be stick on the glass plate for thermal insulation applications. Here the main advantage of the SnO_2 doped with antimony nanoparticles because of their optically transparent nature. But the main challenge to make thin-film form is the homogeneous mixture of the material along with the polymers. But the problem can easily overcome with the use of water-borne polyurethanes (WPU) matrix. Here the required poly(dimethylsiloxane) (PDMS) cast can be made from the leaves of the lotus plant shown in Fig. 6a (Feng, Huang, & Zhong, 2009). The micro-holes of the caste can clearly be seen from the inset of the figure. Due to the presence of these micro-holes hierarchical nanostructures formed for SnO_2-water-borne polyurethanes (WPU) shown in Fig. 6b (Feng et al., 2009). The performance of the SnO_2-water-borne polyurethanes (WPU) thin films are always better than that of the pure water-borne polyurethanes (WPU) films, shown in Fig. 6c (Feng et al., 2009) (Fig. 7).

3 Conclusions

In summary, water-borne polyurethanes (WPU) are a potential replacement for conventional polyurethanes (PU) due to their biodegradable property. Water-borne polyurethanes (WPU) also possess some unique features like retardation of flames, non-toxic nature, and cheap production cost. Water-borne polyurethanes (WPU) can be synthesized in a medium of water with an organic synthesis process. It is possible to remove the hydrophobic groups generally present in the polyurethanes (PU) along with new polymers while the synthesis of water-borne polyurethanes (WPU). The role of the water molecules is to attract the hydrophilic part and separate itself as pre-polymeric water-borne polyurethanes (WPU) in the synthesis. Hence, the water molecules help to emulsify as well as helps to dissolve. Ionic moiety and ionomer are in the general the main constituent of the water-borne polyurethanes (WPU).

A broad range of industrial applications can be observed for the water-borne polyurethanes (WPU) material. In these purposes, a range of nanofillers is in general incorporated inside the matrix of the water-borne polyurethanes (WPU). The role of these nanofillers is in general to reinforce the water-borne polyurethanes (WPU) matrix with enhanced properties.

For instance, iron oxide nanoparticles can be added with the water-borne polyurethanes (WPU) as a material for the shape memory (SMM) applications. The addition of the iron oxide can reduce the recovery time as well as the hyperthermia behavior of the nanoparticles helps in induction heating. 3D scaffolds of Fe_3O_4—water-borne polyurethanes (WPU) material can be obtained by 3D printing by designing in CAD software. The biodegradable flexible 3D structure can be used for potential applications in tissue engineering. For instance, the Fe_3O_4—water-borne polyurethanes

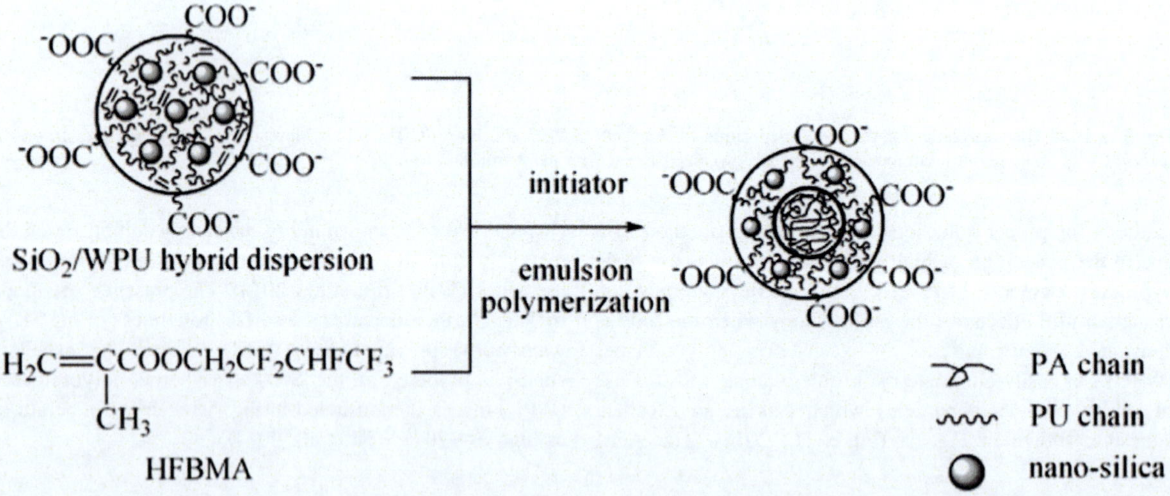

Fig. 6 **a** Schematic illustration of the formation of SiO_2/water-borne polyurethanes (WPU) nanocomposite particle. **b** Three-dimensional images of (a) water-borne polyurethanes (WPU) (b) SiO2/water-borne polyurethanes (WPU)-0 (c) SiO2/water-borne polyurethanes (WPU)-15. **c** Effect of nano-SiO_2 content on the T-peel strength of laminated films (a) at 25 °C(B) Cooking at 120 C and 0.2–0.3 MPa Vapor Atmosphere for 30 min. Reprint with the Permission from Fu et al. (2014). Copyright 2014, Elsevier

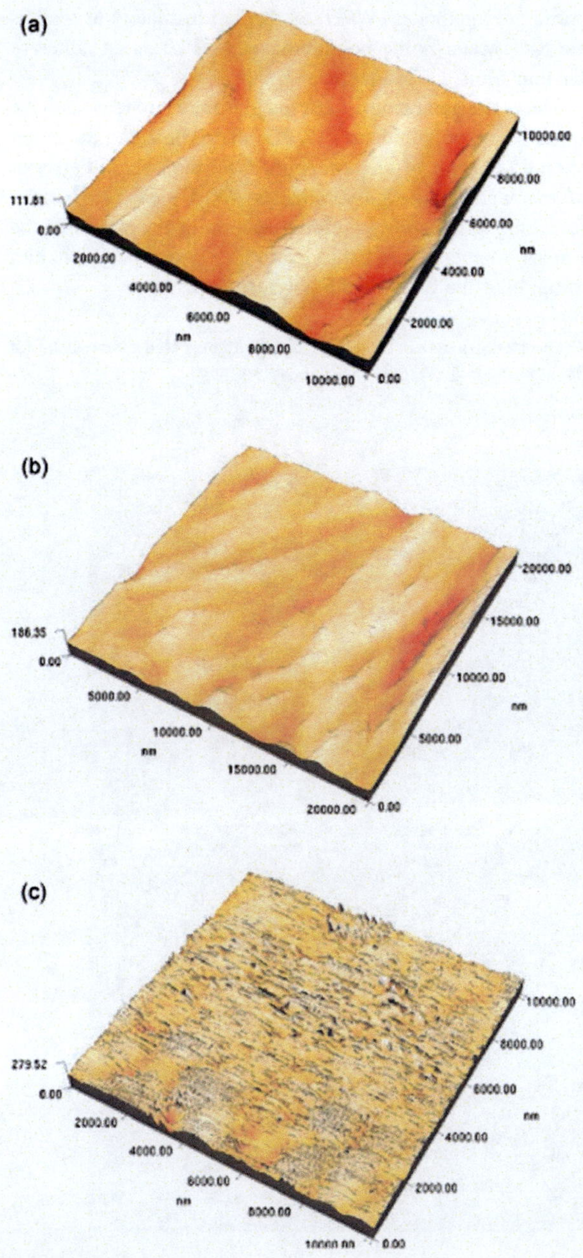

Fig. 6 (continued)

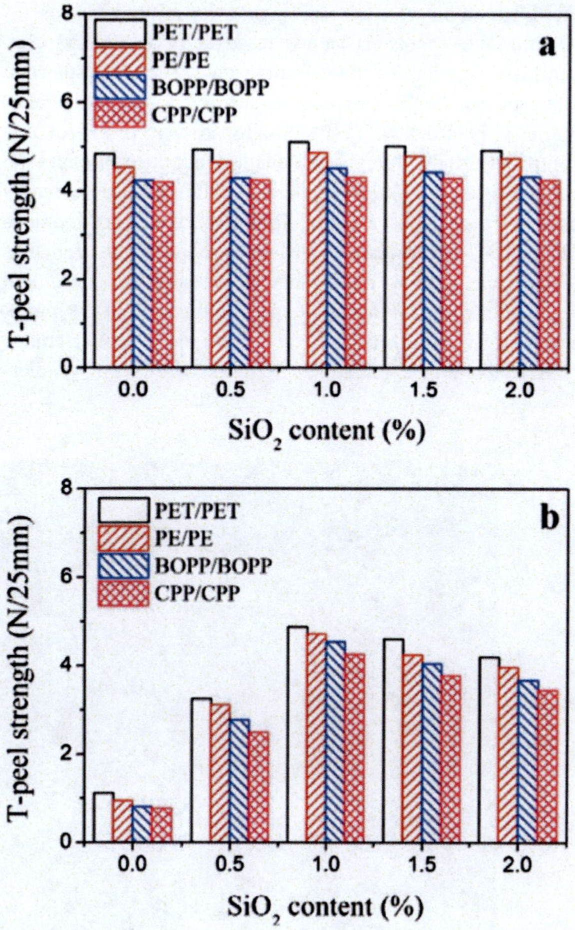

Fig. 6 (continued)

(WPU) material can be used inside the injured tissues for potential treatments as stents, removal of the blood clot, teeth/jaws malpositioning fixing, etc. The hyperthermia effect of the Fe_3O_4 can also be utilized to use the water-borne polyurethanes (WPU) nanoparticles (NP) for the control of drug delivery. Here due to the induction effect the nanoparticles (NP) size expanded and the distance decreases between them considerably. Similarly, other metal oxides like TiO_2 reinforcement in water-borne polyurethanes (WPU) can act as a biomimetic material for stem cell applications. ZnO-water-borne polyurethanes (WPU) matrix successfully utilized as electromagnetic interference (EMI) shielding applications. Reports of use SnO_2-water-borne polyurethanes (WPU) as thermal insulators as well as the SiO_2-water-borne polyurethanes (WPU) as an adhesive for thin films are available.

Here, the interaction between the metal oxides and the water-borne polyurethanes (WPU) matrix leads to the better overall behavior of the compounds. So the overall behavior of the water-borne polyurethanes (WPU) enhances many folds due to the strong ionic interaction of the matrix of the water-borne polyurethanes (WPU) and metal oxides throughout the body of the polymers.

Acknowledgments P. Banerjee sincerely thanks UGC, New Delhi for the grant no.F.30-457/2018 (BSR)-start-up grant.

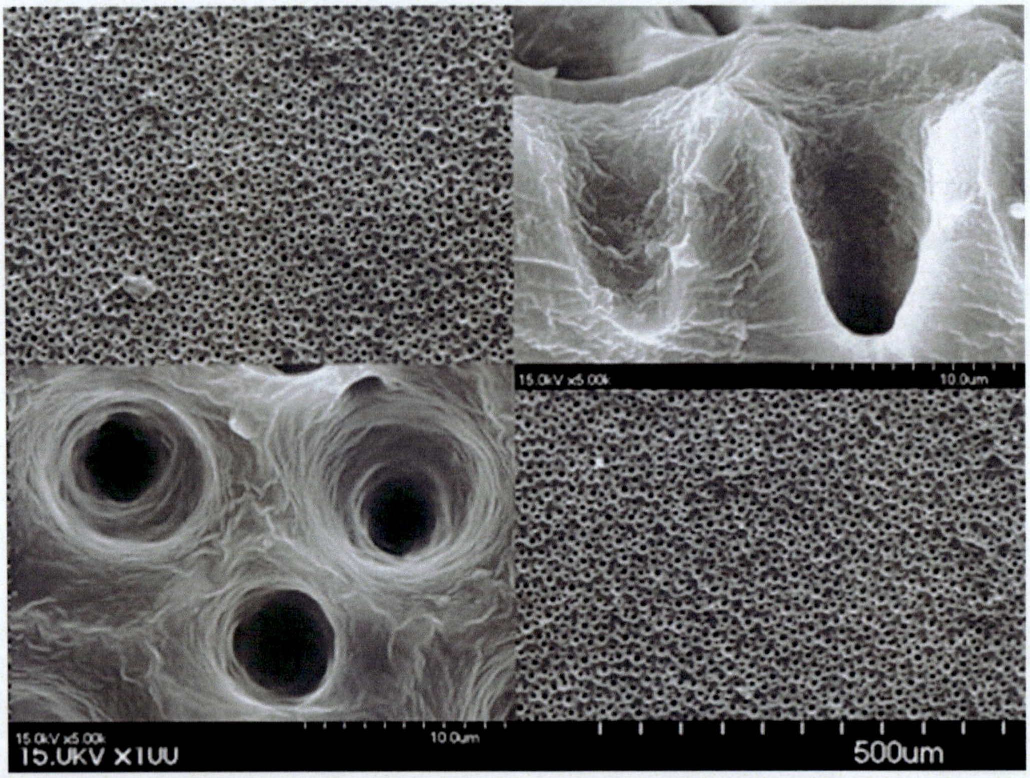

Fig. 7 **a** SEM images of the PDMS stamp replicated from a lotus leaf. The insets were high magnification viewed from section and top. **b** SEM images of the surface microstructures on ATO/PU film (a and b) and pure PU film (c and d). Panels (b) and (d) were magnified images of (a) and (c), respectively. **c** The temperature changes in the polystyrene foam box (a), on the ATO/PU, coated glass plates, (b) and on the ATO/PU films (c) via infrared irradiation time. Reprint with the permission from Feng et al. (2009). Copyright 2009, Elsevier

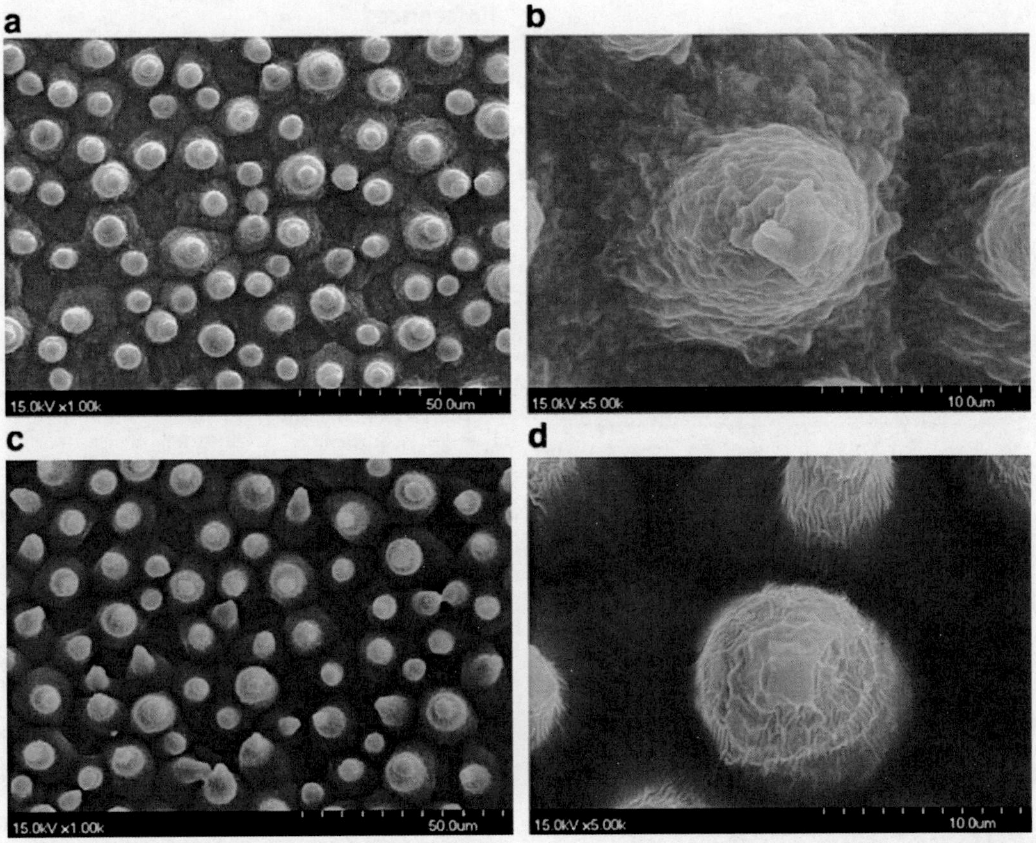

Fig. 7 (continued)

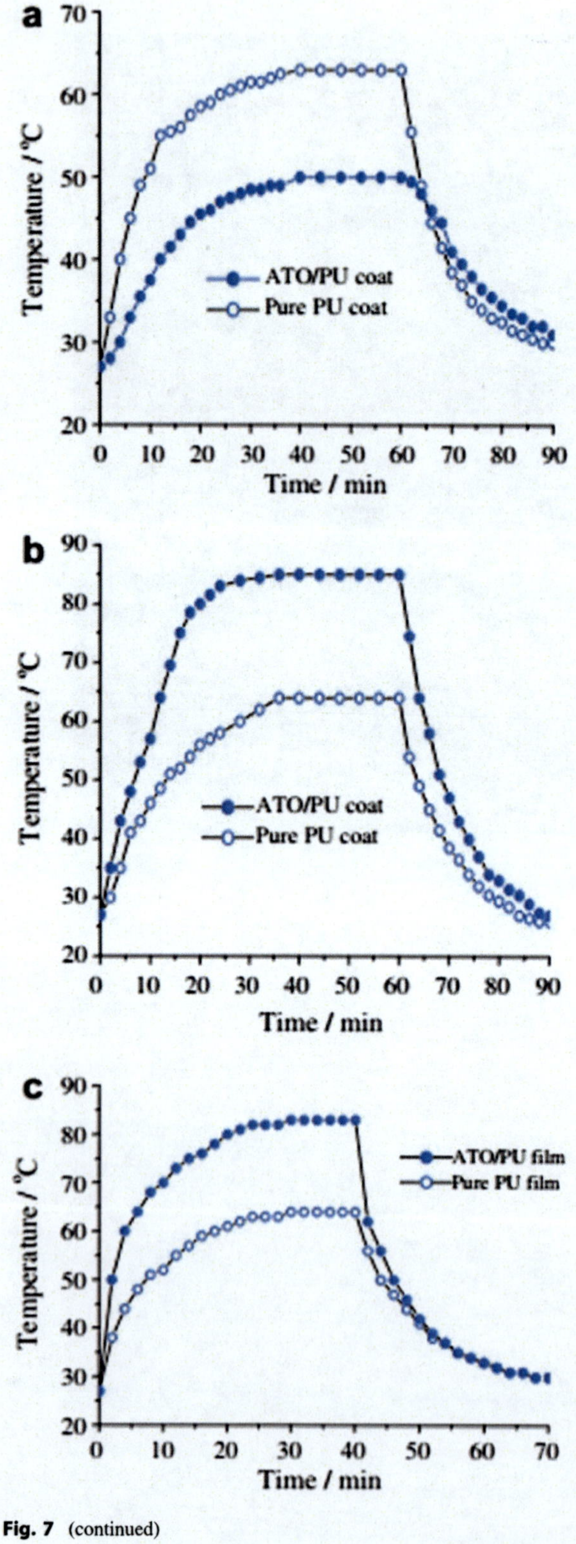

Fig. 7 (continued)

References

Bahadur, A., Saeed, A., Shoaib, M., Iqbal, S., & Anwer, S. (2019). Modulating the burst drug release effect of waterborne polyurethane matrix by modifying with polymethylmethacrylate. *Journal of Applied Polymer Science, 136*(13), 47253.

Bahadur, A., Shoaib, M., Iqbal, S., Saeed, A., ur Rahman, M. S., & Channar, P. A. (2018). Regulating the anticancer drug release rate by controlling the composition of waterborne polyurethane. *Reactive and Functional Polymers, 131*, 134–141.

Banerjee, P., & Franco Jr., A. (2016) Enhanced dielectric and magnetic properties in multiferroic $Bi_{0.99}Y_{0.01}Fe_{0.99}Ni_{0.01}O_3$ ceramic. *Materials Letters, 184*, 17–20.

Banerjee, P., & Franco Jr., A. (2017). Substitution-induced near phase transition with Maxwell–Wagner polarization in $SrBi_2$ (Nb1−xAx) 2O9 ceramics [A= W, Mo and x= 0, 0.025]. *Physica Status Solidi (A), 214*(10), 1700067.

Banerjee, P., & Franco Jr., A. (2019). Role of higher valent substituent on the dielectric and optical properties of Sr0. 8Bi2. 2Nb2O9 ceramics. *Materials Chemistry and Physics, 225*, 213–218.

Banerjee, P., Ghosh, G., & Biswas, S. K. (2007). A simple technique for the measurement of the permittivity of medium loss samples using cavity perturbation method. In *2007 IEEE Applied Electromagnetics Conference (AEMC) 2007* (pp. 1–3). IEEE.

Banerjee, P., Ghosh, G., & Biswas, S. K. (2008). Measurement of the dielectric constant of medium loss cylindrical-shaped samples using cavity perturbation method. In *2008 International Conference on Recent Advances in Microwave Theory and Applications 2008* (pp. 124–125). IEEE.

Banerjee, P., Ghosh, G., & Biswas, S. K. (2009). A system to measure dielectric constant and loss of liquids at microwave frequencies. In *2009 Applied Electromagnetics Conference (AEMC)* (pp. 1–2). IEEE.

Banerjee, P., Ghosh, G., & Biswas, S. K. (2010). Measurement of dielectric properties of medium loss samples at X-band frequencies. *Journal of Metallurgy and Materials Science, 52*(3), 247–255.

Banerjee, P., Ghosh, G., & Biswas, S. K. (2011). A simple method to determine the dielectric constant of small-sized medium-loss samples at X-band frequencies. *International Journal of Electromagnetics and Application* 12–15.

Benhamou, K., Kaddami, H., Magnin, A., Dufresne, A., & Ahmad, A. (2015). Bio-based polyurethane reinforced with cellulose nanofibers: A comprehensive investigation on the effect of interface. *Carbohydrate Polymers, 122*, 202–211.

Bhattarchya, S., De, D., Ghosh, S., Banerjee, P., Saha, S., Mitra, M., Nag, B., Pal, M., Biswas, S. K., & Ghatak, K. P. (2010). Influence of external light waves on the thermoelectric power under strong magnetic field in ultrathin films, quantum wires and quantum dots of optoelectronic materials. *Journal of Computational and Theoretical Nanoscience, 7*(6), 1066–1084.

Cao, X., Chang, P. R., & Huneault, M. A. (2008). Preparation and properties of plasticized starch modified with poly (ε-caprolactone) based waterborne polyurethane. *Carbohydrate Polymers, 71*(1), 119–125.

Cao, X., Dong, H., & Li, C. M. (2007). New nanocomposite materials reinforced with flax cellulose nanocrystals in waterborne polyurethane. *Biomacromolecules, 8*(3), 899–904.

Cao, X., Zhang, L., Huang, J., Yang, G., & Wang, Y. (2003). Structure–properties relationship of starch/waterborne polyurethane composites. *Journal of Applied Polymer Science, 90*(12), 3325–3332.

Chang, P. R., Ai, F., Chen, Y., Dufresne, A., & Huang, J. (2009). Effects of starch nanocrystal-graft-polycaprolactone on mechanical

properties of waterborne polyurethane-based nanocomposites. *Journal of Applied Polymer Science, 111*(2), 619–627.
Chen, G., Wei, M., Chen, J., Huang, J., Dufresne, A., & Chang, P. R. (2008). Simultaneous reinforcing and toughening: New nanocomposites of waterborne polyurethane filled with low loading level of starch nanocrystals. *Polymer, 49*(7), 1860–1870.
Cui, G., Fan, H., Xia, W., Ai, F., & Huang, J. (2008). Simultaneous enhancement in strength and elongation of waterborne polyurethane and role of star-like network with lignin core. *Journal of Applied Polymer Science, 109*(1), 56–63.
De, D., Kumar, A., Adhikari, S. M., Pahari, S., Islam, N., Banerjee, P., Biswas, S. K., Bhattacharya, S., & Ghatak, K. P. (2010). Influence of quantum confinement on the photoemission from superlattices of optoelectronic materials. *Superlattices and Microstructures, 47*(3), 377–410.
Dong, Y. M., Jin, Y. W., & Zhang, Z. J. (2011). Synthesis and application of polyurethane hydrogel carrier on nitrobacteria immobilization. In *Advanced Materials Research 2011* (Vol. 152, pp. 1533–1536). Trans Tech Publications Ltd.
Feng, J., Huang, B., & Zhong, M. (2009). Fabrication of superhydrophobic and heat-insulating antimony doped tin oxide/polyurethane films by cast replica micromolding. *Journal of Colloid and Interface Science, 336*(1), 268–272.
Franco Jr, A., Banerjee, P., & Romanholo, P. L. (2018a) Effect of composition induced transition in the optical band-gap, dielectric and magnetic properties of Gd doped $Na_{0.5}Bi_{0.5}TiO_3$ complex perovskite. *Journal of Alloys and Compounds, 764*, 122–127.
Franco Jr, A., Banerjee, P., & Lima, R. J. (2018b). Dielectric and magnetic properties of three-layers laminated ceramic composite, $K_{0.5}Na_{0.5}NbO_3/CoFe_2O_4/K_{0.5}Na_{0.5}NbO_3$. *Journal of Materials Science: Materials in Electronics. 2018;29*(5):4357–4364.
Fu, H., Yan, C., Zhou, W., & Huang, H. (2013). Preparation and characterization of a novel organic montmorillonite/fluorinated waterborne polyurethane nanocomposites: Effect of OMMT and HFBMA. *Composites Science and Technology, 85*, 65–72.
Fu, H., Yan, C., Zhou, W., & Huang, H. (2014). Nano-SiO_2/fluorinated waterborne polyurethane nanocomposite adhesive for laminated films. *Journal of Industrial and Engineering Chemistry, 20*(4), 1623–1632.
Gao, Z., Peng, J., Zhong, T., Sun, J., Wang, X., & Yue, C. (2012). Biocompatible elastomer of waterborne polyurethane based on castor oil and polyethylene glycol with cellulose nanocrystals. *Carbohydrate Polymers, 87*(3), 2068–2075.
Gogoi, S., & Karak, N. (2014). Biobased biodegradable waterborne hyperbranched polyurethane as an ecofriendly sustainable material. *ACS Sustainable Chemistry & Engineering., 2*(12), 2730–2738.
Gogoi, S., Kumar, M., Mandal, B. B., & Karak, N. (2015). High performance luminescent thermosetting waterborne hyperbranched polyurethane/carbon quantum dot nanocomposite with in vitro cytocompatibility. *Composites Science and Technology, 118*, 39–46.
Guo, W. J., Fu, H. Q., Li, F. Y., Huang, H., & Chen, H. Q. (2009). Study on the mechanical and adhesive properties of waterborne polyurethane with multiple modifications. *Journal of Chemical Engineering of Chinese Universities, 2*, 246–251.
Huang, Y., Yu, H., & Xiao, C. (2006). Effects of Ca2+ crosslinking on structure and properties of waterborne polyurethane-carboxymethylated guar gum films. *Carbohydrate Polymers, 66* (4), 500–513.
Huang, J., Zou, J. W., Chang, P. R., Yu, J. H., & Dufresne, A. (2011). New waterborne polyurethane-based nanocomposites reinforced with low loading levels of chitin whisker. *Express polymer letters, 5*(4).
Ji, X., Zhou, Y., Zhang, B., Hou, C., & Ma, G. (2013). Polydimethylsiloxane and castor oil comodified waterborne polyurethane. *ISRN Polymer Science*.
Lai, X., Shen, Y., & Wang, L. (2011). Preparation and properties of self-crosslinkable polyurethane/silane hybrid emulsion. *Journal of Polymer Research, 18*(6), 2425–2433.
Lee, I. H., Bae, S. Y., Kim, & B. K. (2018). Biodegradable waterborne polyurethane with modified hydroxyethylcellulose. *Express Polymer Letters, 12*(9).
Lee, D. S., Hur, P., & Kim, B. K. (2017). Chemical hybridization of waterborne polyurethane with β-cyclodextrin by sol-gel reaction. *Progress in Organic Coatings, 111*, 107–111.
Lee, H. T., Hwang, J. J., & Liu, H. J. (2006). Effects of ionic interactions between clay and waterborne polyurethanes on the structure and physical properties of their nanocomposite dispersions. *Journal of Polymer Science Part a: Polymer Chemistry, 44* (19), 5801–5807.
Lee, S. J., & Kim, B. K. (2012). Covalent incorporation of starch derivative into waterborne polyurethane for biodegradability. *Carbohydrate Polymers, 87*(2), 1803–1809.
Lee, T. J., Kwon, S. H., & Kim, B. K. (2014). Biodegradable sol–gel coatings of waterborne polyurethane/gelatin chemical hybrids. *Progress in Organic Coatings, 77*(6), 1111–1116.
Lee, H. T., Tsou, C. H., Li, C. L., Gu, J. H., Wu, C. L., Hwang, J. J., & Suen, M. C. (2018). Preparation and characterization of biodegradable polyurethane composites containing attapulgite nanorods. *Advances in Polymer Technology, 37*(1), 208–220.
Lei, W., Zhou, X., Fang, C., Song, Y., & Li, Y. (2019). Eco-friendly waterborne polyurethane reinforced with cellulose nanocrystal from office waste paper by two different methods. *Carbohydrate Polymers, 209*, 299–309.
Li, X., Hu, J., Sun, D., & Zhang, Y. (2014a). Nanosilica reinforced waterborne siloxane-polyurethane nanocomposites prepared via "click" coupling. *Journal of Coatings Technology and Research, 11*(4), 517–531.
Li, X. J., Sun, D. X., & Zhang, Y. H. (2014b). The Cu (I)-mediated azide–alkyne cycloadditions as a facile and efficient route for functionalization of waterborne polyurethane. *Journal of Polymer Research, 21*(1), 320.
Li, S. C., Wang, L. W., & Wang, J. Y. (2012). Study on properties of modified nano-ZnO/water-borne polyurethanes (WPU) hybrid material. In *Advanced Materials Research 2012* (Vol. 510, pp. 742–746). Trans Tech Publications Ltd.
Li, M., Xiao, Y., Chen, Y., Ni, H., Cai, J., Wang, X., Chang, P. R., Anderson, D. P., & Chen, Y. (2016). Soy protein-modified waterborne polyurethane biocomposites with improved functionality. *RSC Advances, 6*(16), 12837–12849.
Liang, H., Feng, Y., Lu, J., Liu, L., Yang, Z., Luo, Y., Zhang, Y., & Zhang, C. (2018). Bio-based cationic waterborne polyurethanes dispersions prepared from different vegetable oils. *Industrial Crops and Products, 122*, 448–455.
Liu, H., Cui, S., Shang, S., Wang, D., & Song, J. (2013). Properties of rosin-based waterborne polyurethanes/cellulose nanocrystals composites. *Carbohydrate Polymers, 96*(2), 510–515.
Liu, H., Li, C., & Sun, X. S. (2017b). Soy-oil-based waterborne polyurethane improved wet strength of soy protein adhesives on wood. *International Journal of Adhesion and Adhesives, 73*, 66–74.
Liu, C. H., Lee, H. T., Tsou, C. H., Gu, J. H., Suen, M. C., & Chen, J. K. (2017a). In Situ Polymerization and characteristics of biodegradable waterborne thermally-treated attapulgite nanorods and polyurethane composites. *Journal of Inorganic and Organometallic Polymers and Materials., 27*(1), 244–256.
Liu, K., Miao, S., Su, Z., Sun, L., Ma, G., & Zhang, S. (2016). Castor oil-based waterborne polyurethanes with tunable properties and excellent biocompatibility. *European Journal of Lipid Science and Technology, 118*(10), 1512–1520.

Ma, L., Li, J., & Luo, Y. (2006). Comparison of shape memory properties of cross-linking and linear water-based polyurethane. *Chemical Industry and Engineering Progress., 25*(1), 78.

Ma, J., Xu, Q., Gao, D., Zhou, J., & Zhang, J. (2012). Blend composites of caprolactam-modified casein and waterborne polyurethane for film-forming binder: Miscibility, morphology and properties. *Polymer Degradation and Stability, 97*(8), 1545–1552.

Miskolczi, N., Sedlarik, V., Kucharczyk, P., & Riegel, E. (2018). Enhancement of the mechanical properties of a polylactic acid/flax fiber biocomposite by water-borne polyurethanes (WPU), water-borne polyurethanes (WPU)/Starch, and TPS polyurethanes using coupling additives. *Mechanics of Composite Materials, 53*(6), 791–800.

Naz, F., Zuber, M., Zia, K. M., Salman, M., Chakraborty, J., Nath, I., & Verpoort, F. (2018). Synthesis and characterization of chitosan-based waterborne polyurethane for textile finishes. *Carbohydrate Polymers, 200*, 54–62.

Nikje, M. M., & Tehrani, Z. M. (2010). Synthesis and characterization of waterborne polyurethane-chitosan nanocomposites. *Polymer-Plastics Technology and Engineering, 49*(8), 812–817.

Omrani, I., Babanejad, N., Shendi, H. K., & Nabid, M. R. (2017). Fully glutathione degradable waterborne polyurethane nanocarriers: Preparation, redox-sensitivity, and triggered intracellular drug release. *Materials Science and Engineering: C, 70*, 607–616.

Ou, C. W., Su, C. H., Jeng, U. S., & Hsu, S. H. (2014). Characterization of biodegradable polyurethane nanoparticles and thermally induced self-assembly in water dispersion. *ACS Applied Materials & Interfaces, 6*(8), 5685–5694.

Pereira, L. G., e Silva, R. L., Banerjee, P., Franco Jr., A. (2020). Role of Gd3+ ions on the magnetic hyperthermic behaviour of anisotropic $CoFe_2O_4$ nanoparticles. *Physica B: Condensed Matter* 412140.

Pessoni, H. V., Alves, T. E., Banerjee, P., & Júnior, A. F. (2019). Functional properties of Ho3+ substituted cobalt ferrite in the context of the reduced mass model. *Physica B: Condensed Matter, 575*, 411676.

Rad, F. H., Sharifan, A., & Asadi, G. (2017). Miscibility and morphology of kefiran/waterborne polyurethane blend films. *International Journal of Food Properties, 20*(sup3), S2764–S2775.

Ratkovski, D. R., Ribeiro, P. R., Machado, F. L., Banerjee, P., Franco Jr., A. (2018). On the magnetic properties of the multiferroic ceramics $Bi_{0.99}Y_{0.01}Fe_{1-x}Ni_xO_3$ ($0.01 \leqslant x \leqslant 0.05$). *Journal of Magnetism and Magnetic Materials, 451*, 620–624.

Ren, L., Guo, X., Zhao, Y., & Qiang, T. (2019). Synthesis and properties of waterborne polyurethane incorporated with phenolic acid grafted oligochitosan. *Progress in Organic Coatings, 135*, 410–416.

Saeedi, S., Omrani, I., Bafkary, R., Sadeh, E., Shendi, H. K., & Nabid, M. R. (2019). Facile preparation of biodegradable dual stimuli-responsive micelles from waterborne polyurethane for efficient intracellular drug delivery. *New Journal of Chemistry, 43* (47), 18534–18545.

Saetung, A., Rungvichaniwat, A., Tsupphayakorn-ake, P., Bannob, P., Tulyapituk, T., & Saetung, N. (2016). Properties of waterborne polyurethane films: Effects of blend formulation with hydroxyl telechelic natural rubber and modified rubber seed oils. *Journal of Polymer Research, 23*(12), 264.

Shang, S., Chiu, K. L., Yuen, M. C., & Jiang, S. (2014). The potential of cuttlebone as reinforced filler of polyurethane. *Composites Science and Technology, 93*, 17–22.

She, Y., Zhang, H., Song, S., Lang, Q., & Pu, J. (2013). Preparation and characterization of waterborne polyurethane modified by nanocrystalline cellulose. *BioResources, 8*(2), 2594–2604.

Shin, E. J., & Choi, S. M. (2018). Advances in waterborne polyurethane-based biomaterials for biomedical applications. In *Novel Biomaterials for Regenerative Medicine 2018* (pp. 251–283). Springer.

Song, Y. Q., Gao, Y. L., Pan, Z. C., Zhang, Y., Li, J. H., Wang, K. J., Li, J. S., Tan, H., & Fu, Q. (2016). Preparation and characterization of controlled heparin release waterborne polyurethane coating systems. *Chinese Journal of Polymer Science, 34*(6), 679–687.

Sun, D., Li, R., Li, X., Wang, W., & Hu, J. (2012). Fabrication of functional waterborne polyurethane/montmorillonite composites by click chemistry method. *Combinatorial Chemistry & High Throughput Screening, 15*(7), 522–528.

Tian, H., Wang, Y., Zhang, L., Quan, C., & Zhang, X. (2010). Improved flexibility and water resistance of soy protein thermoplastics containing waterborne polyurethane. *Industrial Crops and Products, 32*(1), 13–20.

Tsou, C. H., Lee, H. T., Hung, W. S., De Guzman, M., Chen, S. T., Suen, M. C., & Wicaksono, S. T. (2017). Effects of different metals on the synthesis and properties of waterborne polyurethane composites containing pyridyl units. *Polymer Bulletin, 74*(4), 1121–1143.

Valentin, C. B., e Silva, R. L., Banerjee, P., & Franco Jr., A. (2019). Investigation of Fe-doped room temperature dilute magnetic ZnO semiconductors. *Materials Science in Semiconductor Processing, 96*, 122–126.

Wan, T., & Chen, D. (2018). Mechanical enhancement of self-healing waterborne polyurethane by graphene oxide. *Progress in Organic Coatings, 121*, 73–79.

Wang, Y. J., Jeng, U. S., & Hsu, S. H. (2018b). Biodegradable water-based polyurethane shape memory elastomers for bone tissue engineering. *ACS Biomaterials Science & Engineering, 4*(4), 1397–1406.

Wang, W., Lu, L., Xie, Y., Yuan, W., Wan, Z., Tang, Y., & Teh, K. S. (2020). A highly stretchable microsupercapacitor using laser-induced Graphene/NiO/Co_3O_4 electrodes on a biodegradable waterborne polyurethane substrate. *Advanced Materials Technologies, 5*(2), 1900903.

Wang, Y., Lue, A., & Zhang, L. (2009). Rheological behavior of waterborne polyurethane/starch aqueous dispersions during cure. *Polymer, 50*(23), 5474–5481.

Wang, Y., & Zhang, L. (2008). High-strength waterborne polyurethane reinforced with waxy maize starch nanocrystals. *Journal of Nanoscience and Nanotechnology, 8*(11), 5831–5838.

Wang, N., Zhang, L., & Gu, J. (2005). Mechanical properties and biodegradability of crosslinked soy protein isolate/waterborne polyurethane composites. *Journal of Applied Polymer Science, 95* (2), 465–473.

Wang, N., Zhang, L., & Lu, Y. (2004). Effect of the particle size in dispersions on the properties of waterborne polyurethane/casein composites. *Industrial & Engineering Chemistry Research, 43*(13), 3336–3342.

Wang, J., Zhang, H., Miao, Y., Qiao, L., Wang, X., & Wang, F. (2018a). Microphase separation idea to toughen CO_2-based waterborne polyurethane. *Polymer, 138*, 211–217.

Wang, Y., Tian, H., & Zhang, L. (2010). Role of starch nanocrystals and cellulose whiskers in synergistic reinforcement of waterborne polyurethane. *Carbohydrate Polymers, 80*(3), 665–671.

Wang, N., & Zhang, L. (2005). Preparation and characterization of soy protein plastics plasticized with waterborne polyurethane. *Polymer International, 54*(1), 233–239.

Wu, G. M., Chen, J., Huo, S. P., Liu, G. F., & Kong, Z. W. (2014). Thermoset nanocomposites from two-component waterborne polyurethanes and cellulose whiskers. *Carbohydrate Polymers, 105*, 207–213.

Wu, Q., & Zhang, L. (2001). Preparation and characterization of thermoplastic starch mixed with waterborne polyurethane. *Industrial & Engineering Chemistry Research, 40*(2), 558–564.

Xiaojuan, L., Xiaorui, L., Lei, W., & Yiding, S. (2010). Synthesis and characterizations of waterborne polyurethane modified with 3-aminopropyltriethoxysilane. *Polymer Bulletin, 65*(1), 45–57.

Xu, X., Shang, S., Song, Z., Cui, S., Wang, H., & Wang, D. (2011). Preparation and characterization of rosin-based waterborne polyurethane from maleopimaric acid polyester polyol. *BioResources, 6* (3), 2460–2470.

Xu, Y., Yang, Y., Yan, D. X., Duan, H., Zhao, G., & Liu, Y. (2019). Flexible and conductive polyurethane composites for electromagnetic shielding and printable circuit. *Chemical Engineering Journal, 360*, 1427–1436.

Yeh, J. M., Yao, C. T., Hsieh, C. F., Yang, H. C., & Wu, C. P. (2008). Preparation and properties of amino-terminated anionic waterborne-polyurethane–silica hybrid materials through a sol–gel process in the absence of an external catalyst. *European Polymer Journal, 44*(9), 2777–2783.

Yu, S. H., Mi, F. L., Shyu, S. S., Tsai, C. H., Peng, C. K., & Lai, J. Y. (2006). Miscibility, mechanical characteristic and platelet adhesion of 6-O-carboxymethylchitosan/polyurethane semi-IPN membranes. *Journal of Membrane Science, 276*(1–2), 68–80.

Zeng, M., Zhang, L., Wang, N., & Zhu, Z. (2003). Miscibility and properties of blend membranes of waterborne polyurethane and carboxymethylchitin. *Journal of Applied Polymer Science, 90*(5), 1233–1241.

Zeng, M., Zhang, L., & Zhou, Y. (2004). Effects of solid substrate on structure and properties of casting waterborne polyurethane/carboxymethylchitin films. *Polymer, 45*(10), 3535–3545.

Zhang, M., Song, F., Wang, X. L., & Wang, Y. Z. (2012a). Development of soy protein isolate/waterborne polyurethane blend films with improved properties. *Colloids and Surfaces B: Biointerfaces, 100*, 16–21.

Zhang, F., Wang, R., He, Y., Lin, W., Li, Y., Shao, Y., Li, J., Ding, M., Luo, F., Tan, H., & Fu, Q. (2018). A biomimetic hierarchical structure with a hydrophilic surface and a hydrophobic subsurface constructed from waterborne polyurethanes containing a self-assembling peptide extender. *Journal of Materials Chemistry B, 6*(26), 4326–4337.

Zhang, L., Zhang, H., & Guo, J. (2012b). Synthesis and properties of UV-curable polyester-based waterborne polyurethane/functionalized silica composites and morphology of their nanostructured films. *Industrial & Engineering Chemistry Research, 51* (25), 8434–8441.

Zhang, G., Zou, M., Wang, S., & Zhang, Z. (2006). Radiation-induced dispersion polymerization of methyl methacrylate using waterborne polyurethane as stabilizer. *Polymer International, 55*(9), 1099–1103.

Zhao, Z., Jing, Z., Qiu, F., Dai, Y., Xu, J., Yu, Z., & Yang, P. (2017). Preparation, thermal property and morphology analysis of waterborne polyurethane-acrylate. In *AIP Conference Proceedings 2017* (Vol. 1794, No. 1, p. 020026). AIP Publishing LLC.

Zhong, S. L., & Cai, B. Z. (2009). Exploiting and application of environment-friendly waterborne polyurethane filmlaminated adhesives for soft packaging. *China Adhesives, 5*.

Zhou, X., Li, Y., Fang, C., Li, S., Cheng, Y., Lei, W., & Meng, X. (2015). Recent advances in synthesis of waterborne polyurethane and their application in water-based ink: A review. *Journal of Materials Science & Technology, 31*(7), 708–722.

Zhu, Q., Li, X., Fan, Z., Xu, Y., Niu, H., Li, C., Dang, Y., Huang, Z., Wang, Y., & Guan, J. (2018). Biomimetic polyurethane/TiO_2 nanocomposite scaffolds capable of promoting biomineralization and mesenchymal stem cell proliferation. *Materials Science and Engineering: C, 85*, 79–87.

Zia, F., Zia, K. M., Zuber, M., Kamal, S., & Aslam, N. (2015). Starch based polyurethanes: A critical review updating recent literature. *Carbohydrate Polymers, 134*, 784–798.

Zou, J., Zhang, F., Huang, J., Chang, P. R., Su, Z., & Yu, J. (2011). Effects of starch nanocrystals on structure and properties of waterborne polyurethane-based composites. *Carbohydrate Polymers, 85*(4), 824–831.

Waterborne Polyurethanes for Biomedical Applications

Charles Oluwaseun Adetunji, Olugbemi Tope Olaniyan, Abel Inobeme, John Tsado Mathew, and Mohd Imran Ahamed

Abstract

Over the years, there are numerous benefits derived from water borne polyurethane which entails medical devices (vascular prostheses, catheters, pacemakers), reconstructive surgery materials (breast implants, wound dressings, maxillofacial prosthetics, several other tissue implants) obstetrics and gynecology plus reproductive biology (contraceptive sponges, condoms, and assisted reproductive technology devices) medical materials (surgical drapes and hospital sheets). Also, they are extensively utilized in wound and plaster sectors because of the flexible, barrier protective, antimicrobial and soft properties. Therefore, this chapter intends to provide detailed information on the application of water borne polyurethane most especially in the medical field such as drug delivery with nanocapsules plus nanoparticles, bone regeneration, catheters polyurethane, contact lenses, tissue engineering scaffolds.

C. O. Adetunji (✉)
Applied Microbiology, Biotechnology and Nanotechnology Laboratory, Department of Microbiology, Edo State University Uzairue, Iyahmo, Edo State, Nigeria
e-mail: adetunji.charles@edouniversity.edu.ng

O. T. Olaniyan
Laboratory for Reproductive Biology and Developmental Programming, Department of Physiology, Edo State University Uzairue, Iyahmo, Edo State, Nigeria

A. Inobeme
Department of Chemistry, Edo State University Uzairue, Iyahmo, Nigeria

J. T. Mathew
Department of Chemistry, Ibrahim Badamasi University, Lapai, Niger State, Nigeria

M. I. Ahamed
Department of Chemistry, Faculty of Science, Aligarh Muslim University, Aligarh, Uttar Pradesh 202002, India

Keywords

Water borne polyurethane • Drug delivery • Vascular prostheses • Catheters • Breast implants • Wound dressings • Maxillofacial prosthetics • Tissue implants • Bone regeneration

1 Introduction

Water borne polyurethane (WPU) basically refers to a polyurethane that has reacted completely in aqueous medium. There are several health implications and challenges associated with the fact that organic solvents synthesized polyurethanes are not environmentally friendly with the substitution with water-based polyurethane products. Several remarkable investments have been put in with a view to enhancing technological advancement in the area of water borne polyurethane products. The broad applications of WPU are attributed to the remarkable features associated with it such as resistance to abrasion, high mechanical impact strength, and flexibility with regards to temperature. WPU is utilized in various areas ranging from elastomers, coating, adhesive, and sealants. The major limiting factor in the production of WPU is the cost that is involved in the process of manufacturing which is very high and the demand for highly technical skills in the formulation (Ajorlou et al., 2016) (Table 1).

Polyurethane has found remarkable applications in the medical field. This wide usage has been attributed to some of the unique properties of this group of compounds which include high biocompatibility, biodegradability, and biostatic characteristics. These properties are improved by using suitable monomers for the synthesis and preparation of the composites (Brzeska, 2015).

The use of polyurethane in different areas of medicine has been investigated by various researchers. They have been assessed for potential utilization in the development of

Inamuddin et al. (eds.), *Sustainable Production and Applications of Waterborne Polyurethanes*, Advances in Science, Technology & Innovation, https://doi.org/10.1007/978-3-030-72869-4_11

Table 1 Studies on medical application of WPU

Application	Properties	Authors
Implanting devices	High flexibility and mechanical strength	(Ajorlou et al., 2016)
Artificial hearts	High resistance to strains	(He et al., 2014; Zdrahala, 1996; Miyamoto et al., 1999; Pan et al., 2008)
Scaffold for tissue growth	High flexibility and biocompatibility	(He et al., 2014)
Support for aiding healing wounds	Biocompatibility and biodegradability	(Brzeska, 2015; Halima & Sharif, 2018)
Growth of osteogenic cells	Biocompatibility and mechanical strength	(Sartori et al., 2014; Hsu et al., 2018)
Drug delivery	Ease modification and haemocompatibility	(Shin & Choi, 2018; Liu et al., 2019)
Antimicrobial agents	Antimicrobial properties and haemocompatibility	(He et al., 2014)

membranes for the dressing and treatment of wounds, for the treatment of lower stage meniscal loss, nanocarriers of drugs in endovascular usage, moderated discharge for membrane delivery of ketoprofen, as polyurethane that is biologically stable for materials used in bone replacement. Also, the shaped memory polyurethane is employed as a wire in orthodontic application which aids the effective alignment of teeth. The incorporation of chitosan into water borne polyurethane has also been investigated as antimicrobial agent in manufacturing acrylic textiles that can be used in carpets and blankets that are used in hospitals (Bankoti et al., 2017).

One of the major remarkable applications of PUR in the medical field is the production of implanting devices used in the treatment of heart diseases. This usage is based on some unique properties of PUR which involve toughness, ease of introduction of various additives, high mechanical strength, high flexibility without the need for modifying agents, and high haemocompatibility which results in the modification of the surface (Ajorlou et al., 2016).

Studies have shown the high resistance of PURs to environmental effects and strains hence were utilized for the production of the first artificial heart. This has given birth to the pursuit of possible development of biocompatible polyurethane that can be used for the reconstruction of tissues. The utilization of monomeric materials that have high susceptibility to environmental agents makes it possible for the production of PURs that are biodegradable which have promising applications in creating scaffolds that are used for growing living cells or as temporally implanting tools. The soft segment is constructed by the incorporation of ester groups that are sensitive to hydrolysis, mainly with oligomerols (He et al., 2014).

The polyurethane elastomers are a group of compounds that possess hard and soft segments which form a structure of separated microphase. There is enhancement of mechanical properties as a result of the microphase separation (He et al., 2014). At present Pus are widely employed in medicine in aiding wound healing and heart supporting devices as a result of their high compatibility with biological systems. The adjustment of various chemical structures makes it possible for it to be used in suiting different purposes. When in comparison with the conventional polyurethane, the water borne polyurethane is biodegradable and environmentally friendly and is easily dispersed in water through ionic groups as nanoparticles (Brzeska, 2015).

Also, the presence of ionic groups may be regarded as physical crosslinks. The ionic groups bring about the restriction of the mobility of the hard segment and this results in a higher level of microphase separation. Specifically, the presence of anionic groups such as carboxyl structure has proven low inflammatory reaction. Few studies exist in the utilization of PU in antiadhesive barriers. However, in one of the studies, Rehman investigated the antiadhesive potential of the biodegradable PU. Findings from the studies revealed that the peritoneal adhesion was influenced by the compositions of the polyurethane in which case the aliphatic ones were less effective when compared to the aromatic, the various soft segments did not exhibit any significant difference in terms of anti-adhesion effect (Travinskaya et al., 2014).

In the area of tissue bone technology, the elastomeric polyurethanes help in avoiding shear forces that exist between the interface implant and the bone; this, therefore, supports the growth of osteogenic cells. It is also possible to develop soft tissue effectively using water bone polyurethanes and this is remarkably promising in the construction of muscles constructs as well as cartilage, nerve tissues, heart, and blood vessels (Sartori et al., 2014).

One of the most known synthetic elastomers that is widely used in the medical field is water borne polyurethane which is due to its remarkable hemocompatibility and biocompatibility. These materials are also known to comply with the body in healing processes hence can be mediated through the modification of the chemical compositions. Bulk modification is also possible through the incorporation of biomolecules like anticoagulants and other biorecognisable agents. Modifications of such have been done in other to bring about the improvement of implant acceptance. Due to this, the traditional polyurethane that is based on solvent has been widely used in various medical devices such as the dressing of wounds, tubing, and artificial heart (Shin & Choi, 2018). Consequently, polyurethane that is water borne is undoubtedly indispensable in its use in the medical sector. As a result of this, there are many studies currently being carried out with regards to the production of more environmentally friendly polyurethanes.

The commercialization of polyurethane in medical field first occurred in 1960. It was used as an elastomer known as biomer in cardiovascular usage. This progressed in 1991 as a result of the development of cracks in the surface of the polymer. In spite of the limitations, more researches were carried out in comprehending the chemistry of water borne polyurethane in their use in the medical field.

Therefore, this chapter intends to provide detailed information on the application of water borne polyurethane most especially in medical field such as Drug delivery with nanocapsules plus nanoparticles, Bone Regeneration, Catheters polyurethane, Contact Lenses, Tissue engineering scaffolds.

2 Studies on Application of Polyurethane in Medical Field

Shin and Choi (2018) in their work opined that water borne polyurethane has wide application in the medical sector and attributed this to some of the unique properties which include their high hemocompatibility and unique compatibility with biological systems. They further identified some other basic properties of water borne polyurethane which include their elasticity, resistance to fatigue, and ability to tolerate the body during the process of healing.

Lu and Chang (2020) in their work developed an antimicrobial substance based on mono (hyroxyethoxylethyl) using a combination of different metals such as calcium, manganese, zinc, and lead. The other starting materials used include phthalic anhydride, acetates of divalent metals such as lead and manganese. They also synthesized water borne polyurethane oil using linseed oil by acetone processing technique. They observed that the addition of the metallic-based antimicrobial agent brought about improvement of the stability of the material to heat however the gloss of the films was reduced. Additionally, there is so no significant impact in some of the properties such as resistance to impact, mass retention, and adhesion.

Hsu et al. (2018) in their work aimed at improving the regenerated bone membranes, synthesized polycaprolactone derived polyurethanes. The bone membranes targeted were those of remarkable tensile strength as well as elongation. The preparation was done through the polymerization process involving polycaprolactone diols alongside hexamethylene diisocyanate at varying ratios, during which the chain extender used was glycol. They concluded that the membrane of electrospun obtained has a high tensile strength of 19.84 MPa while the elongation at break was found to be 62.7%.

Lee et al. (2018) worked on the preparation of azomethine diols using condensation reaction involving aldehyde of terephthalic acid, ethanol amine, and polyether diamine. The azomethine prepared was then used for the introduction of azomethine groups into the polyurethane backbones. They concluded from their work that it is promising to combine the mechanical properties and self healing of polyurethanes for varying the design and concentration of azomethine diols.

Liu et al. (2019) in a novel work fabricated water borne polyurethane based on antibiofilm property and auranofin realizing antibacterial. This was then assessed for potential usage in various medical areas. The auranofin is a drug that is used in the treatment of rheumatism and has unique anti-microbial properties. The antibacterial agent is surrounded by the carrier of the drug, polyurethane. This also aids the release of the drugs as well as improving its long time antibiofilm and antibacterial potential. The further observed that the coatings were highly hemocompatible with various human cells and tissues.

Hsu et al. (2010) in a similar study synthesized successfully an elastomer that is biodegradable in nature in the form of a nanocomposite of polyurethane. The synthesis was based on the reaction of polyester diols in a water-based green process. They concluded that the nanocomposites and their respective assembling represent a unique biodegradable elastomer that has potential applications in medicine.

He et al. (2014) developed highly cross-linked polyurethane that is water borne. This novel compound with remarkable stability for the long term was prepared using quaternary ammonium salt, polyethylene glycol, isophorone, and I-lysine. Further studies revealed the high antibacterial efficiency towards various strains of bacteria.

Zdrahala (1996) reported that water borne polyurethane is useful in the construction of vascular tissues specifically the development of vascular access, coronary artery as well as peripheral pathway indicators. He further documented that these groups of materials are also used in the formation of conduits that have several layers and features of the vascular vessels. Some of the properties further identified include plasticity and the healing potentials of these compounds.

Miyamoto et al. (1999) in their work investigated the possible development of a novel approach for incorporation of a microporous into caliber vascular prosthesis that is generated by water borne polyurethane. They used a solution made up of dimethylformamide and tetrahydrofuran combined in 1:1 proportion. A varying quantity of calcium trioxocarbonate (IV) was then covered with a mandril of glass and dipped in distilled water for a period of 24 h. The prosthesis obtained was later placed through implantation into carotid and femoral arteries as well as the abdominal aorta of dogs.

Pan et al. (2008) carried out a study in which the embedding effects of synthetic vascular grafts containing recombinant fibrinolytic enzyme factor was investigated. They prepared polyurethane artificial vascular grafts of 4 mm diameter through the leaching and dipping technique. Observations were then made on the morphology and micropore sizes using SEM. It was concluded that the immobilization process could aid the fibrinolytic activity and also bring about the inhibition of thromboembolic formation.

Okoshi et al. (1993) in a similar study fabricated polyurethane–polydimethylsiloxane of spongy nature, with a vascular graft of 1.5 millimeter internal diameter. Unique handling features were shown by grafts of low porosity with lower hydraulic permeability. They deduced from their findings that it was possible to obtain patency as well as a remarkable degree of endothelialization in the rat aorta using the replacement model.

Hsu et al. (2018) carried out the preparation of nanocomposites using water borne polyurethane based on polyester. The substrate contained varying concentrations of silver nanomaterials. Unique dispersion of the nanoparticles was recorded up to 30 ppm of the silver nanoparticle and the confirmation of this was through the use of transmission electron microscopy (TEM). They reported that the adhesion of the microbes (*Escherichia coli* and *Bacillus spp*) the water borne polyurethane silver nanocomposite was low significantly at the concentrations investigated.

Hsu et al. (2014) revealed that recently, progress has been made in the development of an anti-adhesion material for an individual having problems with joint motion and tendon gliding. The authors showed that waterborne biodegradable polyurethane films are capable of producing biocompatible, non-cytotoxic, and good anti-adhesion material based on their mechanical plus physicochemical properties such as thickness change, gelatination, plus water absorption. From their results, it was demonstrated using a biochemical, and histological assessment that waterborne biodegradable polyurethane films showed immense potential as anti-adhesion properties when compared with others. They, therefore, concluded that waterborne biodegradable polyurethane films have a low inflammatory response, appropriate degradation period, and better mechanical activity.

Halima and Sharif (2018) explained that the effectiveness of any material is based on its biocompatibility. It is generally known that immunological reactions may develop to any external material when they interact with tissues of the body. The authors, therefore, suggested that in order to overcome these challenges, the polymerization conditions and selection of monomers for polyurethane should possess good physicochemical activity. They noted that through appropriate tissue engineering, polyurethane can be tailored or designed to meet specific application conditions. Serious attention must be given to polymerization conditions when designing polyurethane for biomedical applications. Owing to the fact that fatigue resistance and high flexural endurance properties are the major requirements in vascular tissue engineering, polyurethane material is recognition more consideration in the design of biomedical applications. In the past few years, notable changes and expansion have been witnessed in tissue engineering using polyurethane material for biomedical applications with other emerging areas.

Yong et al. (2015) demonstrated that the solvent-borne systems are the conventional polyurethane-ureas and polyurethane with hydrophobicity, hence in the production of waterborne polyurethane plus polyurethane-urea the use of internal emulsifier which is environmentally friendly due to less production of volatile organic molecules is encouraged. Studies have shown that polyurethanes are chemically synthesized copolymers made up of urethane bonds in their dynamic chemical structure. In their structure, there are three monomers; polyol, chain extender, and diisocyanate facilitating the synthesis of large polyurethanes with great physico-chemical properties basically dependent on the polyols conformation. The structural integrity of the polyols such as chain length, ether or ester groups, aromatic or aliphatic unit, and hydroxyl functionality determines the flexibility or rigidity of the polyurethanes. The utilization of polyurethanes in biomedical industries represents an important and dynamic area of tremendous advantage. Many studies have highlighted the importance of utilization of waterborne polyurethanes as medical or biomedical implants in the time past, but in recent years, the utilization of polyurethanes has extended tremendous to other medical areas such as tissue engineering, wound dressing, and drug delivery. In the area of wound dressing, medical and health workers find waterborne polyurethane materials very highly compatible biologically (Lu et al., 2005).

3 Wound Dressings and Plasters

The biological compatibility of polyurethanes such as absorbency of wound exudates, hydrophilicity, wound humidity control, dynamic mechanical characteristics, antimicrobial properties, and permeability in speeding up the healing process have to make polyurethanes a good candidate for use. Serious attention has been given to the production of polyurethanes from industrial perspectives such as replacement of the raw material from fossil fuel sources to renewable natural resources. Recent research has focused attention on the polymerization reactants as natural renewable materials derived from various sources such as food and non-food sources (sunflower oil, canola oil plus soybean oil, algae, liquefied wood plus *Physaria tenella*). Others include chitin, glucomannans, heparin, collagen, alginate, chitosan, starch, plus lignin with tremendous medical benefits (Ninan et al., 2013).

Vermette et al. (2001) revealed the utilization of polyurethane ranges from diverse areas of medical fields such as in medical devices (vascular prostheses, catheters, pacemakers), reconstructive surgery materials (breast implants, wound dressings, maxillofacial prosthetics, several other tissue implants) obstetrics and gynecology plus reproductive biology (contraceptive sponges, condoms, and assisted reproductive technology devices) medical materials (surgical drapes and hospital sheets). In the utilization of any synthetic material, biocompatibility is key in order not to generate any immunological response from proteins or chloride ions to the material. Also, in addition, silicone rubbers are sometimes used in biomedical fields but unlike polyurethane, silicone rubbers are prone to more wear and tear which often limits their usage. Studies have revealed that polyurethane-urea polymers are very hemocompatible based on soft and hard segments making these materials biomedically relevant till today in different areas such as intra-aortic balloon catheters, artificial hearts, hemodialysis membranes and heart valves (Xu et al., 2016).

Polyurethanes are extensively utilized in wound and plaster sectors because of their flexible, barrier protective, antimicrobial and soft properties. The chemistry of the commonly used polyurethane are perforated films, semi-permeable adhesive films, and foam plus hydrocolloids nature. Many of the protective properties show transparent and thin films with no or little exudates secretion during or post-operative wounds. They also show no permeability to pathogens and water but to gases make wounds to respire. In the production of polyurethane wound dressing, different bioactive compounds are present like antibacterial agents, glycerine, local anesthetics, zinc or collagen, absorbents, antifungal agents, gel-forming agents, and analgesic agents. Yari et al. (2012) have contributed significantly to the development of novel polyurethane membrane by synthesizing new design by amine reaction from epoxy-polyurethane prepolymer utilized for wound care with antimicrobial properties and wound exudates absorbent ability.

Another study by Ozkaynak et al. (2005) demonstrated the ability of linseed oil to generate polyester polyurethane membrane for wound dressing. The authors revealed that the membrane showed good acid resistance and mechanical properties. Subsequently, Gultekin et al. (2009) produced many polyurethane films for medical coatings by employing glycerol and linoleic acid through the esterification technique. The authors carried out the toxicity and biocompatibility test to ascertain the physicochemical properties. They revealed that polyurethane surface prepared without catalyst or cross-linked showed the best fibroblast growth in wound dressing.

4 Tissue Engineering Scaffolds

In tissue engineering, many physicochemical properties of biomaterials are considered to ascertain their usefulness. Interconnected pores, high degree of porosity, pore dimensions, are majorly considered (2016). Studies have shown that in liver regeneration, the pore dimension of 20 μm should be allowed to facilitate the growth of the liver hepatocytes. For bone and skin, the range of 200 and 400 μm, 20–150 μm are considered for the pore size respectively with interconnection to enhance cell and tissue growth. The techniques utilized for the production of porous or fibrous scaffolds are either advanced or conventional methods. Generally, the conventional methods are thermal-induced segment separation, particle leaching/solvent casting, gas foaming plus melt molding whereas advanced methods are electro-spinning, 3D printing plus inter alia. Polyurethanes in tissue engineering are utilized for vascular grafts, artificial skin, bone grafts, repair of cartilage, and neural connections. Also Hung et al. (2016) recently described the water-based polyurethane in 3D printed scaffolds for regulated cartilaginous tissue engineering. The authors discovered that prepared scaffolds promote the self-aggregation of mesenchymal stem cells.

Halima and Sharif (2018) revealed that serious work on improving the biocompatibility, biodegradability, fatigue resistance, and mechanical properties of polyurethanes in tissue engineering utilization has been ongoing. Typical examples of its biomedical applications include bone regeneration, fibrocartilage repair, and cardiac patches. Hence to achieve an efficient waterborne polyurethane material, serious attention must be placed on mimicking the extracellular plus intracellular fluid conditions to generate appropriate tissue response.

5 Drug Delivery with Nanocapsules Plus Nanoparticles

In recent times, polymeric nanocapsules and nanoparticles have received considerable attention as promising candidate for drugs delivery plus others agents. Thus the production of biocompatible polyurethanes has received a tremendous boost in pharmaceutical, medical, and cosmetics industries due to positive physicochemical properties such as biodegradability and biocompatibility. Major progress has been made in the formation, characterization, and design of polyurethane nanoparticles. Various techniques and approaches are utilized when producing polyurethane particles for drug delivery such as spontaneous emulsification joined with interfacial polycondensation, suspension-polyaddition, suspension-polycondensation, scattering in the organic solvent, mini emulsion technique, and supercritical CO_2. Studies have indicated that the physicochemical properties of these polymers enable them to be a suitable candidate for nucleic acids binding, antibodies, and peptides binding with other biomolecules for biomedical applications. Cusco et al. (2016) have recently invented a novel drug delivery mechanism for cancer therapy built using polyurethane/polyurea nanocapsules which could defend the drug from untimely stimulation and enhancement of direct release and target of tumor cells. Also Rosenbauer et al. (2010) recently developed a stimulus–response polyurethane nanocapsule made up of a polymeric shell and aqueous core facilitating compressed fluorescent dye to release upon UV-light, temperature, or pH stimulation. Halima and Sharif (2018) also revealed that the uniqueness of waterborne polyurethane makes it suitable for drug delivery, hence it was suggested that the high biocompatibility with blood makes it's the appropriate candidate for in vivo drug delivery utilization.

6 Bone Regeneration

Halima and Sharif (2018) demonstrated that in bone regeneration, a number of physiological processes are involved such as differentiation, proliferation, plus migration of osteogenic progenitors, with the generation of extracellular fluids. The authors suggested that synthesis of biodegradable waterborne polyurethane by different authors has restructured that landscape of bone tissue engineering with significant achievement.

7 Catheters Polyurethane

Chandy et al. (2009) also showed that catheters medical instruments are utilized for delivering fluid to different areas in the body. They revealed that there are different forms or types such as central or intravenous vascular catheters which are implanted in the body like cavities or canals and must be biocompatible with the body tissue otherwise severe immunological reaction will develop.

8 Contact Lenses

Halima and Sharif (2018) revealed that contact lenses must have good physicochemical properties such as mechanical strength, permeability to ions, efficient water absorption capacity, plus high transparency. In the past few years, the use of waterborne polyurethane has been seen.

Wang et al. (2010) revealed that polyurethane is a common synthetic elastomer with exceptional biocompatibility widely utilized in biomedical science. In the preparation, certain factors are considered like elasticity, fatigue resistance, durability, plus compliance. Today, in the manufacturing processes a lot of additives in form of biomolecules are combined to improve the biological response, acceptability, and effectiveness. Some of these biomolecules are anti-inflammatory agents or anticoagulants. For high-performance systems, the authors suggested waterborne polyurethane as the best established standard utilized in biomedical fields.

Xu et al. (2012) also contributed to the importance of polyurethane by revealing that waterborne polyurethane is gaining tremendous attention for biomedical utilization owing to the excellent physicochemical properties and application in breast implants, vascular prostheses, pacemaker lead insulation, bioadhesives plus heart valves. The authors highlighted that the surface properties of polyurethane are crucial to its performance, hence medical devices developers pay serious attention to the surface and different surface modifiers plus techniques. Many surface modifiers have been used such as chemical, immobilization of biological compounds, salinization, plus radiation grafting of monomers to improve the quality of polyurethane without altering the entire properties.

9 Conclusion and Future Recommendation to Knowledge

This chapter has provided detailed information on the application of water borne polyurethane in the biomedical field was also highlighted. The techniques utilized for the production of porous or fibrous scaffolds are either advanced or conventional methods. Generally, the conventional methods are thermal-induced segment separation, particle leaching/solvent casting, gas foaming plus melt molding whereas advanced methods are electro-spinning, 3D printing plus inter alia. Polyurethanes in tissue engineering are utilized for vascular grafts, artificial skin, bone grafts, repair of cartilage, and neural connections were also highlighted. Special features of water borne polyurethane on how it improves biocompatibility, biodegradability, fatigue resistance, and mechanical properties of polyurethanes in tissue engineering utilization were highlighted. Typical examples of its biomedical applications include bone regeneration, fibrocartilage repair, and cardiac patches. Also, stimulus–response polyurethane nanocapsules derived from the polymeric shells and aqueous core facilitating compressed fluorescent dye to release upon UV-light, temperature, or pH stimulation were also highlighted. Special highlights on biomedical utilization owing to the excellent physicochemical properties and application in breast implants, vascular prostheses, and pacemaker lead insulation, bioadhesives plus heart valves.

References

Ajorlou, E., Khosroushahi, A. Y., & Yeganeh, H. (2016). Novel water-borne polyurethane nanomicelles for cancer chemotherapy: Higher efficiency of folate receptors than TRAIL receptors in a cancerous Balb/C mouse model. *Pharmaceutical Research, 33*(6), 1426–1439.

Bankoti, K., Rameshbabu, A. P., Datta, S., Maity, P. P., Goswami, P., Datta, P., Ghosh, S. K., Mitra, A., & Dhara, S. (2017). Accelerated healing of full thickness dermal wounds by macroporous waterborne polyurethane-chitosan hydrogel scaffolds. *Materials Science and Engineering C, 81*, 133–143. https://doi.org/10.1016/j.msec.2017.07.018

Brzeska, J. (2015). Synthesis and properties of polyurethanes based on synthetic polyhydroxybutyrate for medical application. In: C. K. Das (Ed.), *Thermoplastic elastomers—Synthesis and applications*. Intech Open. https://doi.org/10.5772/60933. Available from https://www.intechopen.com/books/thermoplastic-elastomers-synthesis-and-applications

Chandy, T., Van Hee, J., Nettekoven, W., & Johnson, J. (2009). Long-term in vitro stability assessment of polycarbonate urethane micro catheters: Resistance to oxidation and stress cracking. *Journal of Biomedical Materials Research Part B: Applied Biomaterials, 89*(2), 314–324. https://doi.org/10.1002/jbm.b.31218

Cusco, C., Soto-Cerrato, V., Rocas, J., & Perez-Tomas, R. (2016). Design and characterization of novel polyurea/polyurethane nanocapsules developed to improve cancer chemotherapy delivery. *European Journal of Cancer, 61*(Suppl. 1), S125–S126. https://doi.org/10.1016/S0959-8049(16)61445-4

Gultekin, G., Atalay-Oral, C., Erkal, S., Sahin, F., Karastova, D., Tantekin-Ersolmaz, B. S., & Guner, S. F. (2009). Fatty acid-based polyurethane films for wound dressing applications. *Journal of Materials Science: Materials in Medicine, 20*(1), 421–431. https://doi.org/10.1007/s10856-008-3572-5

Halima, K., & Sharif, A. (2018). Polyurethane: A versatile scaffold for biomedical applications. *Journal of Bioscience and Bioengineering*, 2(3), SBB.000536. https://doi.org/10.31031/SBB.2018.02.000536.

He, Y., Xie, D., & Zhang, X. (2014). The structure, microphase separated morphology and property of polyurethane and polyureas. *Journal of Material Science, 49*(21), 7339–7352.

Hsu, S., Tseng, H., & Lin, Y. (2010). The biocompatibility and antibacterial properties of waterborne polyurethane-silver nanocomposites biomaterials. https://doi.org/10.1016/j.biomaterials.2010.05.015

Hsu, S.-H., Dai, L.-G., Hung, Y.-M., & Dai, N.-T. (2018). Evaluation and characterization of waterborne biodegradable polyurethane films for the prevention of tendon postoperative adhesion. *International Journal of Nanomedicine, 13*, 5485–5497.

Hsu, S.-h..Hung, K.-C., Lin, Y.-Y., Su, C.-H., Yeh, H.-Y., Jeng, U-S., Lu, C.-Y., Dai, S.A., Fue, W.-E., Lin, J.-C. (2014). Water-based synthesis and processing of novel biodegradable elastomers for medical applications. *Journal of Materials Chemistry B*.

Hung, K.-C., Tseng, C.-S., Dai, L.-G., & Hsu, S.-H. (2016). Water-based polyurethane 3D printed scaff olds with controlled release function for customized cartilage tissue engineering. *Biomaterials, 83*, 156–168.

Lee, S., Wu, S., Chen, H., Tsai, L.-L., Tzeng, J.-J., Lin, C.-H., Lin, Y. (2018). Synthesis and characterization of polycaprolactone-based polyurethanes for the fabrication of elastic guided bone regeneration membrane. *BioMed Research International*, 3240571. https://doi.org/10.1155/2018/3240571

Liu, H., Shukla, S., Vera-González, N., Tharmalingam, N., Mylonakis, E., Fuchs, B. B., & Auranofin, A. S. (2019). Releasing antibacterial and antibiofilm polyurethane intravascular catheter coatings. *Frontiers in Cellular and Infection Microbiology, 9*, 37. https://doi.org/10.3389/fcimb.2019.00037.eCollection

Lu, K.-T., & Chang, J.-P. (2020). Synthesis and antimicrobial activity of metal-containing linseed oil-based waterborne urethane oil wood coatings. *Polymers (Basel), 12*(3), 663. https://doi.org/10.3390/polym12030663

Lu, Y., Tighzert, L., Berzin, F., et al. (2005). Innovative plasticized starch films modified with waterborne polyurethane from renewable resources. *Carbohydrate Polymers, 61*, 174–182.

Miyamoto, K., Sugimoto, T., Okada, M., & Maeda, S. (1999). Experimental studies on application of small-caliber vascular prosthesis produced by polyurethane. *Annals of Thoracic and Cardiovascular Surgery, 5*(3), 174–181.

Ninan, N., Muthiah, M., Park, I.-K., & Grohens, Y. (2013). Faujasites incorporated tissue en.gineering scaffolds for wound healing: in vitro and in vivo analysis. *ACS Applied Materials & Interfaces*, 5(21), 11194−11206. https://doi.org/10.1021/am403436y.

Okoshi, T., Soldani, G., Goddard, M., & Galletti, P. M. (1993). Very small-diameter polyurethane vascular prostheses with rapid endothelialization for coronary artery bypass grafting. *Journal of Thoracic and Cardiovascular Surgery, 105*(5), 791–795.

Ozkaynak, M. U., Atalay-Oral, C., Tantekim-Ersolmaz, S. B., & Guner, S. (2005). Polyurethane films for wound dressing applications. *Macromolecular Symposia, 228*(1), 177–184. https://doi.org/10.1002/masy.200551016

Pan, S., Yin, S., Yao, J., Zheng, H., Yi, W., & Dai, G. (2008). Test of carotid artery replacement by polyurethane artificial vascular graft with recombinant fibrinolytic enzyme factor II-modified lumina in dogs. *Zhongguo Xiu Fu Chong Jian Wai Ke Za Zhi, 22*(7), 832–837.

Rosenbauer, E. M., Wagner, M., Musyanovych, A., & Landfester, K. (2010). Controlled release from polyurethane nanocapsules via pH-UV-Light- or temperature-induced stimuli. *Macromolecules, 43* (11), 5083–5093. https://doi.org/10.1021/ma100481s

Sartori, S., Chiono, V., Tonda-Turo, C., Mattu, C., Gianluca, C. (2014). Biomimetic polyurethanes in nano and regenerative medicine. *Journal of Materials Chemistry B* (32).

Shin, E., & Choi, S. (2018). Advances in waterborne polyurethane-based biomaterials for biomedical applications. *Advances in Experimental Medicine and Biology, 1077*, 251–283. https://doi.org/10.1007/978-981-13-0947-2_14

Travinskaya, T., Savelyev, Y., & Mishchuk, E. (2014). Waterborne polyurethane based starch containing materials: Preparation, properties and study of degradability. *Polymer Degradation and Stability, 101*, 102–108.

Vermette, P., Griesser, J. H., Laroche, G., & Guidoin, R. (2001). *Biomedical applications of polyurethanes (tissue engineering intelligence unit)*. Landes Bioscience. Eurekah.com.

Wang, S., Xiaodong, Z., Tao, L., She, X., & Li, Z. (2010). Cooperatively modified polyvinyl chloride with polyurethane elastomer/nano-silica composite and its mechanical properties. *China Synthetic Rubber Industry, 2*, 5.

Xu, H., Qiu, F., Wang, Y., Yang, D., Wu, W., Chen, Z., & Zhu, J. (2012). Preparation, mechanical properties of waterborne polyurethane and crosslinked polyurethane-acrylate composite. *Journal of Applied Polymer Science, 124*, 958–968.

Xu, J., Li, T., Zhao, W., Li, P., & Wu, Y. (2016). Synthesis and characterization of waterborne polyurethane emulsions based on poly(butylene itaconate) ester. *Designed Monomers and Polymers, 19*(4), 309–318. https://doi.org/10.1080/15685551.2016.1152541

Yari, A., Yeganeh, H., & Bakhshi, H. (2012). Synthesis and evaluation of novel absorptive and antibacterial polyurethane membranes as wound dressing. *Journal of Materials Science: Materials in Medicine, 23*(9), 2187–2202. https://doi.org/10.1007/s10856-012-4683-6

Yong, Q., Nian, F., Liao, B., Huang, L., Wang, L., & Pang, H. (2015). Synthesis and characterization of solvent-free waterborne polyurethane dispersion with both sulfonic and carboxylic hydrophilic chain-extending agents for matt coating applications. *RSC Advances, 5*, 107413–107420.

Zdrahala, R. J. (1996). Small caliber vascular grafts. Part II: Polyurethanes revisited. *Journal of Biomaterials Applications, 11* (1), 37–61. https://doi.org/10.1177/088532829601100102.

Biomedical and Environmental Applications of Waterborne Polyurethane-Metal Oxide Nanocomposites

Jaison Jeevanandam, Sharadwata Pan, and Michael K. Danquah

Abstract

Metal oxide nanoparticles constitute one of the most conjoint subsets of nanosized particles that are under extensive research, attributing to their outstanding abilities to be useful as a catalyst, and in electronics and biomedical application domains. These nanoparticles are fabricated via physical, chemical, and biological approaches, and their properties depend on the selection of the particular synthesis strategy. However, these manufacturing methods possess certain limitations, which hurdles their benefits in large-scale applications. Consequently, metal oxide nanoparticles are blended with biocompatible polymers, to form nanocomposites for maintaining the stability, without compromising their additional significant properties. Contextually, waterborne polyurethanes are developed and incorporated with a wide variety of metal oxide nanoparticles, to form potential polymer-based nanocomposites with high biocompatibility, as well as mechanical and thermal stability. These novel nanocomposites are currently under investigation to be beneficial as coating materials or formulations in biomedical and environmental applications. This chapter aims to illustrate various types of magnetic and non-magnetic waterborne polyurethane-metal oxide nanocomposites, including their characteristics and specific functionalities. Furthermore, the advantages of utilizing these novel nanocomposites in biomedical and environmental applications are discussed, with an eye on their limitations and future outlook.

Keywords

Nanocomposites • Metal oxide nanoparticles • Waterborne polyurethanes • Biocompatibility • Coating materials

1 Introduction

Metal oxide nanoparticles are the most common nanosized particles, which are under extensive research, as they exhibit excellent abilities to be useful as a catalyst in electronic and biomedical fields (Bai & Tang, 2020). These nanoparticles have garnered more applicational significance compared to other types of nanosized particles, including metal nanoparticles, due to their high structural stability, biocompatibility, and the fact that they are comparatively easier to fabricate via ingenuous protocols (Oskam, 2006). Nanosized oxides of metal particles include aluminum, iron, zinc, copper, titanium, magnesium, manganese, zirconium, and cerium oxides (Chavali & Nikolova, 2019). Further, rare earth metals, such as yttrium, strontium, and lanthanum, were also synthesized as metal oxide nanoparticles for exclusive applications with desired properties (Hussein, 1996; Singh et al., 2013). These nanoparticles are fabricated via physical, chemical, and biological approaches, and their properties depend on the rationale of choice of a particular synthesis tactic (Stankic et al., 2016). Physical methods, such as ball milling, vapor deposition, laser ablation, electrospinning (Jeevanandam et al., 2016) and chemical approaches, including sol–gel, hydrothermal, solvothermal, polyol, and precipitation, are the most common synthesis methods of metal oxide nanoparticles (Niederberger & Pinna, 2009). However, while the necessity of sophisticated equipment and clean room facilities are the limitations of

J. Jeevanandam
CQM—Centro de Química da Madeira, MMRG, Universidade da Madeira, Campus da Penteada, 9020-105 Funchal, Portugal

S. Pan
School of Life Sciences Weihenstephan, Technical University of Munich, 85354 Freising, Germany

M. K. Danquah (✉)
Chemical Engineering Department, University of Tennessee, Chattanooga, TN 37403, USA
e-mail: michael-danquah@utc.edu

Inamuddin et al. (eds.), *Sustainable Production and Applications of Waterborne Polyurethanes*, Advances in Science, Technology & Innovation, https://doi.org/10.1007/978-3-030-72869-4_12

physical methods; the usage of toxic chemicals and lack of biocompatibility constitute the drawbacks of chemical approaches (Jeevanandam et al., 2020). Accordingly, biological approaches are introduced to reduce the toxicity of synthesized metal oxide nanoparticles by eliminating the usage of toxic chemicals (Jeevanandam et al., 2016). However, reduction in the stability of the nanoparticles will be a major issue, while utilizing biosynthesis approach for large-scale applications (Hii et al., 2018). Consequently, metal oxide nanoparticles are blended with biocompatible polymers to form nanocomposites, in order to maintain stability as well as biological properties, without compromising on their other significant properties (Prasanna et al., 2019).

Among the polymer-based metal oxide nanocomposites, polyurethanes are recently gaining the focus of researchers, due to their enhanced ability to blend with nanoparticles (Reid et al., 2014). The addition of polyurethanes with metal oxide nanoparticles elevates the physicochemical properties of both components by eliminating their negative aspects (Mallakpour & Khadem, 2017). Generally, polyurethanes are used as a coating material to provide heat resistance, mechanical strength, and flame retardancy to steel or carbon-based metallic structures (Magnin et al., 2020). These polyurethanes can be prepared from vegetable oil or synthetic approaches (Hasirci, 2017; Sawpan, 2018). However, they possess certain toxic organic solvents, which halts their potential to be used in commercial applications (Ahmadi et al., 2020). Later, waterborne polyurethanes are developed and incorporated with a wide variety of metal oxide nanoparticles, to form potential nanocomposites with high biocompatibility, mechanical and thermal stability (Ahmadi & Ahmad, 2020). These novel nanocomposites are currently under research to be useful as coating materials or formulations in biomedical and environmental applications (Christopher et al., 2016). The current chapter expounds on various types of magnetic and non-magnetic waterborne polyurethane-metal oxide nanocomposites, their properties, and their specific functions. It also outlines the benefits of employing these innovative nanocomposites in biomedical and environmental applications, including their shortcomings and future perspectives.

2 Overview of Waterborne Polyurethane

Polyurethanes are generally prepared via flexible and alternating rigid segments, which is responsible for their high-performance capability. The end product of polyurethane is formed by reacting polyol and isocyanate together, which leads to an intermediate, medium molecular weight polymer, which manifests as a thick liquid. Later, the liquid is allowed to react with diamine or diol chain extender, to be transformed into the high molecular weight polyurethane (Hepburn, 1982). It may be noted that most polyurethane systems possess large volatile organic compounds with certain free isocyanates. These polyurethane materials possess mutual properties, such as abrasion, scratch resistance, flexibility, and toughness, in order to be beneficial as coating and adhesives (Engels et al., 2013). However, the existence of organic solvents in polyurethanes may lead to environmental pollution, due to their toxicity. Thus, aqueous polyurethane dispersion systems were introduced as an alternative to the conventional polyurethanes, to avoid their toxicity.

Waterborne (aqueous) polyurethanes are binary colloidal systems, where the polyurethane particles are dispersed in a continuous water medium. These polyurethane materials are fabricated to yield polymers with more hydrophilic groups, to possess enhanced solubility in water. In general, polyurethanes are incompatible with water, and hence, hydrophilic monomers, such as carboxylate, sulfonate, or quaternary ammonium groups with ionic functionality, were incorporated as a backbone to make them water compatible (Honarkar, 2018). The polyurethanes with hydrophilic, ionic functional groups, are commonly referred to as polyurethane ionomers or internal emulsifiers (Dieterich et al., 1970). These waterborne polyurethanes are widely employed as coating materials, compared to conventional polyurethanes, to reduce the atmospheric emission of toxic solvent, to diminish the cost of solvents, and to improve the quality of polymer (Frisch & Klempner, 1998). The backbone and the emulsifier act as the principal structural measures of the waterborne polyurethanes, where the emulsifiers are hydrophilic and amphiphilic polymers to stabilize polyurethane dispersions in water. These emulsifiers are categorized into internal emulsifiers (such as polyethylene oxide as non-ionic centers, and cationic, anionic, and zwitterions as ionic centers) and external emulsifiers (Szycher, 1999). In waterborne polyurethane, the hydrophilic nature of the ionic groups is the leading factor that contributes to the polyurethane dispersion in aqueous medium. For instance, polyurethane is formed by dimethylol-propionic acid as a potential internal emulsifier, where the basic triethyl amine group is used as a neutralizing agent (Honarkar & Barikani, 2014). Waterborne polyurethanes can be subclassified into anionomers, cationomers, and zwitterionomers, depending on the type of charged ionic groups present on their surface. Polyurethane anionomers were fabricated with the help of ionic isocyanate groups in combination with diols, such as phosphonate, sulfonic groups, and carboxylic acid as well as diisocyanates (Francolini et al., 2010). Likewise, polyurethane cationomers are developed by using diisocyanates with sulfur or nitrogen-containing alkyl diols via a quarternization process (Król et al., 2011). Similarly, polyurethane zwitterionomers are polymers with both negative and positive charges with several charge-neutral atoms. These polymers are also synthesized via quarternization processes with lactones and sultones to modify the *N*-alkyldiols (Hwang et al., 1981).

These waterborne polyurethanes are generally synthesized via three main approaches, namely emulsification, acetone process, and hot-melt approach. Guo et al. (2013) synthesized waterborne polyurethane emulsions using monomers, such as toluene diisocyanates (TDI) and isophorone diisocyanates (IPDI), with a hydrophilic chain extender, namely dimethylol-propionic acid (DMPA). In this study, the emulsions were prepared via a two-step pre-polymerization synthesis approach, with chain extending reactions with monomers and chain extenders. These emulsions are identified to possess paper surface sizing properties and the maximum efficiency with increased water resistance was identified for the polyurethane emulsion prepared with 1.6–1.8 of isocyanate (NCO)/polyol (OH) ratio, 5–7% of DMPA, and 5:1 molar ratio of TDI/IPDI (Guo et al., 2013). Likewise, Sukhawipat et al. (2018) utilized step-growth polymerization approach of hydroxyl telechelic natural rubber, with a molecular weight of 2800 g/mol, and emulsifiers such as N-methyl diethanolamine and toluene-2, 4-diisocyanate for the fabrication of novel cationic waterborne polyurethanes. It has been identified that the chain extender influenced the formation of polyurethane film and increases their thermal and mechanical properties (Sukhawipat et al., 2018). Similarly, Guo et al. (2012) fabricated a novel hybrid waterborne polyurethane with poly (n-butyl acrylate-styrene) via a hybrid emulsification process with a chain extender known as dimethylol-propionic acid. These hybrid polymers were proposed to be beneficial for paper sizing applications, due to their folding resistance, gloss, burst strength, and smoothness (Guo et al., 2012). Further, Shendi et al. (2017) synthesized waterborne polyurethane with the help of a novel internal emulsifier extracted from sunflower oil. Renewable dihydroxy acid was derived from epoxide sunflower oil via saponification process with methanol and was utilized as a chain extender in the emulsification process of polyurethane. The resultant polyurethanes were proposed to be useful as an effective coating material (Shendi et al., 2017). Furthermore, Zhu et al. (2018) developed a waterborne polyurethane-acrylate emulsion with high solid content via a two-step process. Initially, the polyurethanes were allowed to react with the monomer of acrylate via semi-continuous seeded approach-based emulsion polymerization. In addition, dimethylol-propionic acid was used as a reactive emulsifier for the polymer synthesis, which confers advantages such as narrow particle size distribution, small particle size, excellent stability, film performance, and 46% more solid content (Zhu et al., 2018).

Sardon et al. (2010) utilized acetone process for the fabrication of waterborne polyurethane dispersions, using poly (propylene glycol), isophorone diisocyanate, triethylamine, and 1, 4-butane diol and 2-*bis* (hydroxymethyl) propionic acid, via the phase-inversion approach. It was reported that stable polyurethanes can be obtained with 60 wt% of acetone at 30 °C of phase-inversion temperature (Sardon et al., 2011). Further, Fang et al. (2014) fabricated low crystalline waterborne polyurethane by combining prepolymer isocyanate via acetone process. The synthesized polyurethane dispersions showed excellent thermal resistance, which is proposed to be beneficial as water-based ink binders (Fang et al., 2014). Furthermore, Sardon et al. (2010) developed self-curable hybrid waterborne polyurethanes by functionalizing 3-(aminopropyl) triethoxysilane (APTES) at room temperature via acetone process. This study emphasized that the characteristics of polyurethane were influenced by the curing temperature, and the concentration of APTES as the curing agent (Sardon et al., 2010). Moreover, Aizpurua et al. (2020) recently fabricated novel UV-curable waterborne acrylic polyurethane dispersions via acetone process, which contains Diels–Alder moieties. These polyurethanes are proven to be re-amendable and recyclable with self-healing properties, which can be useful as a potential sustainable coating material (Aizpurua et al., 2020). Besides, Liu et al. (2018) fabricated a waterborne polyurethane using a solvent-free and self-catalysis approach. The resultant polyurethanes were identified to be highly stable with a narrow particle size distribution, admirable mechanical potency, and enhanced tensile strength of 84.2 MPa (Liu et al., 2018).

Recently, Zhang et al. (2019) synthesized waterborne polyurethane adhesive via the flame-retardant, reactive, hot-melt approach. These polyurethanes possess enhanced thermal stability, flammability, mechanical properties, and bonding strength, with a slight decrement in the tensile strength (Zhang et al., 2020). Similarly, Yousefi et al. (2020) fabricated waterborne polyurethane-zirconium dioxide nanoparticles via the sol–gel approach. The resultant nanocomposite possesses antistatic property with stable adhesion, rough structure formation, and resin performance modification capability to be useful as a potential homogenous coating film. Likewise, Fuensanta et al. (2020) developed a novel waterborne polyurethane urea with ether-carbonate as copolymer and amino-alcohol as chain extenders via prepolymer approach. The ensuing polyurethane possesses a mean particle size of 51–78 nm and 58–133 mPa.s of viscosity. These polyurethanes are proposed to be useful as an environmentally friendly coating material with elevated tack, enhanced adhesion, and cohesion properties (Fuensanta et al., 2020). Further, Zhu et al. (2020) synthesized waterborne polyurethane with polymeric dyes and anthraquinone chromophores, to realize textile colors and enhance color fastness. These polyurethanes exhibited excellent emulsion stability, superior water resistance, ultra-low migration performance and enhanced thermal stability (Zhu et al., 2020). Besides, Jiang et al. (2020) utilized polymerization process to form anionic waterborne polyurethane using hydroxylated tung oil. These polyurethane materials possess excellent post-polymerization crosslinking capability with soft/hard segment compatibility, and enhanced intermolecular interactions (Jiang et al., 2020). However, waterborne polyurethane has certain limitations in terms of stability and physicochemical properties. Thus, they are blended with metal oxide nanoparticles to enhance their characteristics, especially structural stability, and to provide biocompatibility as well as bioavailability, with less toxicity.

3 Waterborne Polyurethane-Metal Oxide Nanocomposites

Waterborne polyurethane is melded with metal oxide nanoparticles to form nanocomposites to improve the exclusive properties of the polyurethanes. The nanosized oxides of metal particles, which can be blended with polyurethanes, are classified into two major classes: magnetic and non-magnetic nanoparticles.

3.1 Magnetic Metal Oxide Nanoparticles

Generally, waterborne polyurethanes are blended with iron oxide nanoparticles to exhibit magnetic properties. Zhang et al. (2013) fabricated novel nanosized composites by using iron oxide (Fe_2O_4) nanoparticles and waterborne polyurethane via co-precipitation and *in situ* polymerization approaches. In this study, oleic acid was used as a potential modifier to enhance the compatibility of the nanoparticle with the monomer of urethane. The results showed that the resultant nanocomposite possesses enhanced magnetic and electrical properties, with initial thermal decomposition at 234–249 °C, elevated residual rate at 800 °C from 9.4 to 28.7%, and volume and surface resistivity of 9 and 8 orders of electric performance magnitudes. This novel nanocomposite with exclusive magneto-electrical properties was proposed to be beneficial in microwave absorption applications (Zhang et al., 2013). Likewise, Chen et al. (2016) fabricated films using waterborne polyurethane-iron oxide nanocomposites by modifying nanosized oxide of iron particles with 3-(trimethoxysilyl) propyl isocyanate (IPTS) as a cross-linker via the *in situ* polymerization technique. These nanocomposites possess enhanced storage stability, solvent resistance, electrical conductivity, and mechano-magnetic characteristics. Specifically, the resultant nanocomposite possesses 8.43×10^{-4} S/cm of high electrical conductivity and 14.22 emu/cm^3 with 8 wt% of low nanocomposite content. These nanocomposites were proposed to be beneficial as an ecofriendly polyurethane material for microwave absorption application (Chen et al., 2016). Similarly, dos Santos et al. (2018) demonstrated the synthesis of waterborne polyurethane-iron oxide nanoparticle blended synthetic talc gel-based nanocomposites. In this study, the synthetic iron oxide nanoparticle talc gel was used as a filler, with several silicon oxide and magnesium oxide, as well as forming a hydroxide group by binding with hydrogen and water molecule via polar interactions. This shows the hydrophilic property of the resultant nanocomposite to improve the dispersion of the filler inside an aqueous matrix. Additionally, the study emphasized that the nanocomposite possesses ferromagnetic and superparamagnetic properties below and above 120 K of Curie temperature, respectively, with exclusive mechanical characteristics (dos Santos et al., 2018).

Recently, Cheng and Hsu (2017) synthesized a novel nanocomposite by encapsulating hydrophobic vitamin K3 or 9-(methylaminomethyl) anthracene and superparamagnetic iron oxide nanoparticles in amphiphilic, biodegradable, anionic waterborne polyurethane nanoparticles. The polyurethanes fabricated via waterborne process were 35 nm in size with unique chemical properties and helps to encapsulate drugs and magnetic nanoparticles via excellent mechanical compliance and a low glass transition temperature. The superparamagnetic nanoparticles in polyurethane are beneficial as a high contrast agent in magnetic resonance imaging applications and possess cell hyperthermic ability to inhibit cancer cells. Further, these novel nanocomposites are proposed to be a multimodal carrier to deliver drugs at target sites with 95% of high encapsulation efficacy. It may be noted that the drug secretion occurs here via diffusion, and not via burst release (Cheng & Hsu, 2017). Furthermore, Gogoi and Karak (2016) demonstrated the synthesis of an exclusive waterborne polyurethane nanocomposite with hyperbranches, to include nickel ferrite ($NiFe_2O_4$) and graphene oxide nanohybrid via in situ polymerization and hydrothermal approach. The resultant nanocomposite possesses excellent stability towards high temperature, with 335 °C as initial degradation temperature. Additionally, it demonstrates an enhanced mechanical characteristic with 24.23 MPa of tensile strength, 280% of elongation at break, and 28.35 MJ m^{-3} of toughness. Besides, these nanocomposites exhibited shape memory behavior with multi-stimuli responses upon being exposed to sunlight, 300 W of microwave, and high-temperature contact. These novel nanocomposites are proposed to be useful as a smart, non-contact triggered, high performing material for selected shape memory applications (Gogoi & Karak, 2016). Moreover, waterborne polyurethane nanocomposites were prepared as elastomers with magnetic iron oxide nanoparticles and insulin, to improve their sustainability and stability towards temperature. The resultant nanocomposite was beneficial in avoiding agglomeration of insulin and nanoparticles in the body fluid while elevating their half-life durability. It may be noted that the encapsulation of insulin using nanocomposite elastomer provides enhanced biocompatibility, bioavailability, and degradability to the magnetic nanoparticles, facilitating the delivery of insulin at the target sites (Farajollah et al., 2015).

In recent times, several novel and complex magnetic nanocomposites have been prepared by incorporating iron oxide nanoparticles. Hoseini and Nikje (2018) developed a unique magnetic nanocomposite with water-dispersible polyurethane and modified iron oxide (Fe_3O_4) nanoparticles via 3-aminopropyltriethoxysilane (APTS) and chemical synthesis via 2, 2 bis (hydroxymethyl) propionic acid, hydroxyl-terminated polybutadiene, and 1,5-naphthalene diisocyanate, by using the in situ polymerization approach. The resultant nanocomposite possesses an elevated melting point, enhanced physical strength, stability towards high temperature, and an enhanced superparamagnetic property (Hoseini & Nikje, 2018). Similarly, Dai et al. (2019) demonstrated the fabrication of magnetic aligned iron oxide-reduced nanocomposites, with graphene oxide and waterborne polyurethane, via a novel chemical reduction method. The resultant nanocomposites were identified to have enhanced complex permittivity, microwave absorption capacity, and tangent dielectric loss with − 67.8 dB of minimum reflection loss value, and a rigid three-dimensional structural orientation. These nanocomposites were proposed to be beneficial as a potential microwave absorption material (Dai et al., 2019). Likewise, Hoseini and Nikje (2018) utilized the sol–gel approach for the effective synthesis of novel iron oxide-APTS magnetic clay, to be blended with waterborne polyurethane, for the formation of nanocomposites. The nanocomposites were formed by blending clay-iron oxide-APTS nanoparticles with 1,5-naphthalene diisocyanate and polypropylene glycol in the matrix of polyurethane, via the in situ polymerization approach, and was proved to possess improved thermal, magnetic, and mechanical properties. Besides, the clay-based nanocomposite exhibited good dispersion ability, due to their covalent bond formation between chains of polymers and functionalized magnetic nanoparticles, even at 3.5 parts by weight of low quantity (Hoseini & Nikje, 2018). In addition, Salahuddin et al. (2020) recently synthesized superparamagnetic nanocomposites with a model cancer drug candidate named Norfloxacin, iron oxide nanoparticles and coated with waterborne polyurethane urea via polyethylene glycol, isophorondiisocyanate along with water as a polymer chain extender. The results showed that the nanocomposite follows the Fickian diffusion mechanism to release the encapsulated drug under magnetic field at 37 °C and follows a first-order reaction at 45 °C Further, the fabricated nanocomposites possess antimicrobial properties with significant anticancer activity against the breast cancer MCF-7 cell line, and has moderate cytotoxicity towards the human lung WI38 fibroblast cell line (Salahuddin et al., 2020).

Nikje et al. (2015) fabricated novel magnetically rigid foams that are reinforced with iron oxide nanoparticles, dipodal silane, APTS, and γ-glycidoxypropyltrimethoxysilane (GPTS) via the sol–gel approach. The synthesized nanocomposites were proven to have enhanced thermomechanical and superparamagnetic features (Nikje et al., 2015). Further, Chen et al. (2015a, 2015b) also developed a novel, well-dispersed, superparamagnetic nanocomposite with waterborne polyurethane and nickel-zinc ferrite, via the *in situ* polymerization method. In this study, the nickel-zinc ferrite nanoparticles were synthesized using a solvothermal approach with 3-(trimethoxysilyl) propyl isocyanate (IPTS), for enhancing their compatibility with monomers of urethane. The resultant nanocomposites were proven to possess excellent magnetic properties with 16.58 emu/cm^3 of maximum saturation magnetization, and are proposed to be useful in microwave absorption applications (Chen et al., 2015a). Furthermore, Nikje et al. (2013) developed novel rigid foams of waterborne polyurethanes, by incorporating superparamagnetic iron oxide-silicon dioxide nanoparticles via the facile one-shot method. The sol–gel approach was utilized to form the core–shell structure of nanoparticles, and 3% of the magnetic nanoparticles were used up in the formulation with foams. The resultant magnetic nanocomposite foams were proven to possess enhanced magnetic, thermal, and mechanical characteristics, compared to the conventional pristine type of foams (Nikje et al., 2013). Besides, Chen et al. (2015a, 2015b) synthesized an exclusive ultraviolet light curable hyperbranched waterborne polyurethane nanocomposite films, with magnetic iron oxide nanoparticles. These nanocomposite films were fabricated via the in situ polymerization approach, where the magnetic nanoparticles were synthesized using a solvothermal method, with vinyltriethoxysilane as a modifier, to enhance their compatibility with polyurethane. The resultant nanocomposite films possess enhanced glass transition temperature, solvent resistance, thermal stability, and hardness, due to the incorporation of UV cured magnetic nanoparticles. The superior magnetic property of the nanocomposite film was proposed to be helpful as a microwave absorption material (Chen et al., 2015b). Moreover, Khatoon and Ahmad (2019) utilized safflower oil polyurethane for the fabrication of nanocomposites with vanadium pentoxide enwrapped in polydiphenylamine. In general, vanadium oxide in the nano-size range, are magnetic in nature (Demishev et al., 2011). However, the nanocomposite formulation exhibited high corrosion resistance performance to be beneficial as a potential anticorrosive agent (Khatoon & Ahmad, 2019). It is noteworthy that most of these magnetic waterborne polyurethane nanocomposites possess microwave absorption, as well as critical mechanical or thermal applications. Some studies also reveal their biomedical applications in cancer management. However, biocompatibility of these magnetic nanocomposites is a major restraint, which confines their biomedical and environmental applications.

3.2 Non-magnetic Metal Oxide Nanoparticles

Several non-magnetic metal oxide nanoparticles were also blended with waterborne polyurethane to fabricate exclusive nanocomposites for various applications. Christopher et al. (2015) fabricated an innovative, highly dispersive, waterborne polyurethane nanocomposite with zinc oxide, which are surface modified with oleic acid. The addition of surface-modified zinc oxide has led to an increment in the water contact angle of the polyurethane, compared to the standalone polymer. Further, these nanocomposites are proposed to be beneficial as a corrosion protection agent that can be coated on mild steel, with exclusive electrochemical resistance, potentiodynamic polarization, high hydrophobicity, and uniform dispersion, without agglomeration as well as exclusive anti-settling behavior (Christopher et al., 2015). Furthermore, Ma and Zhang (2009) fabricated zinc oxide nanowhiskers with flower-like morphology via simple hydrothermal approach and evaluated the effect of adding them with waterborne polyurethane on the thermal, mechanical and antibacterial properties of the resultant nanocomposite. The fabricated flower-like zinc oxide nanoparticle was modified with γ-aminopropyltriethoxysilane to exhibit good dispersion and water resistance properties. These novel nanocomposite films possess excellent antibacterial activity against *Staphylococcus aureus* and *Escherichia coli* (Ma & Zhang, 2009). Likewise, Awad et al. (2011) fabricated an exclusive free volume nanocomposite using 20 nm-sized zinc oxide nanoparticles, which are dispersed in waterborne polyurethane. These nanocomposites possess high and low glass transition temperatures, due to their hard microdomains in the polar region and soft aliphatic chains, along with electrostatic, van der waals force, and hydrogen bonds. Further, the addition of zinc oxide nanoparticles has improved the crosslinking density of the polymer with microphase hard and soft segment separation via exponential function (Awad et al., 2011). Zhou et al. (2014) utilized the sol–gel approach for the fabrication of exclusive nanocomposites with antimony doped tin oxide nanoparticles, with waterborne polyurethane dispersions. The coating efficiency of these nanocomposites was evaluated and the results revealed 35% of their enhanced low near-infrared transmittance and 75% of high visible light transmittance with hearing insulation effect (Zhou et al., 2014).

Halloysite is an aluminosilicate clay composite mineral with 55.7% of oxygen, 21.7% of silicon, 20.9% of aluminum, and 1.56% of hydrogen in the form of $Al_2Si_2O_5(OH)_4$. These novel clay minerals have been fabricated in the form of nanotubes, to load an antibacterial and essential thyme oil, named carvacrol. Further, the carvacrol-loaded halloysite nanotubes were blended with waterborne polyurethanes as an exclusive nanocomposite by Hendessi et al. (2016), who evaluated their ability to be an effective coating material to release the antibacterial essential oil in a sustained mechanism. The resultant nanocomposite films exhibited effective antibacterial activity against *Aeromonas hydrophila,* and coating them on the surfaces inhibited the bacterial growth for two days. Thus, these novel nanocomposite films were proposed to be beneficial as a potential anti-biofilm surface coating agent to prevent bacterial growth on the surface of materials, without or less toxicity towards the environment (Hendessi et al., 2016). Similarly, attapulgite is a clay soil mineral found in the southeastern United States, which is a magnesium aluminum phyllosilicate with $(Mg, Al)_2SiO_{10}(OH)0.4(H_2O)$ as empirical formula. Peng et al. (2011) blended these clay minerals with waterborne polyurethane to form nanosized composites. The resultant organically modified clay minerals were identified to be homogeneously dispersed in the polyurethane with improved tensile strength, stability towards high temperature, and elongation at breaking of nanocomposites (Peng et al., 2011). Further, Sow et al. (2010) fabricated a nanocomposite formulation using nanosized alumina and silica with waterborne polyurethanes. The nanoparticles were incorporated with waterborne polyurethane and acrylate resin and a photo-initiator to absorb ultraviolet (UV) rays. It has been reported that the polymerization with the photo-initiator and nanoparticles were less efficient, however, the addition of only nanosized functionalized silica was better than nano alumina (Sow et al., 2010). Moreover, Deyab et al. (2017) fabricated a novel nanocomposite with acidic $Rb_2Co(H_2P_2O_7)_2 . 2H_2O$ pyrophosphate and waterborne polyurethane. These nanocomposites were evaluated for its effectiveness as coating material with 3.5% of sodium chloride solution to protect corrosion in carbon steel. This study emphasized that the nanocomposite with waterborne polyurethane possess enhanced mechanical and corrosion resistance properties (Deyab et al., 2017).

Recently, Jiang et al. (2019) fabricated a novel nanocomposite using copper, zinc oxide, and reduced graphene oxide with waterborne polyurethane. These nanocomposites were evaluated to be a coating material for controlling the adhesion of bacterial colonies and corrosion of metal in the circulating cooling water system. The resultant nanocomposite exhibited superior antibacterial activity against gram-positive *Staphylococcus aureus* and gram-negative *Escherichia coli* bacterial strains, with 94.3% of antibacterial activity, 5 H of hardness, and 93.3% of corrosion inhibition efficiency (Jiang et al., 2019). Further, Nosrati et al. (2020) also developed an exclusive complex nanocomposite by blending titanium dioxide, polyaniline, halloysite nanotube, carbon nanotubes, and protected by waterborne polyurethane. These nanocomposites are found to possess excellent hydrophilicity, which effectively decolorizes dye contaminants with high corrosion resistance, potential, and lower corrosion current as a coating additive material with anticorrosive and self-cleaning benefits (Nosrati et al., 2020). Furthermore, Ye et al. (2019) developed a unique zinc oxide array via seed growth approach using zinc nitrate and hexamethylenetetramine and blended with waterborne polyurethane via the electrospinning method. These nanocomposites exhibited excellent structural and morphological stability, mechanical characteristics, wettability, and infrared properties to be highly beneficial as an anticorrosive coating material to protect stainless steel substrates. Additionally, the incorporation of hexagonal wurtzite crystal structure of zinc oxide arrays has improved the hydrophilicity of the nanocomposite with exclusive corrosion resistance (Ye et al., 2019). Moreover, Kale et al. (2019) synthesized a complex nanocomposite using graphene oxide functionalized with epoxy resin, silicon dioxide functionalized with amine and waterborne polyurethane via the physical mixing approach. These nanocomposite films were proven to possess enhanced thermal and mechanical properties with hydrolysis and water resistance capability. Also, these nanocomposites were proposed to be beneficial as leather coating agents via the screen printing technique with excellent flexibility, adhesion, and abrasion resistance properties (Kale et al., 2019). In addition, Guo et al. (2018) prepared a novel $[(1-x)$ manganese dioxide-multiwalled carbon nanotube] nanocomposite by protecting them with waterborne polyurethane via the in situ polymerization approach. The resultant nanocomposites exhibited superior microwave absorption property with a minimum reflection loss value of -28.7 dB with 2.5 mm of thickness at 6.4 GHz and excellent thermal stability and tensile strength (Guo et al., 2018). All these studies showed that the non-magnetic waterborne polyurethane-metal oxide nanocomposites are under extensive research or widely useful in the context of microwave absorption, as well as other biomedical applications.

4 Biomedical Applications

Waterborne polyurethane-metal oxide nanocomposites are generally less toxic towards living organisms and are widely used or under research to be advantageous as a potential nanocomposite material for biomedical applications. These nanocomposites are proposed to be beneficial as antimicrobial agents and drug delivery systems, especially for the treatment of cancer. Nguyen et al. (2019) fabricated a novel nanocomposite with acrylic polyurethane, iron oxide, and silver hybrid nanoparticles of 10–20 nm in size. These nanocomposites were proven to possess enhanced coating hardness with high mechanical stability, as well as an enhanced antibacterial efficacy against *Escherichia coli*. Also, the nanocomposite possesses infrared and relative hardness properties after curing the coating for 6 days with elevated adhesion of 2.21 N/mm^2 and 125 kg/cm of impact resistance (Nguyen et al., 2019). Later, Nguyen et al. (2020) developed a similar nanocomposite via green-based photo-crosslinking agents. These nanocomposites also possess an enhanced swelling degree, high relative hardness, and increase gel fraction, while coating them on a material surface. The curing of nanocomposite with UV rays improved their abrasion resistance to 126.54 lite/mil and exhibited an effective antibacterial property against *E. coli*, which is inversely proportional to the density of the cross-linkers (Nguyen et al., 2020). Similarly, Li et al. (2019) fabricated a novel polyurethane-acrylate-based nanocomposite with the support of reduced graphene oxide and titanium dioxide as photo-initiator, via the in situ polymerization approach. The resultant nanocomposite exhibited improved electrical conductivity, UV-curing potential, synergistic antibacterial efficacy against *S. aureus* due to graphene oxide and titania and promoted electron–hole pair separation to generate reactive oxygen species. These nanocomposites were proposed to be useful in high-performance antibacterial coating in biomedical applications (Li et al., 2019). Likewise, Malkappa et al. (2018) developed hybrid organic inorganic nanosized colloids of titanium dioxide as core and silicon dioxide as shell, that are blended with waterborne dispersible polyurethanes. The pristine nanocomposite films with stable nanosized colloids, after curing, exhibited enhanced mechanical and thermal properties, with high young's modulus and tensile strength. Besides, these nanocomposites also exhibited effective protection against both *S. aureus* and *E. coli* with high zone of inhibition, and are proposed to be beneficial as a potential antibacterial coating material (Malkappa et al., 2018).

In general, polyurethanes are widely used in drug delivery applications to be favorable for nanoparticle systems, stimulus-response systems, as well as shape memory, and covalently conjugated system. The reaction of diisocyanates in polyurethanes with polyols leads to the carbamate linkage formation, which is often combined with diamines as chain extenders and diols, to enhance their mechanical properties (Basu et al., 2016). Particularly, metal oxide nanoparticle blended polyurethane composites are used to fabricate smart materials with shape memory, which will be useful in the development of biocompatible implants, such as self-adjusting orthodontic wires and intravenous cannula (Athimoolam & Moorthy, 2012). Soto et al. (2018) developed a novel nanocomposite using polyurethane and nanosized Fe_3O_4 magnetite particles, via a simple suspension casting approach. The results revealed that the nanocomposites possess shape memory ability with a superparamagnetic behavior at 300 K. Even after exposure to temperature and an alternate magnetic field, these composites with 3 nominal wt% of magnetic nanoparticles, retained their original shape with a swift and complete recovery rate, via remote activation of the deformed samples (Soto et al., 2018). Later, shape-memory polyurethane-based nanocomposites were synthesized using nanosized platelets of ferromagnetic iron oxide and graphene. In this study, the addition of nanoplates as fillers has been proven to improve the rheological, thermal, shape memory, and mechanical properties of polyurethane. Further, polyurethane matrix with an increment in the filler content altered the viscosity of the synthesized nanocomposite, and increased their magnetic property dependent shape memory characteristics (Urban & Strankowski, 2017). Also, Rahman (2020) recently emphasized that polyurethane-zinc oxide nanocomposites possess noteworthy features to be widely used in biomedical applications. In the article, the author listed that adhesion, self-healing, and particle size of zinc oxide particles are the contributing factors for the biological properties of the nanocomposites. These nanocomposites are reported to be beneficial as effective antibacterial coating agents, scaffolds, membranes, foams, and packaging films. Further, they also possess high mechanical strength and biocompatibility with antibacterial properties, which has been proposed to be beneficial in injection equipment, tubing, hospital bedding, implants, surgical drapes, and wound dressing applications (Rahman, 2020).

5 Environmental Applications

Currently, waterborne polyurethane-metal oxide nanocomposites are used as a significant antifouling agent to protect the materials from damage via microbial growth, which are to be immersed in fresh or sea water for a long time, especially in boats and ships. Even though, certain chemically synthesized metal oxide nanoparticles exhibited toxic reactions towards the marine environment and its intrinsic flora and fauna (Baker et al., 2014). It is apparent that blending them with ecofriendly, non-toxic biocompatible polyurethane will reduce their adverse toxic reactions (Honarkar, 2018). Xiao-min et al. (2015) evaluated the performance of polymerizable, nanosized silicon dioxide by modifying UV-curable films made up of waterborne polyurethane. These nanocomposites were identified to be highly stable towards elevated temperatures, show maximum elongation break at 10.303 MPa and 2.418% of water absorptivity, without affecting the surface flatness of the film. Also, the resultant nanocomposite exhibited enhanced water resistance and improved mechanical properties, which will be beneficial as a potential antifouling film (Luo et al., 2015). Similarly, Nosrati et al. (2019) synthesized nanocomposite using waterborne polyurethane, titanium dioxide, polyaniline, carbon nanotube, and halloysite for environmental protection applications. The anticorrosion study of these materials via Tafel plots showed that they possess high corrosion resistance with high electrical conductivity. Further, the electrochemical impedance spectroscopy revealed that the barrier properties of halloysite, titanium dioxide nanoparticles, and carbon nanotubes, are the contributing factors in the anticorrosive nature of the nanocomposite. Furthermore, the hydrophilic nature of the nanocomposite was demonstrated by the sessile drop approach, where the contact angle of the water with the material is evaluated by exposing them towards UV lamps and light-emitting diode (LED) lights. The results showed that the nanocomposite possesses an enhanced hydrophilic property with a water contact angle of 77.37°, which will be beneficial for self-cleaning applications. These are actually altered after exposing them towards light, due to the electron–hole pair transfer in titanium dioxide nanoparticles. Moreover, the nanocomposites were exposed toward the LED light and their degradation and decolorization properties were evaluated towards the toxic Malachite green dye, which showed 59.28% of efficiency in decolorizing the dye after 5 h of light irradiation. Thus, these novel nanocomposites are highly beneficial in environmental remediation applications to eliminate toxic dyes in water, without a noticeable toxicity towards other organisms (Nosrati et al., 2020).

Recently, Zheng et al. (2019) fabricated a unique nanocomposite by blending an emulsion of waterborne polyurethane and inorganic silica (silicon dioxide) fluoride nanoparticles. The spraying coating approach with low surface energy was used to provide the superhydrophobic surface property with enhanced optical transparency, aging resistance, abrasion resistance, and strong adhesion to the organic–inorganic composite material. The self-cleaning property of the nanocomposite was evaluated by immersing them in toluene solution dyed with Sudan 1 for 1–2 h. The results showed that the nanocomposite coated surface exhibited self-cleaning properties by maintaining superhydrophobicity, even after 250 cycles of abrasion, soil pollution, oil contamination, and ultraviolet light irradiation, to indicate its superior antifouling efficacy (Zheng et al., 2019). Similarly, Xiao-dong et al. (2007) evaluated the adhesion property of marine bacteria on the nanocomposite coated surface, which are prepared with waterborne polyurethane, polytetrafluoroethylene, and titanium dioxide nanoparticles. The results emphasized that the nanocomposite coating impeded the adhesion of marine bacteria to exhibit an antifouling property, and protect the surface from environmental degradation (Huang et al., 2007). Likewise, standalone metal oxide nanoparticles are either widely used or under extensive research, to act as a potential material for environmental applications, such as eco-remediation (Liu et al., 2017), wastewater treatment (Shanker et al., 2016), plant growth (Rizwan et al., 2019), biofertilizers (Kalia & Kaur, 2019) and biopesticides (Lade & Gogle, 2019). It is worthy to note that both magnetic and non-magnetic nanosized oxides of metal particles can be used as a potential material for a wide range of environmental applications (Gebre & Sendeku, 2019; Magro & Vianello, 2019). In particular, iron oxide nanoparticles exhibited the ability to sense toxic gases in the environment, which will be beneficial to eliminate or reduce the toxic gases or particles, in addition to monitoring bioremediation processes (Li et al., 2016; Rahman et al., 2017). Besides, the addition of rare earth metals and metallic ions as dopants in metal oxide nanoparticles,has elevated their mechanical and thermal properties, to sustain in the harsh environment, monitor the level of pollutants, and maintain environmental safety (Adeleye et al., 2018). However, most of these chemically synthesized metal oxide nanoparticles are toxic and hazardous towards the environment (Jeevanandam et al., 2018). Thus, the addition of non-toxic, environmentally compatible waterborne polyurethane will reduce the toxicity of the metal oxide nanoparticles, and elevate their physicochemical properties, thereby being highly beneficial for environmental applications. Table 1 is the summary of various biomedical and environmental applications of waterborne polyurethane-metal oxide nanocomposites.

6 Future Perspectives

Generally, waterborne polyurethanes are intermixed with metal oxide nanoparticles, in order to elevate their physicochemical properties, ensure biocompatibility and reduce their toxicity, thereby posing as an effective agent facilitating a wide range of biomedical and environmental applications (Zhang et al., 2019). However, these nanocomposites are not yet commercially available in the market for actual applications. This can be attributed to certain challenges, which must be addressed, to boost their large-scale manufacturing. The first constraint is the inconvenience in fabricating the nanocomposite in the actual nano-size range, i.e., 10–100 nm. When several nanoparticles are embedded or encapsulated in a polymer to form nanocomposites, the composite dimension will expand, which due to the swelling phenomenon, could even result in the manifestation of a microparticle. The swelling of the composite will reduce the unique properties of the nanosized particles, and thus, may lead to a decrement in their application efficacy (Aydınoğlu et al., 2014). Thus, smaller nanoparticles, such as quantum dots, can only be encapsulated with polyurethanes to form nanocomposites (Qiang et al., 2020). However, nanoparticles can be embedded in the matrix of waterborne polyurethane, to overcome the challenge of swelling mechanism in the next generation of composite fabrication strategies (Chou et al., 2006). Another inadequacy associated with the waterborne polyurethane-metal oxide nanoparticles lies in the context of the employed precursor in the synthesis approach. Most of the monomer precursors used for the synthesis of polyurethane may boast of a chemical origin, which can be cytotoxic towards humans, animals, plants, or microbes while utilizing them in biomedical or environmental applications (Singh & Jain, 2009). These chemically originated polyurethanes will add on to the cytotoxicity of chemically synthesized metal oxide nanoparticles, when they are blended together as nanocomposites (Jeevanandam et al., 2019). Thus, it has been proposed that 'green' or biological approaches to fabricate polyurethane-metal oxide nanocomposites, will be highly effective, especially for biomedical applications, to drastically reduce their toxic behavior (Kumar et al., 2015; Natarajan & Suuki). However, biological synthesis using biomolecular extracts may affect the stability of the resultant nanocomposites (Jeevanandam et al., 2016). Hence, hybrid strategies, which utilize a combination of both chemical and biological approaches, are highly recommended for waterborne polyurethane-metal oxide nanocomposite synthesis, to yield highly stable, biocompatible, and less toxic particles (Zhang et al., 2008).

The viscosity of the polymer plays a major role in the formation of waterborne polyurethane composite with metal oxide nanoparticles. These polymers can be fabricated into nanofibers via the electrospinning approach, where the metal oxide nanoparticles will be amalgamated or embedded on the surface of the fibers, to form composites (Meng et al., 2010). Such electrospun polyurethane-metal oxide nanocomposite fibers will be useful in producing antimicrobial textiles for medical purposes, sutures, bandages, and scaffolds, especially for implant applications (Lee, 2009). These polymeric nanocomposites can additionally be formulated into dendrimer or liposomal structures in the future, to encapsulate potential drug candidates for their enhanced and controlled delivery at the target sites (Feast et al., 2003; Yin et al., 2020). Although these nanocomposites are beneficial in several aspects, the research is still in its infancy, as various physicochemical properties corresponding to these are yet to be revealed and/or adequately characterized. Hence, extensive investigation is necessary for the large-scale production of waterborne polyurethane-metal oxide nanocomposites, for future commercial applications in biomedical and environmental domains.

Table 1 Biomedical and environmental applications of water polyurethane-metal oxide nanocomposites

Nanocomposite	Characteristics	Applications	References
Biomedical applications			
Acrylic polyurethane-iron oxide-silver hybrid nanoparticles	10–20 nm in size with coating hardness, mechanical stability, impact resistance, infrared, relative hardness property, and high adhesion	Antibacterial activity against *E. coli*	Nguyen et al. (2019), Nguyen et al. (2019)
Acrylic polyurethane-iron oxide-silver hybrid nanoparticles	Fabricated via green-based photo-crosslinking agents, high swelling degree relative hardness, gel fraction, and abrasion resistance	Antibacterial activity against *E. coli*	Nguyen et al. (2020), Nguyen et al. (2020)
	Polyurethane-acrylate-graphene oxide-titanium dioxide	High electrical conductivity and UV-curing potential	Synergistic antibacterial activity against *S. aureus*
Li et al. (2019), Li et al. (2019)			
Waterborne polyurethane-titanium dioxide-silicon dioxide nano-colloids	High mechanical, thermal property, young's modulus, and tensile strength	Antibacterial activity against *E. coli* and *S. aureus*	Malkappa et al. (2017), Malkappa et al. (2018)
Waterborne polyurethane-magnetite	Superparamagnetic property at 300 K	Shape memory ability to be used as implants	Soto et al. (2018), Soto et al. (2018)
Polyurethane-iron oxide-graphene nanoplatelets	Ferromagnetic material with high thermal, mechanical, and rheological property	Shape memory ability to be used as implants	Urban and Strankowski (2017), Urban and Strankowski (2017)
Polyurethane-zinc oxide	Self-healing, high mechanical strength, and biocompatibility	Antibacterial agent to be used in injection equipment, tubing, hospital bedding, implants, surgical drapes, and wound dressing applications	Rahman (2020), Rahman (2020)
Environmental applications			
Waterborne polyurethane-silicon dioxide films	Thermal stability, water absorptivity, maximum elongation break, water resistance, and high mechanical property	Antifouling agent	Xiao-min et al. (2015), Luo et al. (2015)
Waterborne polyurethane-titanium	dioxide-polyaniline-carbon nanotube-halloysite	High corrosion resistance, electrical conductivity, hydrophilic, high water contact angle, self-cleaning property	Environmental degradation and decolorization of toxic Malachite green dye (59%)
Nosrati et al. (2019), Nosrati et al. (2020)			
Waterborne polyurethane-silicon dioxide (silica) fluoride	Low surface energy, superhydrophobic surface property, optical transparency, aging resistance, abrasion resistance, and strong adhesion	Self-cleaning property even after immersing them in toluene dyed Sudan 1 and antifouling activity	Zheng et al. (2019), Zheng et al. (2019)
Waterborne			

polyurethane-polytetrafluoroethylene-titanium dioxideImpede adhesion of marine bacteriaAntifouling propertyXiao-dong et al. (2007), Huang et al. (2007)

7 Conclusions

This chapter emphasizes the advantages of blending metal oxide nanoparticles with waterborne polyurethane to form novel nanocomposites. Various magnetic and non-magnetic nanocomposites are prepared by using waterborne polyurethanes and metal oxide nanoparticles, to either directly act as a mechanically stable composite, or encapsulate significant molecules for targeted delivery in biomedical and environmental applications. Nevertheless, several shortcomings persist, such as swelling of particles and toxicity, which impede their large-scale and commercial utilization. Thus, it is necessary to address and overcome these limitations via advantages in the phytochemistry, biosynthesis of nanoparticles, and coating methods, to utilize them as an important commercial material to advance the biomedical industry.

References

Adeleye, A. S., Pokhrel, S., Mädler, L., & Keller, A. A. (2018). Influence of nanoparticle doping on the colloidal stability and toxicity of copper oxide nanoparticles in synthetic and natural waters. *Water Research, 132*, 12–22.

Ahmadi, Y., & Ahmad, S. (2020). Recent progress in the synthesis and property enhancement of waterborne polyurethane nanocomposites: Promising and versatile macromolecules for advanced applications. *Polymer Reviews, 60*(2), 226–266.

Ahmadi, Y., Kim, K.-H., Kim, S., & Tabatabaei, M. (2020). Recent advances in polyurethanes as efficient media for thermal energy storage. *Energy Storage Materials*

Aizpurua, J., Martin, L., Fernández, M., González, A., & Irusta, L. (2020). Recyclable, remendable and healing polyurethane/acrylic coatings from UV curable waterborne dispersions containing Diels-Alder moieties. *Progress in Organic Coatings, 139*, 105460. https://doi.org/10.1016/j.porgcoat.2019.105460

Alavi Nikje, M. M., Farahmand Nejad, M. A., Shabani, K., & Haghshenas, M. (2013). Preparation of magnetic polyurethane rigid foam nanocomposites. *Colloid and Polymer Science, 291* (4), 903–909. https://doi.org/10.1007/s00396-012-2808-6

Athimoolam, M., & Moorthy, T. V. (2012). Polymer nanocomposite Materials and shape memory applications—A Review. *Procedia Engineering, 38*, 3399–3408.

Awad, S., Chen, H., Chen, G., Gu, X., Lee, J. L., Abdel-Hady, E. E., & Jean, Y. C. (2011). Free volumes, glass transitions, and cross-links in zinc oxide/waterborne polyurethane nanocomposites. *Macromolecules, 44*(1), 29–38. https://doi.org/10.1021/ma102366d

Aydınoğlu, D., Akgül, Ö., Bayram, V., & Şen, S. (2014). Polymer nanocomposite hydrogels with improved metal adsorption capacity and swelling behavior: Influence of spirulina immobilization onto montmorillonite clay. *Polymer-Plastics Technology and Engineering, 53*(16), 1706–1722.

Bai, C., & Tang, M. (2020). Toxicological study of metal and metal oxide nanoparticles in zebrafish. *Journal of Applied Toxicology, 40* (1), 37–63.

Baker, T. J., Tyler, C. R., & Galloway, T. S. (2014). Impacts of metal and metal oxide nanoparticles on marine organisms. *Environmental Pollution, 186*, 257–271.

Basu, A., Farah, S., Kunduru, K. R., Doppalapudi, S., Khan, W., & Domb, A. J. (2016). 8—Polyurethanes for controlled drug delivery. In S. L. Cooper & J. Guan (Eds) *Advances in Polyurethane Biomaterials*. Woodhead Publishing, pp. 217–246.

Chavali, M. S., & Nikolova, M. P. (2019). Metal oxide nanoparticles and their applications in nanotechnology. *SN Applied Sciences, 1* (6), 607.

Chen, S., Chen, S., Zhao, G., & Chen, J. (2015a). Fabrication and properties of novel superparamagnetic, well—Dispersed waterborne Polyurethane/Ni–Zn ferrite nanocomposites. *Composites Science and Technology, 119*, 108–114. https://doi.org/10.1016/j.compscitech.2015.09.020

Chen, S., Zhang, S., Li, Y., & Zhao, G. (2015b). Synthesis and properties of novel UV—curable hyperbranched waterborne polyurethane/Fe_3O_4 nanocomposite films with excellent magnetic properties. *RSC Advances, 5*(6), 4355–4363. https://doi.org/10.1039/C4RA13683G

Chen, S., Zhang, S., Jin, T., & Zhao, G. (2016). Synthesis and characterization of novel covalently linked waterborne polyurethane/Fe_3O_4 nanocomposite films with superior magnetic, conductive properties and high latex storage stability. *Chemical Engineering Journal, 286*, 249–258. https://doi.org/10.1016/j.cej.2015.10.089

Cheng, K.-W., & Hsu, S.-H. (2017). A facile method to prepare superparamagnetic iron oxide and hydrophobic drug-encapsulated biodegradable polyurethane nanoparticles. *International Journal of Nanomedicine, 12*, 1775–1789. https://doi.org/10.2147/IJN.S120290

Chou, C.-W., Hsu, S.-H., Chang, H., Tseng, S.-M., & Lin, H.-R. (2006). Enhanced thermal and mechanical properties and biostability of polyurethane containing silver nanoparticles. *Polymer Degradation and Stability, 91*(5), 1017–1024.

Christopher, G., Anbu Kulandainathan, M., & Harichandran, G. (2015). Highly dispersive waterborne polyurethane/ZnO nanocomposites for corrosion protection. *Journal of Coatings Technology and Research, 12* (4) (2015) 657–667. https://doi.org/10.1007/s11998-015-9674-3

Christopher, G., Kulandainathan, M. A., & Harichandran, G. (2016). Biopolymers nanocomposite for material protection: Enhancement of corrosion protection using waterborne polyurethane nanocomposite coatings. *Progress in Organic Coatings, 99*, 91–102.

Dai, M., Zhai, Y., Wu, L., & Zhang, Y. (2019). Magnetic aligned Fe_3O_4-reduced graphene oxide/waterborne polyurethane composites with controllable structure for high microwave absorption capacity. *Carbon, 152*, 661–670. https://doi.org/10.1016/j.carbon.2019.05.067

Demishev, S. V., Chernobrovkin, A. L., Glushkov, V. V., Grigorieva, A. V., Goodilin, E. A., Ohta, H., Okubo, S., Fujisawa, M., Sakurai, T., Sluchanko, N. E., Samarin, N. A., & Semeno, A. V. (2011). Magnetic properties of vanadium oxide nanotubes and nanolayers. *Physical Review B, 84*(9), 094426. https://doi.org/10.1103/PhysRevB.84.094426

Deyab, M. A., Essehli, R., El Bali, B., & Lachkar, M. (2017). Fabrication and evaluation of $Rb_2Co(H_2P_2O_7)_2 \cdot 2H_2O$/waterborne polyurethane nanocomposite coating for corrosion protection aspects. *RSC Advances, 7*(87), 55074–55080. https://doi.org/10.1039/C7RA11212B

Dieterich, D., Keberle, W., & Witt, H. (1970). Polyurethane ionomers, a new class of block polymers. *Angewandte Chemie International Edition in English, 9*(1), 40–50.

dos Santos, L. M., Ligabue, R., Dumas, A., Le Roux, C., Micoud, P., Meunier, J.-F., Martin, F., Corvo, M., Almeida, P., & Einloft, S. (2018). Waterborne polyurethane/Fe_3O_4-synthetic talc composites: Synthesis, characterization, and magnetic properties. *Polymer Bulletin, 75*(5), 1915–1930. https://doi.org/10.1007/s00289-017-2133-9

Engels, H. W., Pirkl, H. G., Albers, R., Albach, R. W., Krause, J., Hoffmann, A., Casselmann, H., & Dormish, J. (2013). Polyurethanes: Versatile materials and sustainable problem solvers for today's challenges. *Angewandte Chemie International Edition, 52* (36), 9422–9441.

Fang, C., Zhou, X., Yu, Q., Liu, S., Guo, D., Yu, R., & Hu, J. (2014). Synthesis and characterization of low crystalline waterborne polyurethane for potential application in water-based ink binder. *Progress in Organic Coatings, 77*(1), 61–71. https://doi.org/10.1016/j.porgcoat.2013.08.004

Farajollah, R., Alavi Nikje, M. M., Dorkoosh, F. A., & Vakili, M. (2015). Preparation and characterization of waterborne magnetic polyurethane nanocomposites conjugated with insulin for increasing the sustainability of the drug. *Journal of Applied Researches in Chemistry (JARC), 8* (4), 5–12.

Feast, W. J., Rannard, S. P., & Stoddart, A. (2003). Selective convergent synthesis of aliphatic polyurethane dendrimers. *Macromolecules, 36*(26), 9704–9706.

Francolini, I., D'Ilario, L., Guaglianone, E., Donelli, G., Martinelli, A., & Piozzi, A. (2010). Polyurethane anionomers containing metal ions with antimicrobial properties: Thermal, mechanical and biological characterization. *Acta Biomaterialia, 6*(9), 3482–3490.

Frisch, K. C., & Klempner, D. (1998). *Advances in Urethane: Science & Technology, 14*. CRC Press.

Fuensanta, M., Khoshnood, A., Rodríguez-Llansola, F., & Martín--Martínez, J. M. (2020). New waterborne polyurethane-urea synthesized with ether-carbonate copolymer and amino-alcohol chain extenders with tailored pressure-sensitive adhesion properties. *Materials,13*(3), 627.

Gebre, S. H., & Sendeku, M. G. (2019). New frontiers in the biosynthesis of metal oxide nanoparticles and their environmental applications: An overview. *SN Applied Sciences, 1*(8), 928.

Gogoi, S., & Karak, N. (2016). Biobased waterborne hyperbranched polyurethane/$NiFe_2O_4$@rGO nanocomposite with multi-stimuli responsive shape memory attributes. *RSC Advances, 6*(97), 94815–94825. https://doi.org/10.1039/C6RA16848E

Guo, M., Jiang, X., & Fu, H. (2018). Preparation and study of novel modified [(1−x)MnO_{2-x}MWCNTs]/waterborne polyurethane composites with microwave absorption properties. *Industrial & Engineering Chemistry Research, 57*(40), 13406–13416. https://doi.org/10.1021/acs.iecr.8b02310

Guo, Y.-H., Guo, J.-J., Li, S.-C., Li, X., Wang, G.-S., & Huang, Z. (2013). Properties and paper sizing application of waterborne polyurethane emulsions synthesized with TDI and IPDI. *Colloids and Surfaces A: Physicochemical and Engineering Aspects, 427*, 53–61. https://doi.org/10.1016/j.colsurfa.2013.03.017

Guo, Y.-H., Li, S.-C., Wang, G.-S., Ma, W., & Huang, Z. (2012). Waterborne polyurethane/poly(n-butyl acrylate-styrene) hybrid emulsions: Particle formation, film properties, and application. *Progress in Organic Coatings, 74*(1), 248–256. https://doi.org/10.1016/j.porgcoat.2011.12.016

Hasirci, N. (2017). Polyurethanes. *High performance biomaterials* (pp. 71–90). Routledge.

Hendessi, S., Sevinis, E. B., Unal, S., Cebeci, F. C., Menceloglu, Y. Z., & Unal, H. (2016). Antibacterial sustained-release coatings from halloysite nanotubes/waterborne polyurethanes. *Progress in Organic Coatings, 101*, 253–261. https://doi.org/10.1016/j.porgcoat.2016.09.005

Hepburn, C. (1982). *Polyurethane elastomers*. Applied Science Publishers Ltd.

Hii, Y. S., Jeevanandam, J., Chan, Y. S., & Danquah, M. K. (2018). Cytotoxicity and biomedical applications of metal oxide nanoparticles synthesized from plants. In *Phytochemistry, Volume 2 Pharmacognosy, Nanomedicine, and Contemporary Issues* (pp. 459–486). Apple Academic Press.

Honarkar, H. (2018). Waterborne polyurethanes: A review. *Journal of Dispersion Science and Technology, 39*(4), 507–516. https://doi.org/10.1080/01932691.2017.1327818

Honarkar, H., & Barikani, M. (2014). Nanoclay dispersion and its effect on properties of waterborne polyurethanes. *Iranian Journal of Polymer Science and Technology (persian), 26*, 393–401.

Hoseini, Z., & Nikje, M. M. (2018). Synthesis and characterization of novel magnetic polyurethane nanocomposites based on 1,5-Naphthalene diisocyanate (NDI) containing Clay-Fe_3O_4@APTS nanoparticles. *Materials Research Express, 6* (1), 016101. https://doi.org/10.1088/2053-1591/aae2f6

Hoseini, Z., & Nikje, M. M. A. (2018). Synthesis and characterization of a novel thermally stable water dispersible polyurethane and its magnetic nanocomposites. *Iranian Polymer Journal, 27*(10), 733–743. https://doi.org/10.1007/s13726-018-0650-5

Huang, X., Zhang, Z., Qi, Y., Sun, X., & Zhang, Z. (2007). Adhesion of marine bacteria on three antifouling coatings based on waterborne polyurethane. *Journal of Dalian Maritime University* (1), 1.

Hussein, G. A. M. (1996). Rare earth metal oxides: Formation, characterization and catalytic activity thermoanalytical and applied pyrolysis review. *Journal of Analytical and Applied Pyrolysis, 37* (2), 111–149.

Hwang, K. K. S., Yang, C. Z., & Cooper, S. L. (1981). Properties of polyether-polyurethane zwitterionomers. *Polymer Engineering & Science, 21*(15), 1027–1036.

Jeevanandam, J., Barhoum, A., Chan, Y. S., Dufresne, A., & Danquah, M. K. (2018). Review on nanoparticles and nanostructured materials: History, sources, toxicity and regulations. *Beilstein Journal of Nanotechnology, 9*, 1050–1074.

Jeevanandam, J., Sundaramurthy, A., Sharma, V., Murugan, C., Pal, K., Kodous, M. H. A., & Danquah, M. K. (2020). Sustainability of one-dimensional nanostructures: Fabrication and industrial applications. *Sustainable nanoscale engineering* (pp. 83–113). Elsevier.

Jeevanandam, J., Chan, Y. S., Pan, S., & Danquah, M. K. (2019). Metal oxide nanocomposites: Cytotoxicity and targeted drug delivery applications. In *Hybrid nanocomposites: Fundamentals, synthesis and applications*. Pan Stanford Publishing, pp. 111–147.

Jeevanandam, J., Chan, Y. S., & Danquah, M. K. (2016). Biosynthesis of metal and metal oxide nanoparticles. *ChemBioEng Reviews, 3*(2), 55–67.

Jiang, G., Li, X., Che, Y., Lv, Y., Liu, F., Wang, Y., Zhao, C., & Wang, X. (2019). Antibacterial and anticorrosive properties of CuZnO@RGO waterborne polyurethane coating in circulating cooling water. *Environmental Science and Pollution Research, 26*(9), 9027–9040. https://doi.org/10.1007/s11356-019-04374-0

Jiang, L., Ren, Z., Liu, W., Liu, H., & Zhu, C. (2020). Synthesis and molecular interaction of Tung oil-based anionic waterborne polyurethane dispersion. *Journal of Applied Polymer Science* 49383.

Kale, M. B., Luo, Z., Zhang, X., Dhamodharan, D., Divakaran, N., Mubarak, S., Wu, L., & Xu, Y. (2019). Waterborne polyurethane/graphene oxide-silica nanocomposites with improved mechanical and thermal properties for leather coatings using screen printing. *Polymer, 170*, 43–53. https://doi.org/10.1016/j.polymer.2019.02.055

Kalia, A., & Kaur, H. (2019). Nano-biofertilizers: Harnessing dual benefits of nano-nutrient and bio-fertilizers for enhanced nutrient use efficiency and sustainable productivity. *Nanoscience for Sustainable Agriculture* (pp. 51–73). Springer.

Khatoon, H., & Ahmad, S. (2019). Vanadium pentoxide-enwrapped polydiphenylamine/polyurethane nanocomposite: High-performance anticorrosive coating. *ACS Applied Materials & Interfaces, 11*(2), 2374–2385. https://doi.org/10.1021/acsami.8b17861

Król, P., Król, B., Pielichowska, K., & Pikus, S. (2011). Comparison of phase structures and surface free energy values for the coatings synthesised from linear polyurethanes and from waterborne polyurethane cationomers. *Colloid and Polymer Science, 289*(15–16), 1757–1767.

Kumar, Y. S., Unnithan, A. R., Sen, D., Kim, C. S., & Lee, Y. S. (2015). Microgravity biosynthesized penicillin loaded electrospun polyurethane–dextran nanofibrous mats for biomedical applications. *Colloids and Surfaces a: Physicochemical and Engineering Aspects, 477*, 77–83.

Lade, B. D., & Gogle, D. P. (2019). Nano-biopesticides: Synthesis and applications in plant safety. *Nanobiotechnology Applications in Plant Protection* (pp. 169–189). Springer.

Lee, S. (2009). Multifunctionality of layered fabric systems based on electrospun polyurethane/zinc oxide nanocomposite fibers. *Journal of Applied Polymer Science, 114*(6), 3652–3658.

Li, W., Mayo, J. T., Benoit, D. N., Troyer, L. D., Lewicka, Z. A., Lafferty, B. J., Catalano, J. G., Lee, S. S., Colvin, V. L., & Fortner, J. D. (2016). Engineered superparamagnetic iron oxide nanoparticles for ultra-enhanced uranium separation and sensing. *Journal of Materials Chemistry A, 4*(39), 15022–15029.

Li, X., Wang, S., Xie, J., Hu, J., & Fu, H. (2019). Polyurethane-acrylate-supported rGO/TiO_2 electrical conductive and antibacterial nanocomposites. *International Journal of Polymeric Materials and Polymeric Biomaterials, 68*(6), 319–327. https://doi.org/10.1080/00914037.2018.1452227

Liu, X., Iocozzia, J., Wang, Y., Cui, X., Chen, Y., Zhao, S., Li, Z., & Lin, Z. (2017). Noble metal–metal oxide nanohybrids with tailored nanostructures for efficient solar energy conversion, photocatalysis and environmental remediation. *Energy & Environmental Science, 10*(2), 402–434.

Liu, Z., Wu, B., Jiang, Y., Lei, J., Zhou, C., Zhang, J., & Wang, J. (2018). Solvent-free and self-catalysis synthesis and properties of waterborne polyurethane. *Polymer, 143*, 129–136. https://doi.org/10.1016/j.polymer.2018.04.010

Luo, X.-M., Liu, R., Zhang, P., Ge, B. H., & Cao, M. (2015). Study on polymerizable nano-SiO_2 modifying UV-curing waterborne polyurethane film performance. *Journal of Functional Materials, 2*, 02109–02114.

Ma, X.-Y., & Zhang, W.-D. (2009). Effects of flower-like ZnO nanowhiskers on the mechanical, thermal and antibacterial properties of waterborne polyurethane. *Polymer Degradation and Stability, 94*(7), 1103–1109. https://doi.org/10.1016/j.polymdegradstab.2009.03.024

Magnin, A., Pollet, E., Phalip, V., Avérous, L. (2020). Evaluation of biological degradation of polyurethanes. *Biotechnology Advances, 39*, 107457.

Magro, M., & Vianello, F. (2019). Bare iron oxide nanoparticles: Surface tunability for biomedical, sensing and environmental applications. *Nanomaterials, 9*(11), 1608.

Malkappa, K., Rao, B. N., Suresh, G., Ramana, C. V., & Jana, T. (2018). Organic/inorganic hybrid nanocolloids of water dispersible polyurethanes with antibacterial activity. *Colloid and Polymer Science, 296* (1), 95–106. https://doi.org/10.1007/s00396-017-4229-z

Mallakpour, S., & Khadem, E. (2017). Hybrid optically active polymer/metal oxide composites: Recent advances and challenges. *Hybrid polymer composite materials* (pp. 379–406). Elsevier.

Meng, J., Han, Z., Kong, H., Qi, X., Wang, C., Xie, S., & Xu, H. (2010). Electrospun aligned nanofibrous composite of MWCNT/polyurethane to enhance vascular endothelium cells proliferation and function. *Journal of Biomedical Materials Research Part A, 95*(1), 312–320.

Natarajan, E., & Suuki, S. (2013). *Removal of E. coli ATCC 25922 (indicator of fecal contamination) from drinking water by biosynthesized silver nanoparticles impregnated polyurethane foam.*In International Conference on Advanced Nanomaterials & Emerging EngineeringTechnologies (pp. 252-255). IEEE.

Nguyen, T. N. L., Do, T. V., Nguyen, T. V., Dao, P. H., Trinh, V. T., Mac, V. P., Nguyen, A. H., Dinh, D. A., Nguyen, T. A., Vo, T. K. A., Tran, D. L., & Le, T. L. (2019). Antimicrobial activity of acrylic polyurethane/Fe_3O_4–Ag nanocomposite coating. *Progress in Organic Coatings, 132*, 15–20. https://doi.org/10.1016/j.porgcoat.2019.02.023

Nguyen, T. V., Do, T. V., Ha, M. H., Le, H. K., Le, T. T., Linh Nguyen, T. N., Dam, X. T., Lu, L. T., Tran, D. L., Vu, Q. T., Dinh, D. A., Dang, T. C., & Nguyen-Tri, P. (2020). Crosslinking process, mechanical and antibacterial properties of UV-curable acrylate/Fe_3O_4-Ag nanocomposite coating. *Progress in Organic Coatings, 139*, 105325. https://doi.org/10.1016/j.porgcoat.2019.105325

Niederberger, M., & Pinna, N. (209). *Metal oxide nanoparticles in organic solvents: Synthesis, formation, assembly and application*. Springer Science & Business Media.

Nikje, M. M. A., Akbar, R., Ghavidel, R., & Vakili, M. (2015). Preparation and characterization of magnetic rigid polyurethane foam reinforced with dipodal silane iron oxide nanoparticles Fe_3O_4@APTS/GPTS. *Cellular Polymers, 34*(3), 137–156. https://doi.org/10.1177/026248931503400302

Nosrati, R., Kiani, G., Karimzad Ghavidel, A., & Rashidi, A. (2020). Improving environmental protection of waterborne polyurethane coating by adding TiO2/polyaniline/HNT/CNT nanocomposite. *Environmental Science and Pollution Research, 27* (6), 6438–6448. https://doi.org/10.1007/s11356-019-07333-x

Oskam, G. (2006). Metal oxide nanoparticles: Synthesis, characterization and application. *Journal of Sol-Gel Science and Technology, 37* (3), 161–164.

Peng, L., Zhou, L., Li, Y., Pan, F., & Zhang, S. (2011). Synthesis and properties of waterborne polyurethane/attapulgite nanocomposites. *Composites Science and Technology, 71*(10), 1280–1285. https://doi.org/10.1016/j.compscitech.2011.04.012

Prasanna, S. R. V. S., Balaji, K., Pandey, S., & Rana, S. (2019). Metal oxide based nanomaterials and their polymer nanocomposites. *Nanomaterials and polymer nanocomposites* (pp. 123–144). Elsevier.

Qiang, T., Han, M., & Wang, X. (2020). Waterborne polyurethane/carbon quantum dot nanocomposite as a surface coating material exhibiting outstanding luminescent performance. *Progress in Organic Coatings, 138*, 105433.

Rahman, M. M. (2020). Polyurethane/zinc oxide (PU/ZnO) composite—Synthesis, protective property and application. *Polymers, 12* (7). https://doi.org/10.3390/polym12071535

Rahman, M. M., Hussain, M. M., & Asiri, A. M. (2017). Fabrication of 3-methoxyphenol sensor based on Fe_3O_4 decorated carbon nanotube nanocomposites for environmental safety: Real sample analyses. *PLoS One, 12* (9), e0177817.

Reid, D. L., Draper, R., Richardson, D., Demko, A., Allen, T., Petersen, E. L., & Seal, S. (2014). In situ synthesis of polyurethane–TiO_2 nanocomposite and performance in solid propellants. *Journal of Materials Chemistry A, 2*(7), 2313–2322.

Rizwan, M., Ali, S., Ali, B., Adrees, M., Arshad, M., Hussain, A., ur Rehman, M. Z., & Waris, A. A. (2019). Zinc and iron oxide nanoparticles improved the plant growth and reduced the oxidative stress and cadmium concentration in wheat. *Chemosphere, 214*, 269–277.

Salahuddin, N., Rehab, A., & Abd-Elghany, S. (2020). In vitro thermo-triggered drug release from magnetic polyurethane-urea nanocomposite. *Journal of Drug Delivery Science and Technology, 56*, 101564. https://doi.org/10.1016/j.jddst.2020.101564

Sardon, H., Irusta, L., Fernández-Berridi, M. J., Luna, J., Lansalot, M., & Bourgeat-Lami, E. (2011). Waterborne polyurethane dispersions obtained by the acetone process: A study of colloidal features. *Journal of Applied Polymer Science, 120*(4), 2054–2062. https://doi.org/10.1002/app.33308

Sardon, H., Irusta, L., Fernández-Berridi, M. J., Lansalot, M., & Bourgeat-Lami, E. (2010). Synthesis of room temperature self-curable waterborne hybrid polyurethanes functionalized with (3-aminopropyl)triethoxysilane (APTES). *Polymer, 51*(22), 5051–5057. https://doi.org/10.1016/j.polymer.2010.08.035

Sawpan, M. A. (2018). Polyurethanes from vegetable oils and applications: A review. *Journal of Polymer Research, 25*(8), 184.

Shanker, U., Jassal, V., & Rani, M. (2016). Catalytic removal of organic colorants from water using some transition metal oxide nanoparticles synthesized under sunlight. *RSC Advances, 6*(97), 94989–94999.

Shendi, H. K., Omrani, I., Ahmadi, A., Farhadian, A., Babanejad, N., & Nabid, M. R. (2017). Synthesis and characterization of a novel internal emulsifier derived from sunflower oil for the preparation of waterborne polyurethane and their application in coatings. *Progress in Organic Coatings, 105*, 303–309. https://doi.org/10.1016/j.porgcoat.2016.11.033

Singh, H., & Jain, A. K. (2009). Ignition, combustion, toxicity, and fire retardancy of polyurethane foams: A comprehensive review. *Journal of Applied Polymer Science, 111*(2), 1115–1143.

Singh, S., Srivastava, P., Kapoor, I. P. S., & Singh, G. (2013). Preparation, characterization, and catalytic activity of rare earth metal oxide nanoparticles. *Journal of Thermal Analysis and Calorimetry, 111*(2), 1073–1082.

Soto, G. D., Meiorin, C., Actis, D. G., Mendoza Zélis, P., Moscoso Londoño, O., Muraca, D., Mosiewicki, M. A., & Marcovich, N. E. (2018). Magnetic nanocomposites based on shape memory polyurethanes. *European Polymer Journal, 109*, 8–15. https://doi.org/10.1016/j.eurpolymj.2018.08.046

Sow, C., Riedl, B., & Blanchet, P. (2010). Kinetic studies of UV-waterborne nanocomposite formulations with nanoalumina and nanosilica. *Progress in Organic Coatings, 67*(2), 188–194. https://doi.org/10.1016/j.porgcoat.2009.10.002

Stankic, S., Suman, S., Haque, F., & Vidic, J. (2016). Pure and multi metal oxide nanoparticles: Synthesis, antibacterial and cytotoxic properties. *Journal of Nanobiotechnology, 14*(1), 1–20.

Sukhawipat, N., Saetung, N., Pilard, J.-F., Bistac, S., & Saetung, A. (2018). Synthesis and characterization of novel natural rubber based cationic waterborne polyurethane—Effect of emulsifier and diol class chain extender. *Journal of Applied Polymer Science, 135*(3), 45715. https://doi.org/10.1002/app.45715

Szycher, M. (1999). *Szycher's handbook of polyurethanes*. CRC press.

Urban, M., & Strankowski, M. (2017). Shape memory polyurethane materials containing ferromagnetic iron oxide and graphene nanoplatelets. *Materials, 10* (9). https://doi.org/10.3390/ma10091083

Ye, X., Wang, Z., Ma, L., Wang, Q., & Chu, A. (2019). Zinc oxide array/polyurethane nanocomposite coating: Fabrication, characterization and corrosion resistance. *Surface and Coatings Technology, 358*, 497–504. https://doi.org/10.1016/j.surfcoat.2018.11.080

Yin, H., Du, B., Chen, Y., Song, N., Li, Z., Li, J., Luo, F., & Tan, H. (2020). Dual-encapsulated biodegradable 3D scaffold from liposome and waterborne polyurethane for local drug control release in breast cancer therapy. *Journal of Biomaterials Science, Polymer Edition* (just-accepted), 1–13.

Yousefi, E., Dolati, A., & Najafkhani, H. (2020). Preparation of robust antistatic waterborne polyurethane coating. *Progress in Organic Coatings, 139*, 105450.

Zhang, C., Zhang, X., Dai, J., & Bai, C. (2008). Synthesis and properties of PDMS modified waterborne polyurethane–acrylic hybrid emulsion by solvent-free method. *Progress in Organic Coatings, 63*(2), 238–244.

Zhang, J., Li, Y., Hu, C., Huang, W., & Su, L. (2019). Anti-corrosive properties of waterborne polyurethane/poly (o-toluidine)-ZnO coatings in NaCl solution. *Journal of Adhesion Science and Technology, 33*(10), 1047–1065.

Zhang, S., Li, Y., Peng, L., Li, Q., Chen, S., & Hou, K. (2013). Synthesis and characterization of novel waterborne polyurethane nanocomposites with magnetic and electrical properties. *Composites Part A: Applied Science and Manufacturing, 55*, 94–101. https://doi.org/10.1016/j.compositesa.2013.05.018

Zhang, W., Xue, W., Zeng, Z., & Li, J. (2020). Synthesis and properties of flame-retardant reactive hot melt polyurethane adhesive. *Journal of Adhesion Science and Technology, 34*(2), 178–191. https://doi.org/10.1080/01694243.2019.1663692

Zheng, H., Pan, M., Wen, J., Yuan, J., Zhu, L., & Yu, H. (2019). Robust, transparent, and superhydrophobic coating fabricated with waterborne polyurethane and inorganic nanoparticle composites. *Industrial & Engineering Chemistry Research, 58*(19), 8050–8060. https://doi.org/10.1021/acs.iecr.9b00052

Zhou, H., Wang, H., Tian, X., Zheng, K., & Cheng, Q. (2014). Preparation and properties of waterborne polyurethane/antimony doped tin oxide nanocomposite coatings via sol–gel reactions. *Polymer Composites, 35* (6), 1169–1175. https://doi.org/10.1002/pc.22764

Zhu, J., Li, J., Sang, C., & Luo, Y. (2020). The synthesis of polymeric dyes based on waterborne polyurethane: A reaction kinetics study using UV absorption spectroscopy. *New Journal of Chemistry, 44* (7), 2930–2940. https://doi.org/10.1039/C9NJ05490A

Zhu, Z., Li, R., Zhang, C., & Gong, S. (2018). Preparation and properties of high solid content and low viscosity waterborne polyurethane—Acrylate emulsion with a reactive emulsifier. *Polymers, 10* (2). https://doi.org/10.3390/polym10020154